Polyamine Drug Discovery

RSC Drug Discovery Series

Editor-in-Chief:
Professor David Thurston, *London School of Pharmacy, UK*

Series Editors:
Dr David Fox, *Pfizer Global Research and Development, Sandwich, UK*
Professor Salvatore Guccione, *University of Catania, Italy*
Professor Ana Martinez, *Instituto de Quimica Medica-CSIC, Spain*
Dr David Rotella, *Montclair State University, USA*

Advisor to the Board:
Professor Robin Ganellin, *University College London, UK*

Titles in the Series:
1: Metabolism, Pharmacokinetics and Toxicity of Functional Groups: Impact of Chemical Building Blocks on ADMET
2: Emerging Drugs and Targets for Alzheimer's Disease; Volume 1: Beta-Amyloid, Tau Protein and Glucose Metabolism
3: Emerging Drugs and Targets for Alzheimer's Disease; Volume 2: Neuronal Plasticity, Neuronal Protection and Other Miscellaneous Strategies
4: Accounts in Drug Discovery: Case Studies in Medicinal Chemistry
5: New Frontiers in Chemical Biology: Enabling Drug Discovery
6: Animal Models for Neurodegenerative Disease
7: Neurodegeneration: Metallostasis and Proteostasis
8: G Protein-Coupled Receptors: From Structure to Function
9: Pharmaceutical Process Development: Current Chemical and Engineering Challenges
10: Extracellular and Intracellular Signaling
11: New Synthetic Technologies in Medicinal Chemistry
12: New Horizons in Predictive Toxicology: Current Status and Application
13: Drug Design Strategies: Quantitative Approaches
14: Neglected Diseases and Drug Discovery
15: Biomedical Imaging: The Chemistry of Labels, Probes and Contrast Agents
16: Pharmaceutical Salts and Cocrystals
17: Polyamine Drug Discovery

How to obtain future titles on publication:
A standing order plan is available for this series. A standing order will bring delivery of each new volume immediately on publication.

For further information please contact:
Book Sales Department, Royal Society of Chemistry, Thomas Graham House, Science Park, Milton Road, Cambridge, CB4 0WF, UK
Telephone: +44 (0)1223 420066, Fax: +44 (0)1223 420247, Email: books@rsc.org
Visit our website at http://www.rsc.org/Shop/Books/

Polyamine Drug Discovery

Edited by

Patrick M. Woster
Medical University of South Carolina, Charleston, SC, USA

Robert A. Casero, Jr.
John Hopkins University, Baltimore, MD, USA

RSCPublishing

RSC Drug Discovery Series No. 17

ISBN: 978-1-84973-190-4
ISSN: 2041-3203

A catalogue record for this book is available from the British Library

Published by The Royal Society of Chemistry,
Thomas Graham House, Science Park, Milton Road,
Cambridge CB4 0WF, UK

Registered Charity Number 207890

For further information see our web site at www.rsc.org

Preface

There have been many significant advances in polyamine research since the field was brought to the forefront of biological and biomedical research in the 1970s. Many of the significant findings since that time, whether in plant research, cell biology, the development of therapeutics, genetics or another related field, have been aided by the availability of synthetic compounds specifically designed to inhibit enzymes in the polyamine pathway or otherwise disrupt polyamine metabolism. For example, a search of PubMed using the terms difluoromethylornithine AND dfmo AND eflornithine produces 1179 references dating to 1980, and including research in diverse areas such as plant biochemistry, cancer cell biology, parasitology, insect biochemistry, synthetic chemistry, drug development and human clinical trials. Despite the diversity of fields of endeavor within polyamine research and the significant impact of modulators of polyamine metabolism, a book dedicated to the discovery and development of synthetic compounds targeting polyamine metabolism as drugs has never been produced. The purpose of this book is to fill that void by presenting an overview of drug-discovery research within the polyamine field.

The impetus for a significant portion of polyamine research has been provided by the availability of synthetic analogs that produce defined effects on polyamine metabolism *in vitro* and *in vivo*. This book begins with a chapter that outlines the synthetic approaches to these analogs, covering areas such as nucleotide synthesis and synthetic routes used to access various polyamine analogs. The structural biology aspects of polyamine drug discovery are detailed, as are efforts to design and discover specific inhibitors of enzymes in the polyamine pathway. Chapters are also included that address the role of polyamine analogs as antiparasitic agents, antineoplastic agents and epigenetic modulators. In addition, the important role played by polyamine oxidation is detailed. Other important areas within polyamine drug discovery research, such

RSC Drug Discovery Series No. 17
Polyamine Drug Discovery
Edited by Patrick M. Woster and Robert A. Casero, Jr.
© Royal Society of Chemistry 2012
Published by the Royal Society of Chemistry, www.rsc.org

as polyamine transport, the development of polyamine–metal complexes as antitumor agents and the design of polyamine-based gene transfer reagents, are discussed. Finally, a chapter is included that describes the promising results from recent human clinical trials involving drugs targeting polyamine metabolism. The result is a broad overview of polyamine drug discovery and the translation of new chemical entities from basic chemistry to studies involving patients.

Those of us who have a long history in polyamine drug discovery research know that it is a cyclic endeavor, with drug discovery successes appearing periodically and clinical successes appearing steadily but infrequently. However, recent clinical research with existing compounds, including DFMO and the bis(ethyl)polyamine analogs PG-11093 and PG-11144, bode well for the future. In particular, DFMO has found utility as a chemopreventative agent in combination with sulindac, and the bis(ethyl)polyamines have produced promising results in combination antitumor studies. Not since the advent of DFMO has the field of polyamine drug discovery research been so close to bringing a drug to market. We hope that after perusing this book, the reader will have gained an appreciation for polyamine drug discovery efforts that are occurring on multiple fronts. We also hope that you will share the confidence inherent in modern polyamine researchers that significant successes in polyamine drug discovery are on the horizon.

Patrick M. Woster
Robert A. Casero, Jr

Contents

RSC Drug Discovery Series No. 17
Polyamine Drug Discovery
Edited by Patrick M. Woster and Robert A. Casero, Jr.
© Royal Society of Chemistry 2012
Published by the Royal Society of Chemistry, www.rsc.org

CHAPTER 1

Polyamine Drug Discovery: Synthetic Approaches to Therapeutic Modulators of Polyamine Metabolism

PATRICK M. WOSTER*

Department of Pharmaceutical and Biomedical Sciences, Medical University of South Carolina, 70 President St., Charleston, SC 29425, USA

1.1 Introduction

In the following chapters, a complete description of the design, bioevaluation and development of modulators of polyamine metabolism is presented. There are numerous synthetic approaches to these inhibitors, and as such a comprehensive review of the chemical literature in this area is beyond the scope of this book. In this chapter, specific examples of synthetic approaches to nucleosides, analogs of the natural polyamines and other agents that affect polyamine metabolism are described. The reader should bear in mind that the literature is replete with alternative strategies for the synthesis of compounds described herein. However, the examples provided will allow the reader to appreciate the vast chemical diversity that is available to medicinal chemists working in the polyamine field.

RSC Drug Discovery Series No. 17
Polyamine Drug Discovery
Edited by Patrick M. Woster and Robert A. Casero, Jr.
© Royal Society of Chemistry 2012
Published by the Royal Society of Chemistry, www.rsc.org

1.2   Polyamine Metabolism as a Drug Target

The mammalian polyamine biosynthetic pathway is shown in Figure 1.1.[1,2] Ornithine is converted to putrescine by the action of the enzyme ornithine decarboxylase (ODC). Mammalian ODC, a dimeric enzyme with a molecular weight of about 80 000, is a typical pyridoxal phosphate-requiring amino acid decarboxylase that has been studied quite extensively.[3] ODC is known to be one of the control points in the polyamine biosynthetic pathway, producing a product that is committed to polyamine biosynthesis. The synthesis and degradation of ODC are controlled by a number of factors including degradation assisted by a specific ODC antizyme, a polyamine-induced protein that binds to ODC and promotes ubiquitin-independent degradation by the 26S proteasome.[4] As a result, ODC has a functional half-life of about 10 min. Putrescine is next converted to spermidine via an aminopropyltransferase known as spermidine synthase, which requires decarboxylated S-adenosylmethionine as a co-substrate.[5] A second closely related but distinct aminopropyltransferase, spermine synthase, then adds an additional aminopropyl group to spermidine to yield spermine, the longest polyamine occurring in mammalian systems.[5] The by-product for the spermidine and spermine synthase reactions is 5'-methylthioadenosine (MTA), a potent product inhibitor for the aminopropyl transfer process.[6] In mammalian systems, MTA is rapidly hydrolyzed by the enzyme MTA-phosphorylase, and the components are converted to adenosine and methionine *via* salvage pathways.[7] The aminopropyl donor for both aminopropyltransferases is decarboxylated S-adenosylmethionine (dc-AdoMet), produced from S-adenosylmethionine (AdoMet) by S-adenosylmethionine decarboxylase (AdoMet-DC).[8] AdoMet-DC, like ODC, is a highly regulated enzyme in mammalian cells, and also serves as a regulatory point in the pathway. However, unlike ODC, AdoMet-DC belongs to a class of pyruvoyl enzymes that do not require pyridoxal phosphate as a cofactor (see below).[9]

Polyamine metabolism is tightly controlled by a combination of inducible enzymes and the import/export of cellular polyamines. In addition to the enzymes mentioned above, intracellular polyamine content is modulated by a pair of acetyltransferases. Spermidine in the cell nucleus is acetylated on the four-carbon end by spermidine-N8-acetyltransferase, possibly altering the compound's binding affinity for DNA.[10,11] A specific deacetylase can then reverse this enzymatic acetylation. Cytoplasmic spermidine and spermine serve as substrates for spermidine/spermine-N[1]-acetyltransferase (SSAT), resulting in acetylation on the three-carbon end of each molecule (Figure 1.1).[5,12] The acetylated spermidine or spermine then acts as a substrate for acetylpolyamine oxidase (APAO),[13] which catalyzes the formation of 3-acetamidopropionaldehyde and either putrescine or spermidine, respectively. Excess acetylated polyamines can also be exported from the cell via the polyamine transport system.[14] More recently, a second polyamine oxidase, the inducible spermine oxidase (SMO) was discovered and characterized.[15,16] Thus, SSAT, APAO and SMO together serve as a reverse route for the interconversion of polyamines. An additional mechanism for control of cellular polyamines is provided by the polyamine

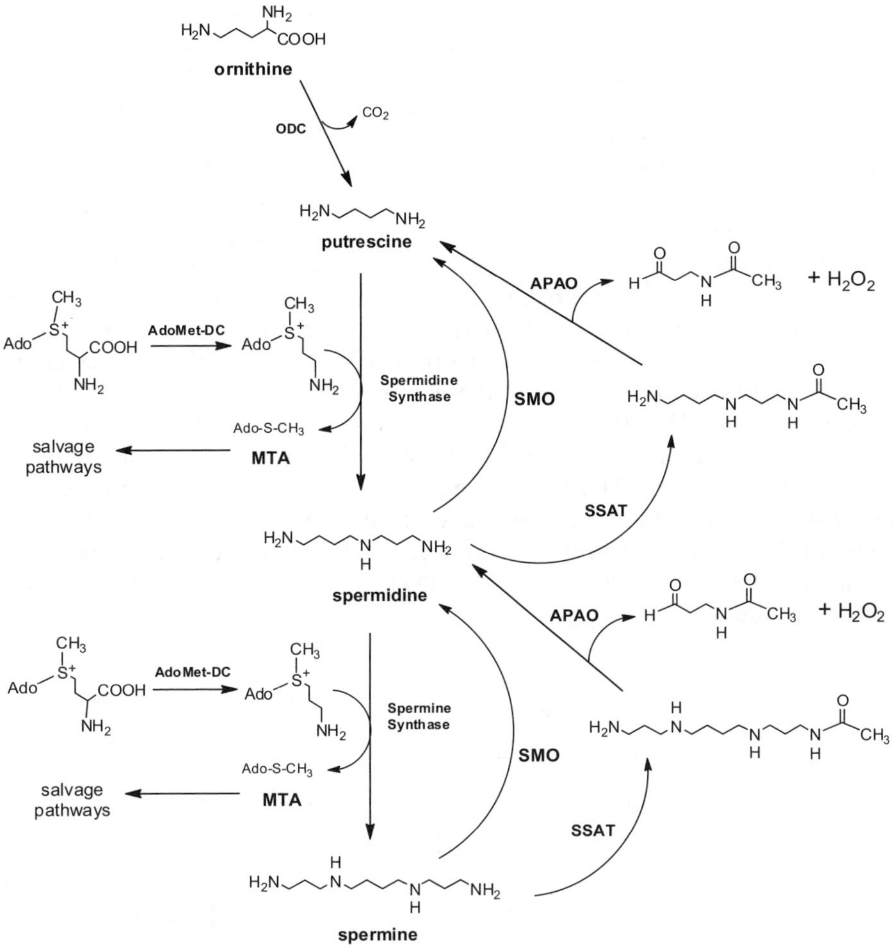

Figure 1.1 Mammalian polyamine metabolic pathway.

transport system, which has been well characterized in some organisms (bacteria, yeast), but has not been well characterized in mammalian organisms.[17] The function of enzymes in polyamine metabolism and the polyamine transport system, and the consequences of modulating their activity, are described in more detail elsewhere in this book.

1.3 Synthetic Approaches to Modulators of Polyamine Metabolism and Function

1.3.1 Ornithine Decarboxylase (ODC)

Mammalian ODC is a highly unstable protein, and cellular levels of ODC depend on rates of synthesis and degradation as outlined above. For this reason, reversible and irreversible inhibitors of ODC have proven to be of limited value, since the synthesis of new protein occurs very rapidly in response to reduced polyamine levels in the cell. The catalytic mechanism of ODC involves the formation of a Schiff's base between the amino group of ornithine and the pyridoxal phosphate cofactor which is tightly bound to ODC. The most useful inhibitor of ODC to date, α-difluoromethylornithine (DFMO, **1**, Scheme 1.1), takes advantage of this aspect of the mechanism, and belongs to a group of rationally designed mechanism-based inactivators specifically targeted to individual amino acid decarboxylases. The chemical synthesis of DFMO is shown in Scheme 1.1.[18] The (bis)benzylidene-protected amino ester **2** is treated with lithium diisopropylamide (LDA) followed by exposure to 1-chloro-2,2-difluoroethane to form the alkylated product **3**. Removal of the benzylidene protecting groups and cleavage of the methyl ester are accomplished simultaneously to afford DFMO **1** in a 60% overall yield. It is noteworthy that the pathway shown in Scheme 1.1 is not used at the industrial scale, and the large-scale production of DFMO is an expensive undertaking. Thus, until recently, the drug has been produced almost exclusively in sufficient quantities for inclusion in commercial preparations such as the the lifestyle drug Vaniqa®. Although DFMO is available commercially in small quantities for research, the cost is prohibitive.

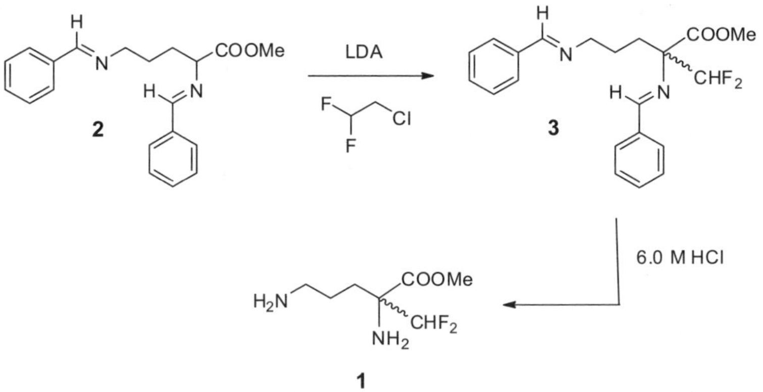

Scheme 1.1 Synthesis of 2,2-difluoromethylornithine (DFMO, **1**).

The mechanism of inactivation of ODC by DFMO is shown in Scheme 1.2. As a substrate analog, DFMO forms a Schiff's base with the pyridoxal phosphate cofactor bound to ODC. The subsequent decarboxylation step results in the generation of a latent electrophile, and ODC is rapidly and irreversibly deactivated by forming a covalent bond with CYS_{360}.[19] The discovery of DFMO provided an enormous stimulus to the field of mammalian polyamine biology. Historically, DFMO has been marketed as a treatment for *Pneumocystis carinii* secondary infections in immunocompromised patients,[5] and has been shown to be effective in curing infections of *Trypanosoma brucei gambiense* (but not *T. brucei rhodesiense*) in limited clinical trials.[20–24]

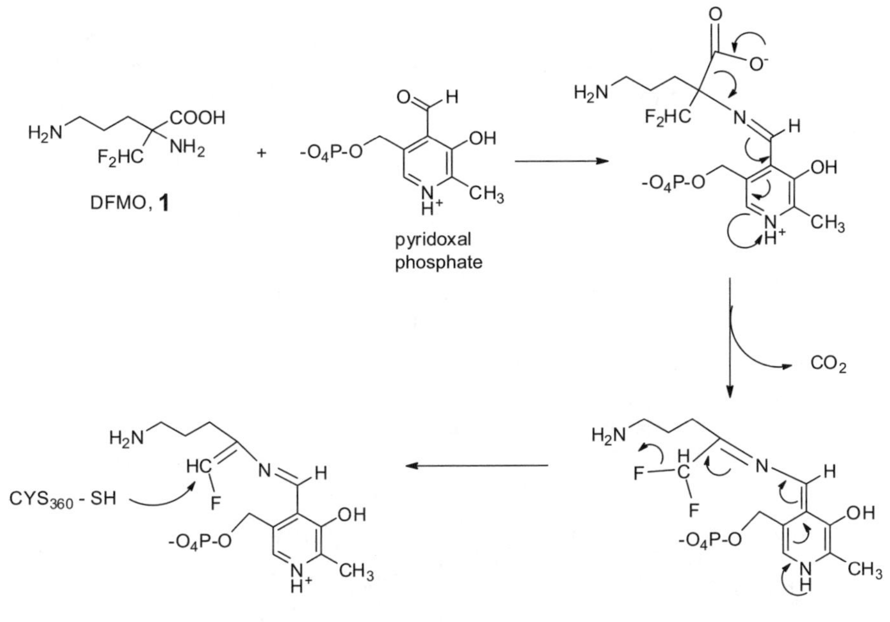

Scheme 1.2 Mechanism of inactivation of ODC by DFMO.

A relatively small number of ODC inhibitors related to **1** have been described, mostly in the mid- to late 1980s, by the highly productive Merrell Dow research group, but none were as successful as **1** in either *in vitro* or *in vivo* studies. The ODC inactivator (2*R*,5*R*)-6-heptyne-2,5-diamine (*R*,*R*-MAP), was shown to possess a K_i of 3 μM, and to penetrate mammalian cells relatively well. This compound was the subject of human clinical trials[25] but was never marketed. Based on the promising activity of **1** and *R*,*R*-MAP, a number of fluorine-containing mechanism based inactivators of ODC were developed, the most potent of which was 2,2-difluoro-5-hexyne-1,4-diamine, **4**.[26] The synthesis of **4** is outlined in Scheme 1.3, and illustrates that reasonably complex syntheses are often required to access simple but specifically designed target molecules. The requisite ester ethyl 2,2-difluoro-4-pentenoate **5** was reduced in quantitative yield to afford the corresponding alcohol **6**. Formation of the triflate followed by addition of phthalimide then gave the protected aminoolefin **7**. Ozonolysis was then used to convert the olefinic linkage to the aldehyde **8**. The acetylene group was added *via* a Grignard reagent (HCC-MgBr), followed by addition of phthalimide using a Mitsunobu reaction to produce **9**. The phthalimide protecting groups were removed (hydrazine) to afford **10**, followed by conversion to the dihydrochloride **4**. Despite early successes, **4** was not developed as a drug following the dissolution of the Merrell Dow research effort in polyamine research. More recently, Gehring *et al.* have described transition-state analog inhibitors of ODC involving the structures of ornithine and pyridoxal phosphate.[27]

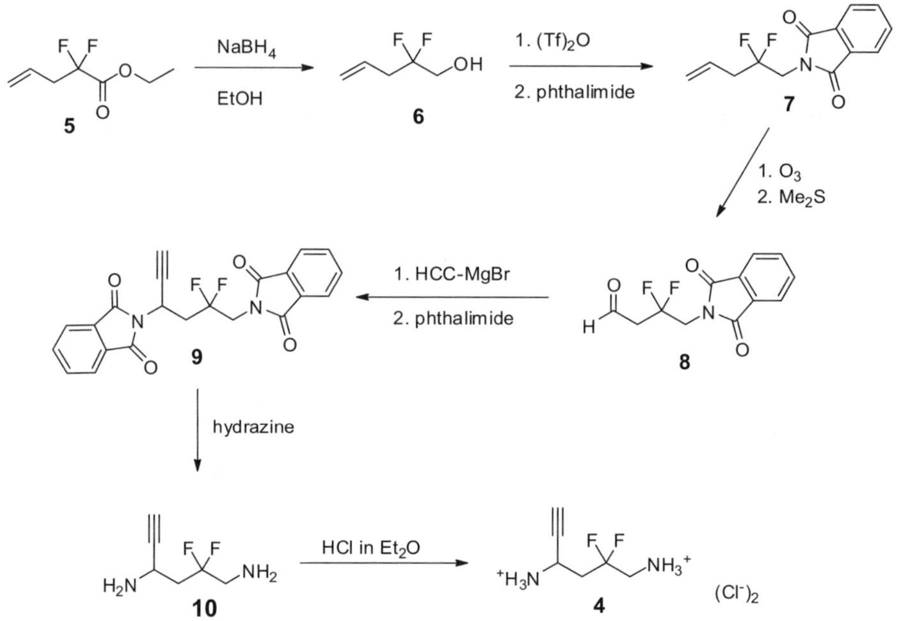

Scheme 1.3   Synthesis of 2,2-difluoro-5-hexyne-1,4-diamine, **4**.

1.3.2   *S*-Adenosylmethionine Decarboxylase (AdoMet-DC)

Mammalian AdoMet-DC is a pyruvoyl enzyme which has two subunits of 32 000 M_r.[28] The pyruvate residue at the N-terminus of one of the subunits serves the function of pyridoxal phosphate, forming a Schiff's base with the primary amino group in AdoMet. AdoMet-DC is an inducible enzyme, and responds dramatically to either polyamine depletion or elevation of spermidine or spermine. AdoMet-DC requires putrescine for activation, which proceeds autocatalytically, as shown in Figure 1.2. In the case of human AdoMet-DC, cleavage occurs between residues Glu67 and Ser68. The Ser68 oxygen attacks the adjacent carbonyl carbon of Glu67 to generate a five-membered oxyoxazolidine intermediate that rapidly rearranges to form an ester intermediate. His243 then abstracts a proton from the alpha carbon of Ser68, thus cleaving the ester and forming two chains. The N-terminal end of the cleavage product becomes the β-chain, and the C-terminal portion becomes the α-chain. The N-terminal dehydroalanine residue resulting from cleavage of the ester tautomerizes to form an imine, which is then hydrolyzed to a pyruvate.[9]

The antileukemic agent methylglyoxal bis(guanylhydrazone) (MGBG, **11**, Figure 1.3) is a potent competitive inhibitor of the putrescine-activated mammalian AdoMet-DC, with a K_i value of less than 1 μM.[28] However, MGBG is of limited use as an inhibitor due to a wide variety of other effects on cells, including induction of severe mitochondrial damage, interference with polyamine transport and induction of SSAT. In an attempt to abrogate these

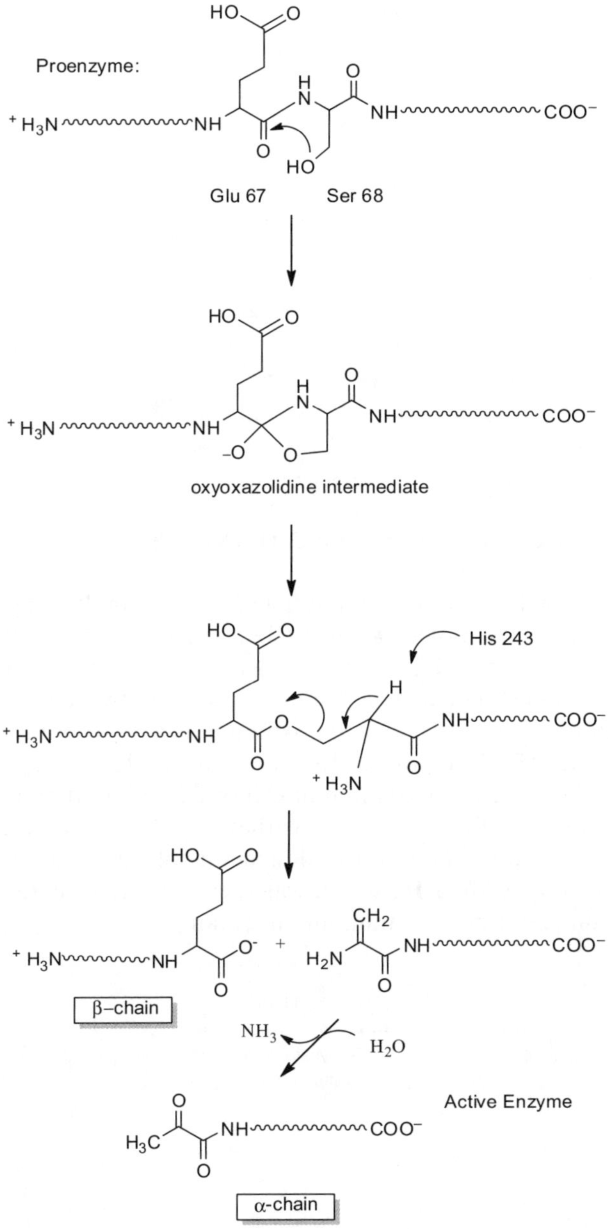

Figure 1.2 Origin of pyruvoyl residue of AdoMet-DC from *Escherichia coli.*

off-target effects, a series of restricted rotation MGBG analogs was synthesized and evaluated. The most promising of these agents were the Ciba-Geigy compounds CBG 48664 (**12**) and CGP 39937 (**13**). Both compounds proved to be nanomolar inhibitors of mammalian AdoMet-DC,[29,30] but CGP 48664 **13** was the more effective antitumor agent. This compound produced growth

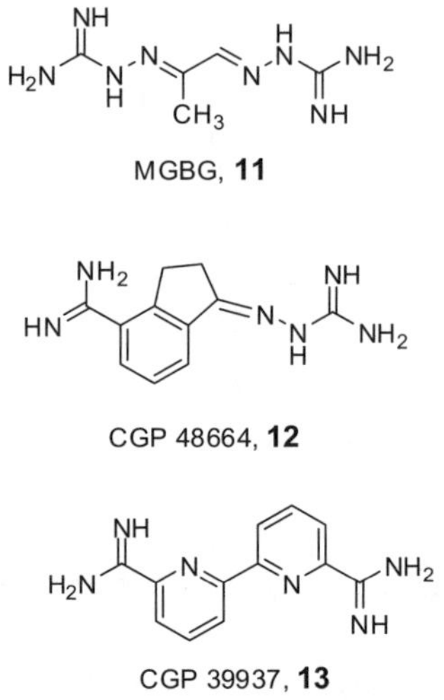

MGBG, **11**

CGP 48664, **12**

CGP 39937, **13**

Figure 1.3 Chemical structures of MGBG **11**, CGP 48664 **12** and CGP 39937 **13**.

inhibition in a panel of tumor cell lines including one multidrug-resistant line, and was 1000 times less potent against Chinese hamster ovary (CHO) cells *in vitro*.[31] The synthesis of **12** is shown in Scheme 1.4.[29] The starting material 1-oxo-2,3-dihydro-1*H*-indene-4-carbonitrile **14** was treated with hydrogen sulfide in pyridine, followed by triethyloxonium tetrafluoroborate to afford the carbimidothioate **15**. Compound **15** was converted to the corresponding amidine **16** in the presence of ammonium chloride, followed by the addition of *N*-aminoguanidine to form **12**. The synthesis of CGP 39937 is detailed in Scheme 1.5.[30] Compound **17** was stirred in methanol with a catalytic amount of sodium methoxide, yielding **18**, which was then converted to **13** via treatment with ammonium chloride and ammonia in ethanol.

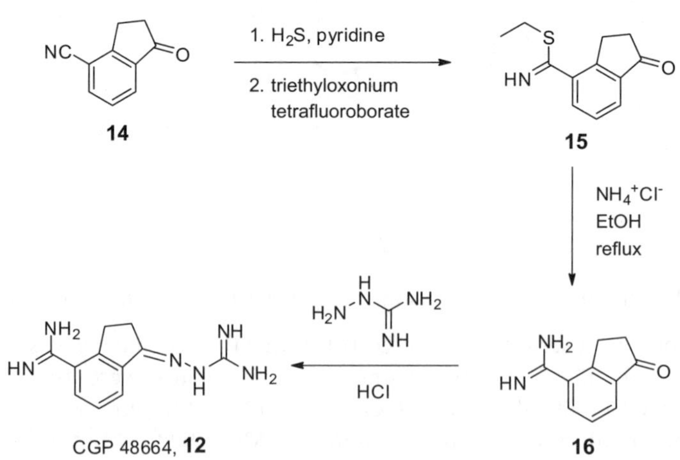

Scheme 1.4 Synthesis of the MGBG analogue CGP 48864.

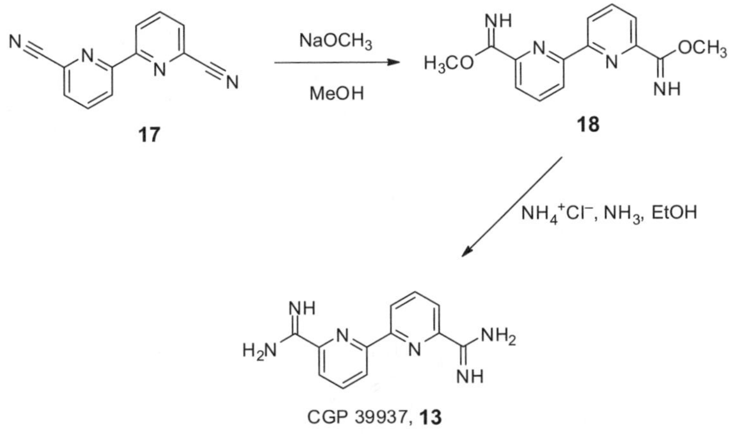

Scheme 1.5 Synthesis of the MGBG analogue CGP 39937.

There are a multitude of known AdoMet-DC inhibitors, the most effective of which are analogs of *S*-adenosylmethionine. Initially, analogs of AdoMet containing a sulfonium center were studied. The α-difluoromethyl analog of AdoMet (analogous to DFMO **1**) has been synthesized but has no activity against AdoMet-DC.[32] *S*-(5′-deoxy-5′-adenosyl)methylthioethylhydroxylamine (AMA) has been shown to act as an irreversible inhibitor of AdoMet-DC in L1210 cells with an IC_{50} of 100 μm.[33] Subsequently, a series of restricted rotation analogs of AdoMet were produced and evaluated as irreversible, enzyme-activated inhibitors of AdoMet-DC (**19–23**, Figure 1.4).[34–37] The first of these analogs, AdoMac **19**, was synthesized as shown in Scheme 1.6.[34] The mixed ester/alcohol **26** was converted to the corresponding phthalimide **27** under Mitsunobu conditions (phthalimide, DEAD, triphenylphosphine), followed by removal of the phthalimide protecting group (hydrazine) to give amine **28**. The free amine **28** was N-Boc protected to yield **29**, the methyl ester was cleaved (LiOH) to form **30**, and this alcohol was mesylated in the presence of lithium chloride to afford the requisite alkyl chloride **31**. Compound **31** was then coupled to thioacetyladenosine **32** (NaOCH₃), and the N-Boc and isopropylidene moieties in the protected adduct **33** were removed under acid conditions. Methylation in the presence of silver perchlorate then afforded AdoMac, **19**, as a mixture of two diastereomers. Subsequently, the synthesis of the four pure diastereomers of AdoMac was undertaken, as shown in Scheme 1.7.[36] The (meso)diacetate **35** was selectively de-esterified at the pro-(S) carbon to produce enantiomerically pure 1*S*,4*R*-**26**. Compound **26** was elaborated as described in Scheme 1.6 to form **33**. The isopropylidene and N-Boc protecting groups were removed, and the mixture of diastereomers was resolved by careful flash chromatography to afford the pure 1*S*,4*S*- and 1*S*,4*R*-diastereomers of **34**, followed by methylation to produce the corresponding 1*S*,4*S*- and 1*S*,4*R*-diastereomers of **19**. Compound **19** was shown to act as a mechanism-based, enzyme-activated inactivator of both the *E. coli* and human forms of

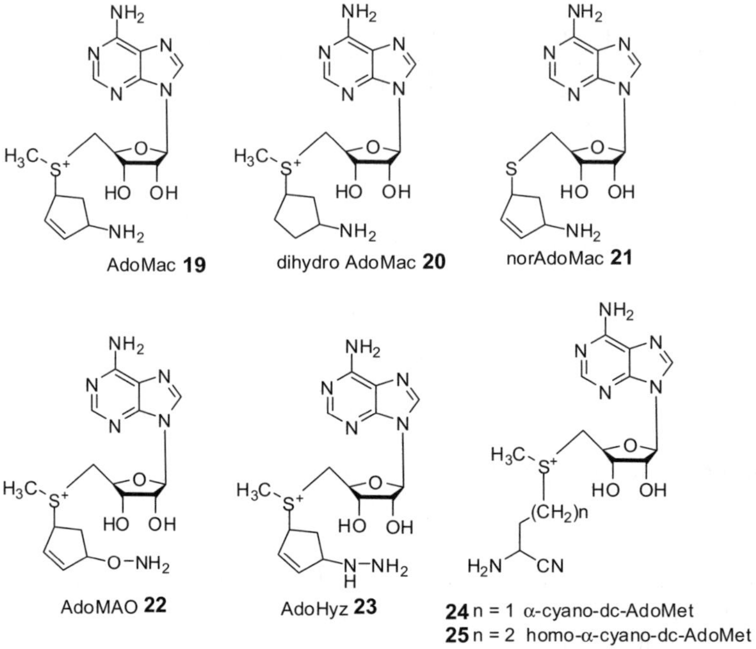

Figure 1.4 Chemical structures of AdoMac **19**, dihydroAdoMac **20**, norAdoMac **21**, AdoMAO **22**, AdoHz **23**, α-cyano-dc-AdoMet **24** and homo-α-cyano-dc-AdoMet **25**.

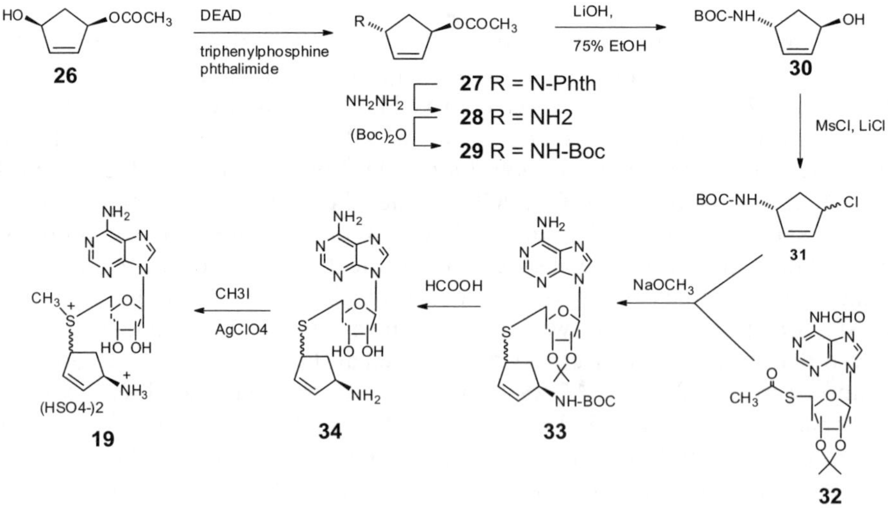

Scheme 1.6 Synthesis of AdoMac **19** as a mixture of diastereomers.

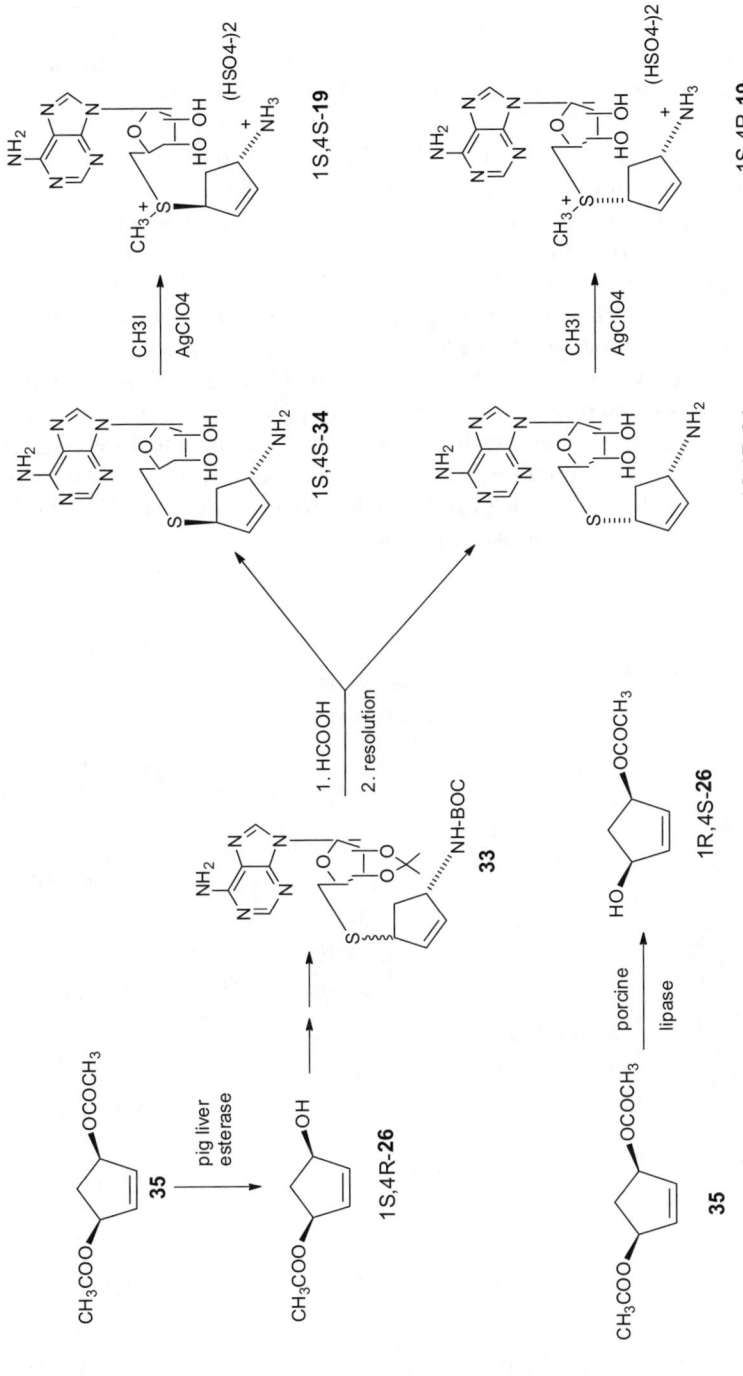

Scheme 1.7 Synthesis of the four pure diastereomeric forms of AdoMac **19**.

S-adenosylmethionine decarboxylase,[38] and optimal inhibitory activity was produced by different pure diastereomers in the case of each isozyme. The dihydro form of AdoMac, **20**, and norAdoMac **21** (Figure 1.4), were also synthesized, and each lacked the driving force for generation of the latent electrophile produced by **19**. Compounds **20** and **21** were both weak, reversible inhibitors of the enzyme. Similar synthetic routes were used to produce the homologs AdoMao **22** and AdoHyz **23** (Figure 1.4) both of which were mechanism-based inhibitors of the enzyme.[37] The α-cyano derivatives of dc-AdoMet **24** and **25** (Figure 1.4) were synthesized as shown in Scheme 1.8. The TBDMS-protected iodoalcohol **35** was coupled with the glycine equivalent **36** (LDA) under phase transfer conditions to afford **37**. The TBDMS group was then removed, and the resultant alcohol was mesylated to form **38**, which was coupled to 5′-thioacetyl-2′,3′-isopropylideneadenosine **32**. The coupled product was deprotected and methylated (CH₃I, AgClO₄) to afford **24** and **25**. Compounds **24** and **25** also acted as mechanism-based inhibitors of AdoMet-DC, with K_i values of 9 and 50 μM, respectively, against the *E. coli* form of the enzyme. Interestingly, the specificity of the two compounds was reversed when evaluated against the human form of AdoMet-DC, with **24** and **25** exhibiting K_i values of 246.6 and 7.2 μM, respectively, against human AdoMet-DC.

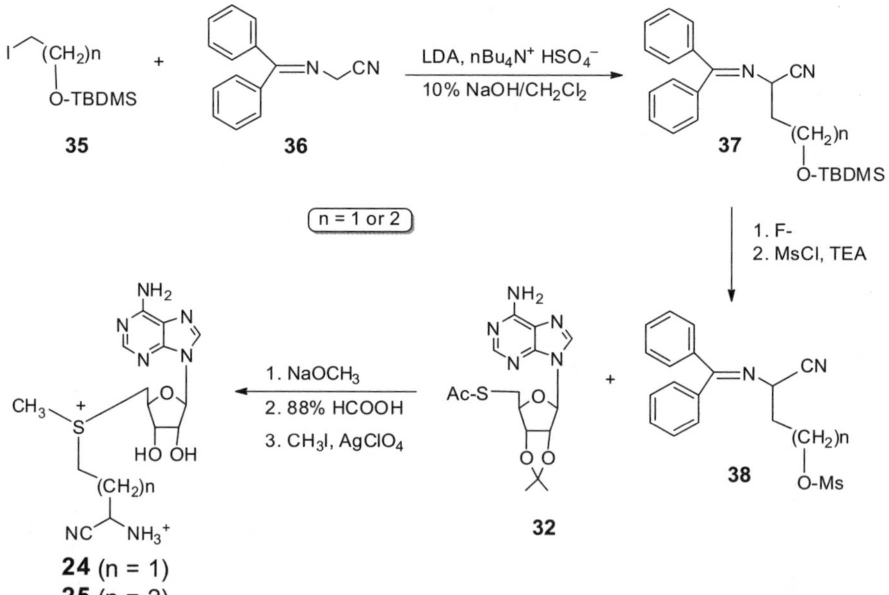

Scheme 1.8   Synthesis of the α-cyano analogues **24** and **25** of dc-AdoMet.

To circumvent stability problems associated with AdoMet analogs containing a 5′-sulfonium center, several groups have synthesized analogs featuring a

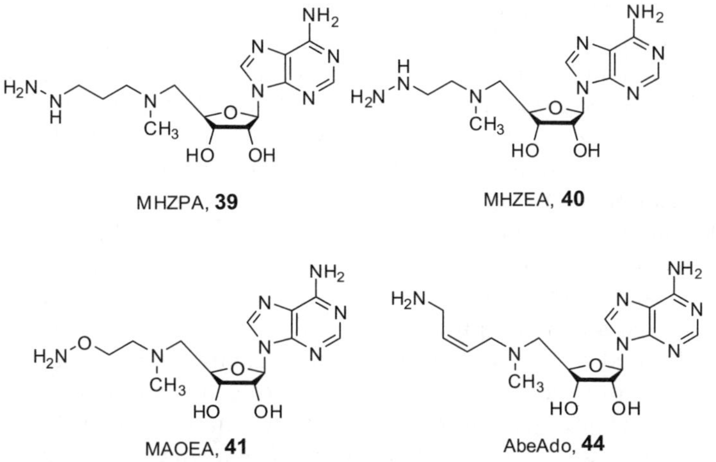

Figure 1.5 Chemical structures of AdoMet-DC inhibitors MHZPA **39**, MHZEA **40** and MAOEA **41**.

nitrogen isosteric replacement at this position. Secrist has described the synthesis of a series of nucleophilic AdoMet analogs (MHZPA **39**, MHZEA **40** and MAOEA **41**, Figure 1.5) as potential inhibitors of AdoMet-DC. These derivatives are potent, irreversible inhibitors of the enzyme, with IC_{50} values as low as 70 nM; however, these analogs appear to be rapidly metabolized in cells, a fact which may limit their usefulness *in vivo*. An example of the original synthesis of these analogs appears in Scheme 1.9. 2′,3′-isopropylidene-5′-tosyladenosine **42** is treated with 2-*N*-(methyl)aminoethanol to produce **43**. Removal of the protecting groups under acid conditions then afforded the desired compound **41**. In later work, it was found that these syntheses proceeded more efficiently in the absence of an isopropylidene protecting group. Originally synthesized in 1987 (prior to the availability of a crystal structure for AdoMet-DC), these compounds fell into disuse until recently, when they were used to aid in the design of C8-substituted homologs through co-crystallization studies with human

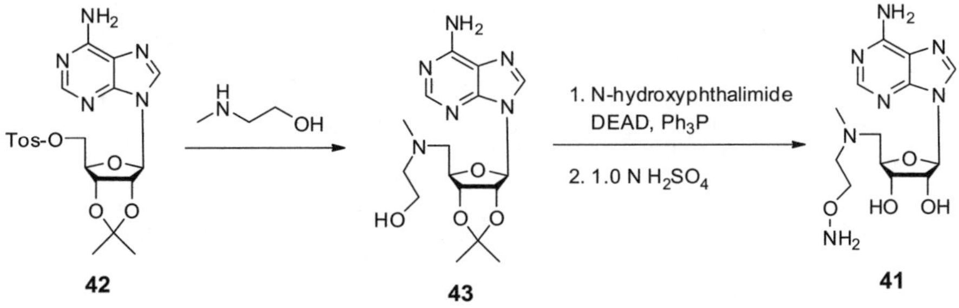

Scheme 1.9 Synthesis of the irreversible AdoMet-DC inhibitor MAOEA.

AdoMet-DC.[39] These novel C8 homologs were synthesized from the corresponding 8-bromo nucleoside in the presence of tetraalkyltin, an alkylamine or another suitable nucleophile. Elaboration of the precursor as described above then resulted in the formation of a series of 8-substituted AdoMet inhibitors. Finally, Casara *et al.* described the synthesis and enzymatic evaluation of 5′-{[(*Z*)-4-amino-2-butenyl]methylamino}-5′-deoxyadenosine (AbeAdo, **44**, Figure 1.5), a potent, irreversible inhibitor of AdoMet-DC.[40] AbeAdo produces a time-dependent loss of enzyme activity, with a K_i of 0.3 µM and a K_{inact} of 3.6 min^{-1}, making it the most promising AdoMet-DC inhibitor to date. The synthesis of **44** appears in Scheme 1.10. (*Z*)-2-butene-1,4 diol **45** was converted to the mono THP derivative **46** in the presence of dihydropyran. Treatment of **46** with phthalimide under Mitsunobu conditions, followed by cleavage of the phthalimide (hydrazine), removal of the THP-protecting group (HCl) and N-Boc protection of the nitrogen afforded the protected aminoalcohol **47**. Compound **47** was converted to the corresponding chloride **48**, which was appended to 5′-(methylamino)-5′-deoxy-2′,3′-*O*-isopropylideneadenosine **49**, and subsequent acid deprotection then yielded AbeAdo **44**. The mechanism of inactivation of mammalian *S*-adenosylmethionine decarboxylase by **44** is shown in Scheme 1.11. Initially, it was thought that **44** rearranged to a Michael acceptor after forming the requisite imino linkage with the terminal pyruvate of AdoMet-DC. However, it was subsequently shown that **44** inactivates AdoMet-DC by transaminating the terminal pyruvate residue to an alanine, as shown in Scheme 1.11.[41] Recently, interest in AdoMet-DC inhibitors featuring a nitrogen rather than a sulfonium center for the treatment of parasitic disease has been revived. A series of analogs related to **44** has recently been synthesized and found to be effective against *Trypanosoma brucei rhodesiense*, one of the causative organisms of human African trypanosomiasis.[42–44]

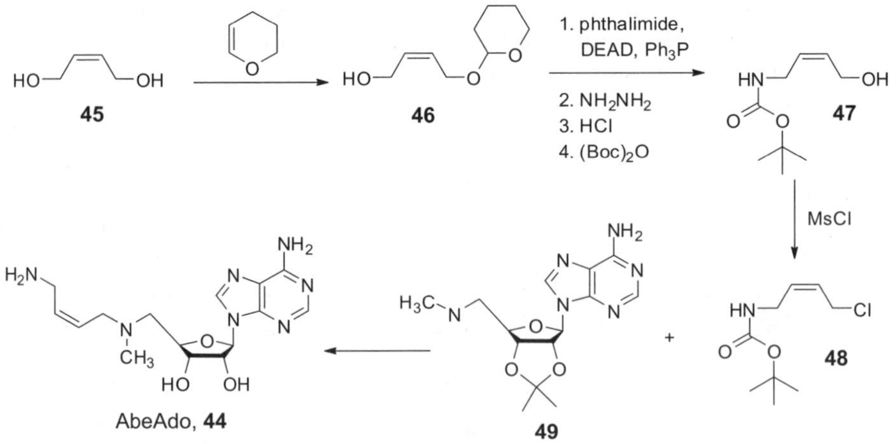

Scheme 1.10 Synthesis of the AdoMet-DC irreversible inactivator AbeAdo, **44**.

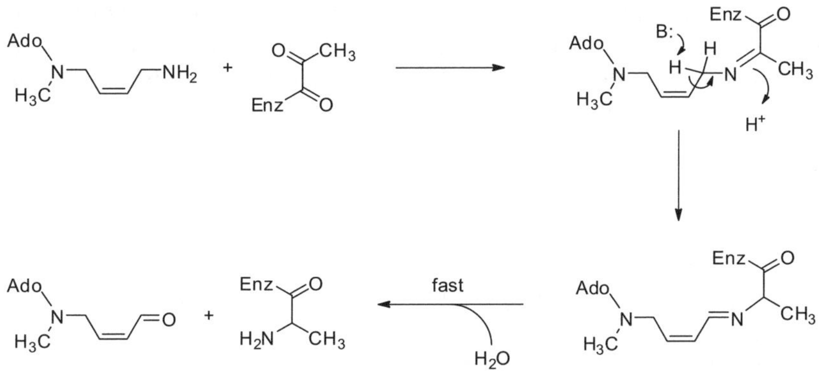

Scheme 1.11   Inactivation of *S*-adenosylmethionine decarboxylase by AbeAdo, **44**.

1.3.3   Spermidine Synthase and Spermine Synthase

Selective inhibition of the individual aminopropyltransferases has proven to be a significant problem, due to the similarity of the reactions catalyzed by the two enzymes. Mammalian spermidine synthase and spermine synthase each consist of two subunits (M_r 35 000 each or 44 000 each, respectively) and require no cofactors.[5] Both mammalian enzymes have been isolated and purified to homogeneity. A variety of MTA analogs have been shown to act as inhibitors for the aminopropyltransferases (5′-ethylthioadenosine, 5′-isobutylthioadenosine (SIBA), 5′-methylthiotubercidin, *etc.*) but these agents cannot distinguish between the two enzymes. An exception to this situation is dimethyl-(5′-adenosyl)-sulfonium perchlorate ($AdoS^+(CH_3)_2$), which appears to act selectively at spermine synthase with an IC$_{50}$ of 8 µM.[45] However, this compound also has inhibitory activity toward AdoMet-DC, limiting its usefulness as a specific inhibitor for biochemical studies. By far the most significant advance in the selective inhibition of the aminopropyltransferases is the development of *S*-adenosyl-1,8-diamino-3-thiooctane (AdoDATO **50**, Scheme 1.12)[45] and *S*-adenosyl-1,12-diamino-3-thio-9-azadodecane (AdoDATAD **51**, Scheme 1.12.),[46] which are multisubstrate analog inhibitors for spermidine synthase and spermine synthase, respectively. AdoDATO **50** inhibits spermidine synthase with an IC$_{50}$ of less than 50 nM while having no significant effect on spermine synthase. Conversely, AdoDATAD **51** inhibits spermine synthase selectively (IC$_{50}$ = 20 nM) while leaving spermidine synthase unaffected. However, **51** appears to be rapidly metabolized in L1210 cells by polyamine oxidase, resulting in the formation of **50** and other cytotoxic metabolites.[47] The synthesis of **50** and **51** is detailed in Scheme 1.12. The syntheses depended on the availability of the fully protected side-chain moieties **53** and **54**, which were produced in multistep syntheses as outlined in the references.[45,46] 5′-Thioacetyl-2′,3′-isopropylidene adenosine **32** was converted to the corresponding thiolate anion **52** in the presence of sodium methoxide, and the appropriate side chain synthon **55** or **56** was added to form the fully protected coupled product. Compound **55** was treated with formic acid to afford AdoDATO **50**. The azido groups in **56** were reduced, the *p*-nitro-Cbz protecting group was removed simultaneously by catalytic hydrogenolysis, and then the isopropylidine moiety was removed with formic acid to yield **51**.

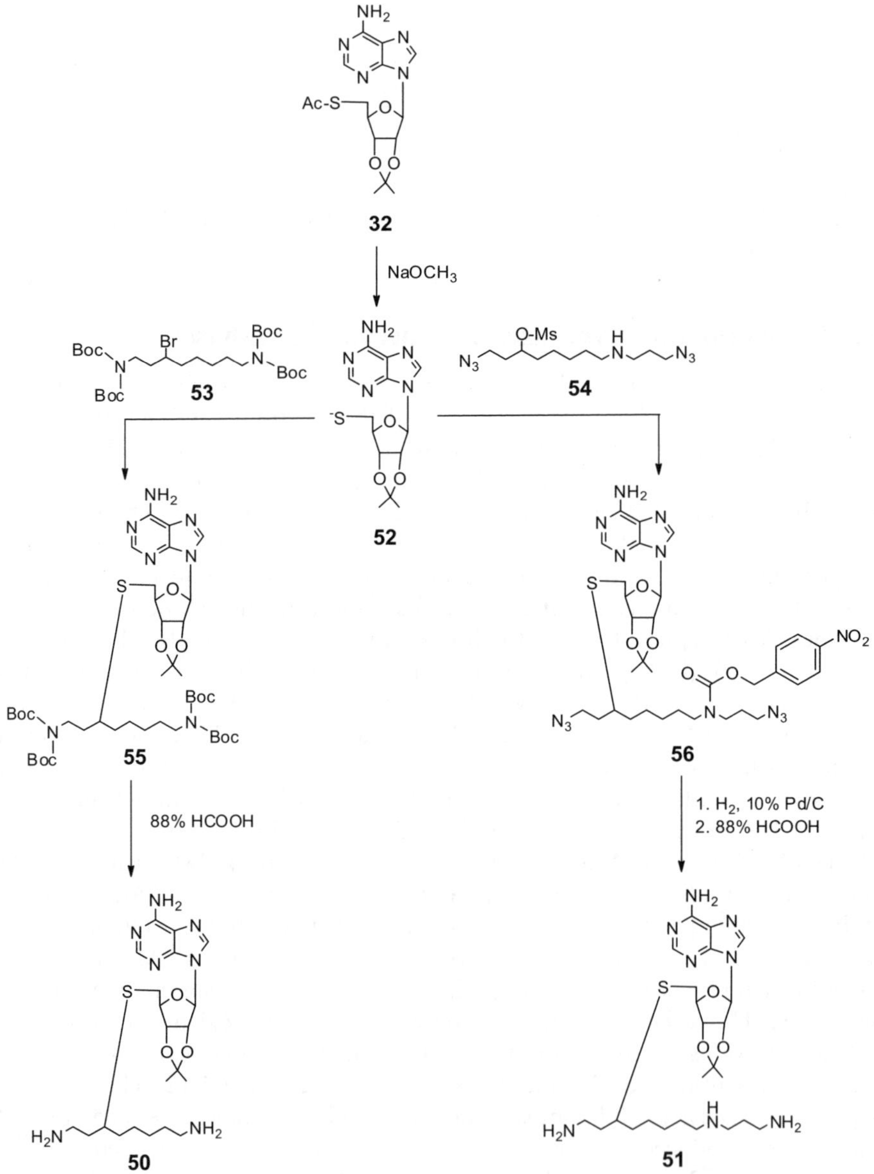

Scheme 1.12 Synthesis of aminopropyltransferase inhibitors AdoDATO **50** and AdoDATAD **51**.

1.3.4 Terminally Alkylated Polyamine Analogs

A more complete description of the biological effects of the terminally alkylated polyamines appears in Chapter 5 of this volume. The chemical rationale for the design of these agents stems from the finding that raising the pK_a of the terminal nitrogens by alkylation produces analogs which disrupt polyamine metabolism. These analogs readily enter cells using the polyamine transport system, and down regulate ODC and AdoMet-DC but do not substitute for the cellular functions of spermidine and spermine.[2,5,48,49] There are a multitude of strategies for the synthesis of these analogs, and the routes shown in this section are meant only to provide the reader with examples of possible approaches. The synthesis of symmetrically alkylated polyamine analogs is straightforward (Scheme 1.13) and depends only on the availability of the corresponding di-, tri- or tetraamine backbone.[50] Treatment of the backbone component, in this case spermine **57**, with the required number of equivalents of tosyl chloride in pyridine yielded the fully protected backbone structure, in this case **58**. Removal of the acidic terminal hydrogens (NaH), followed by addition of a suitable alkyl halide, then afforded the protected bis-alkylated product. Simultaneous removal of the tosyl protecting groups under relatively harsh conditions (sodium in liquid ammonia) then afforded the desired symmetrically substituted polyamine, such as bis(ethyl)spermine **59**. This approach was used by Bergeron to access a large number of symmetrically substituted polyamines,[50–54] a few of which were advanced to human clinical trials.

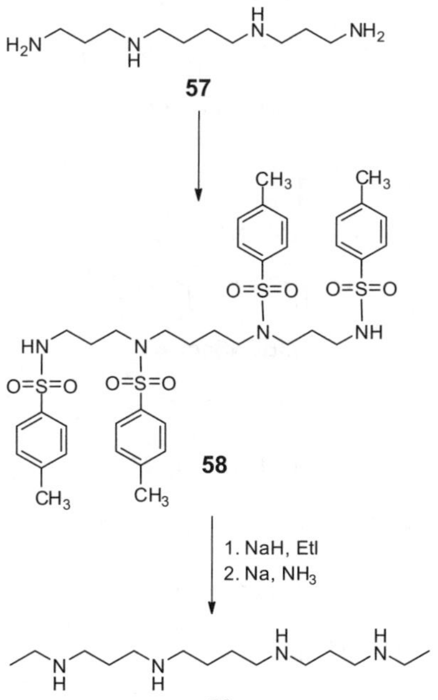

Scheme 1.13 Common synthetic route to symmetrically substituted alkylpolyamine analogues.

The first synthesis of an unsymmetrically substituted polyamine analog was published in 1992,[55] and the synthetic route was designed to eliminate the possibility of side products at the expense of acceptable chemical yields. The first examples of unsymmetrically substituted alkylpolyamines were PENSpm **60** and CPENSpm **61** (Figure 1.6). The biological characterization of **60** and **61** is outlined in Chapter 5. In order to devise a more facile route to unsymmetrically substituted alkylpolyamines such as **62–66** (Figure 1.6), the route shown in Scheme 1.14 was devised.[56] An appropriate diamine **67** is cyanoethylated (acrylonitrile, ethanol, reflux) to form **68**, and the central nitrogens are protected ((Boc)$_2$O) to yield **69**. The nitrile moieties are reduced (Raney Ni, H$_2$, 50 psi) to afford **70**, and the resulting primary amines are protected as the corresponding mesitylates **71**. Addition of a mesityl protecting group[57] to the terminal nitrogen of the alkylpolyamine precursor creates an acidic proton which can be removed using NaH, generating a strong nucleophile. In the subsequent alkylation reaction, the ratio of mono- to disubstitution can be controlled by adjusting the stoichiometry of the electrophile; thus, 1.2 equivalents of alkyl halide results in the formation of mostly the monosubstituted product **72**. The use of 2.2 equivalents of alkyl halide produces mainly the symmetrically disubstituted derivative. Deprotection of **72** (30%

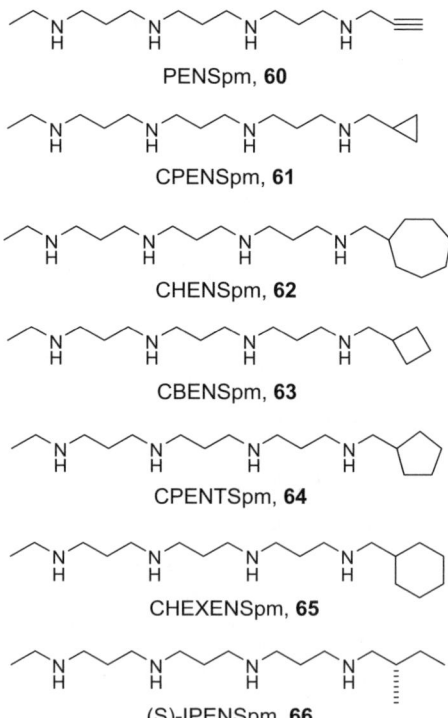

Figure 1.6 Chemical structures of selected unsymmetrically substituted alkylpolyamine analogues **60–66**.

HBr in acetic acid) affords the monosubstituted alkylpolyamine **73**, or **72** can be re-alkylated to form the unsymmetrically substituted precursor **74**, which on deprotection (30% HBr in acetic acid) gives the desired product **75**.

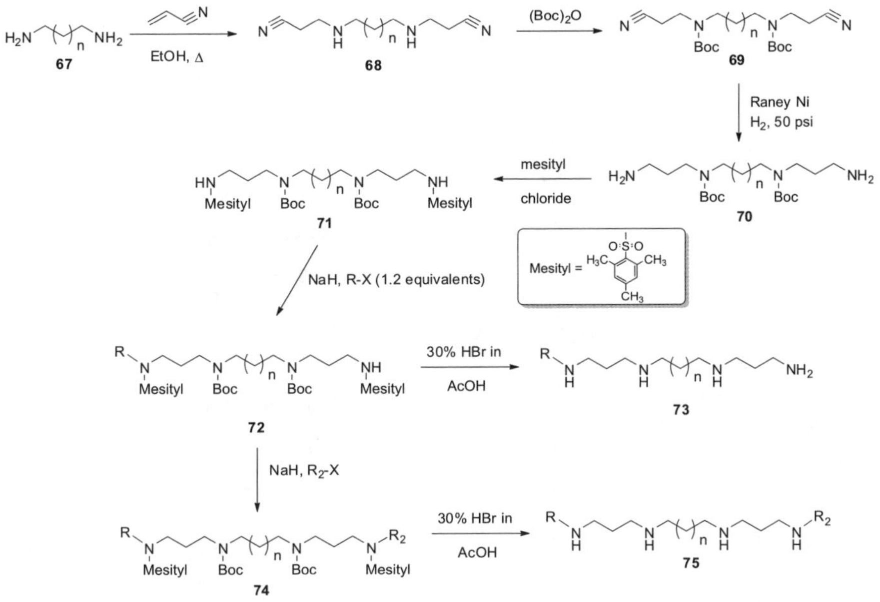

Scheme 1.14 Synthesis of unsymmetrically substituted alkylpolyamine analogues.

In addition to the syntheses described above, there are a number of methods for the stepwise elongation of polyamine analogs. Some common examples are shown in Scheme 1.15. As mentioned above, removal of the acidic hydrogen of a mesityl protected primary amine such as **76**, followed by the addition of a suitable electrophile such as **77**, results in elongation of the chain to form **78**. The use of a bromoalkyl phthalimide like **77** in this reaction affords a product with orthogonal protecting groups that facilitate selective functionalization of the nitrogens in the chain. Using reductive amination, a primary amine precursor such as **79** can be reacted with a variety of commercially available aldehydes **80** to form chain-elongated products such as **81**. Similarly, primary amine **79** can be coupled to a suitable carboxylic acid **82**, followed by reduction of the amide bond to the corresponding methylene and secondary amine **81** in the presence of diborane.

1.3.5 Polyamine-Based Epigenetic Modulators

As described in Section 1.3.4, alteration in the pK_a of the terminal nitrogens of the natural polyamines has been shown to produce modulators of polyamine metabolism. In addition to installing alkyl substituents to form secondary amines, other approaches have been used to access analogs of the linear

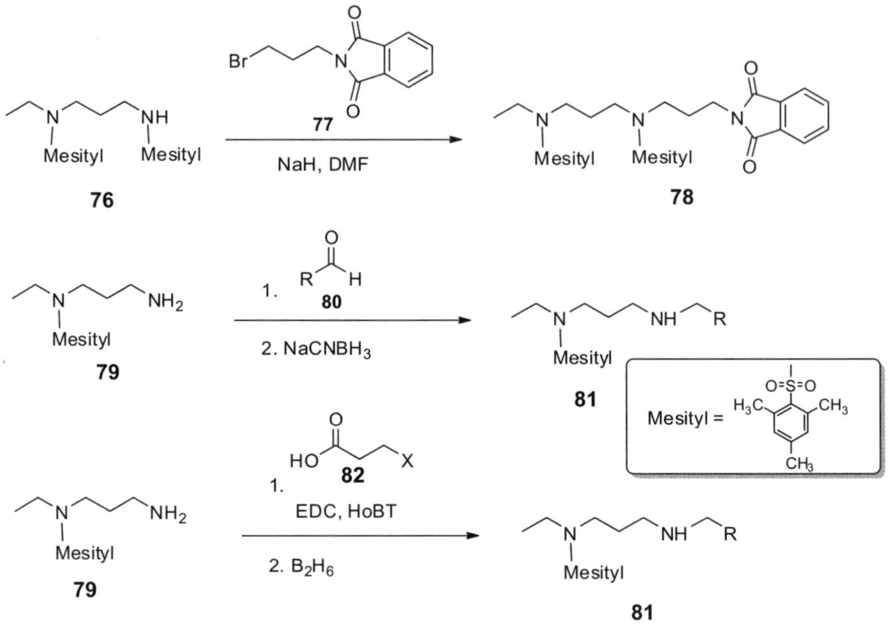

Scheme 1.15   Example methods for the synthetic elongation of alkylpolyamine chains.

polyamines with altered pK_a values at the terminal nitrogen moieties. Poly-aminoguanidines and polyaminobiguanides have been synthesized that have potent antitrypanosomal activity,[58] and more recently these analogs have been shown to inhibit the histone demethylase lysine-specific demethylase 1 (LSD1; see Chapter 5).[59–61] These derivatives are highly effective at promoting the re-expression of aberrantly silenced genes in tumor cells and are also effective at limiting tumor growth *in vivo*. The synthesis of these analogs is outlined in Scheme 1.16. The versatile synthetic derivative **82** (produced as shown in Scheme 1.14 but substituting a mesityl protecting group for N-Boc) is reacted with the N-cyano alkylamine **83** to afford the protected intermediate **84**. Deprotection (HBr) then affords the desired polyaminoguanidines **85**. Likewise, addition of **82** to the N-cyanoalkylguanidine **87** yields the protected intermediate **87**, which is depro-tected (HBr) to form the target polyaminobiguanides **88**. Isosteric analogs of **85** and **88** have recently been synthesized that retain the antiparasitic and epigenetic activities of the parent molecules,[62] and their synthesis is shown in Scheme 1.17. The (bis) primary amine **70** (Scheme 1.14) is allowed to react with the appro-priate isothiocyanate **89** or isocyanate **90** to afford the protected (bis)thioureas **91** or (bis)ureas **92**, respectively. Removal of the N-Boc protecting groups then yields the desired target analogs **93** and **94**.

In addition to epigenetic modulation of gene expression through LSD1, polyamine-containing analogs have been reported that are effective inhibitors of the histone deacetylases (HDACs). A series of polyaminohydroxamic acids (PAHAs) was synthesized as shown in Scheme 1.18.[63,64] The protected,

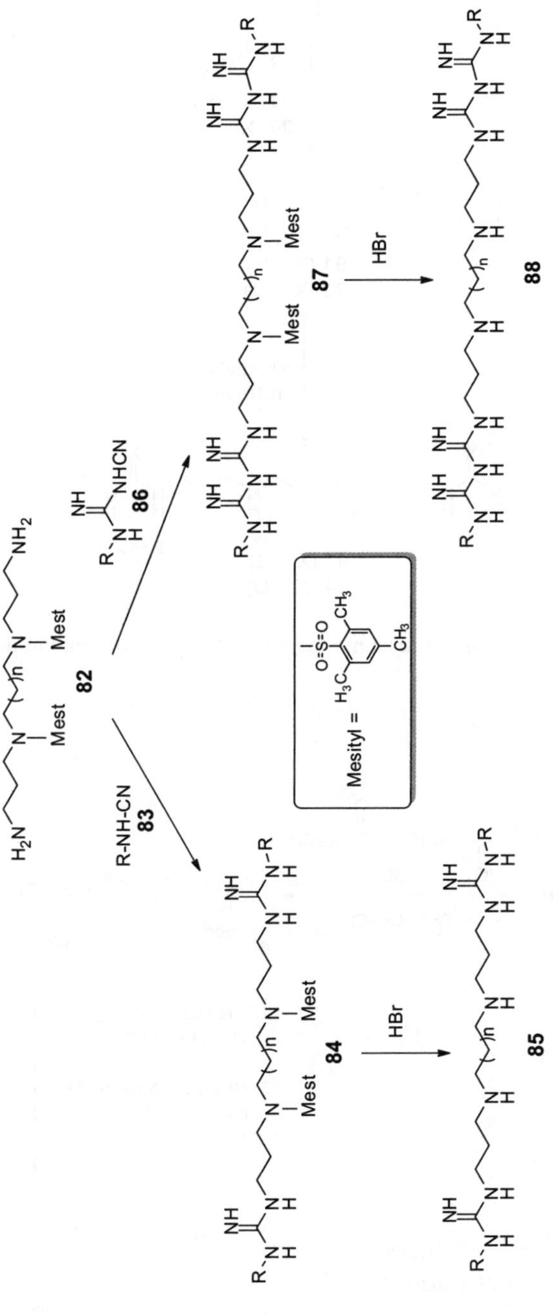

Scheme 1.16 Synthesis of polyaminoguanidines and polyaminobiguanides.

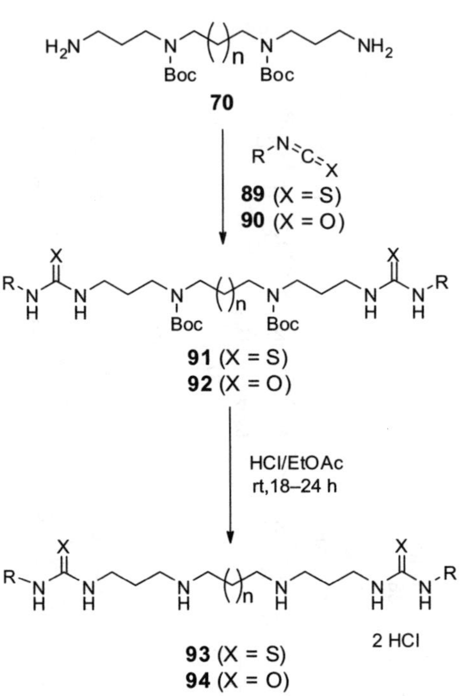

Scheme 1.17 Synthesis of isosteric analogues of the polyaminoguanidines.

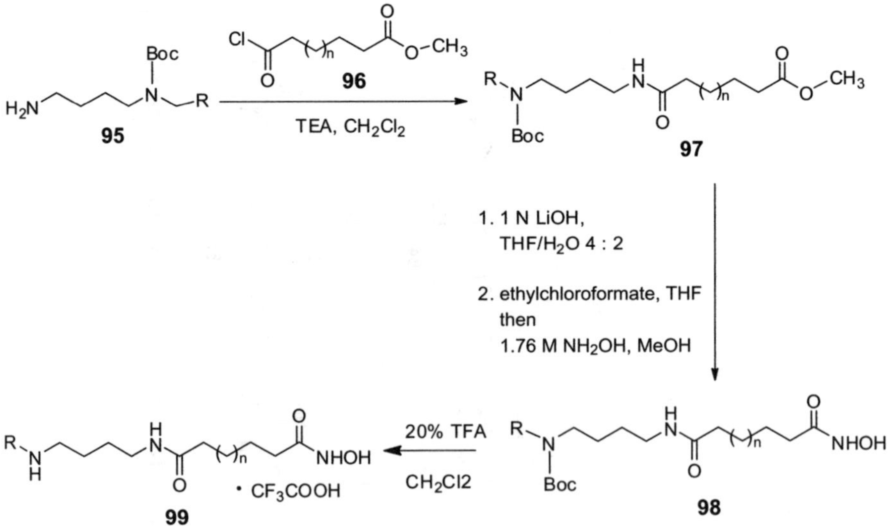

Scheme 1.18 Synthesis of polyaminohydroxamic acid HDAC inhibitors.

monoalkylated putrescine derivative **95** is coupled to the mixed ester/acid chloride **96** to form intermediate **97**. The ester functionality in **97** is then hydrolyzed (LiOH), and the resulting acid is converted to the corresponding hydroxamic acid using ethylchloroformate and hydroxylamine. In a similar fashion, a series of polyaminobenzamide (PABA) analogs has been synthesized (Scheme 1.19) that comprise potent inhibitors of HDAC, and in some cases show selectivity towards individual HDAC isoforms.[64] 2-Aminoaniline **100** is first mono-N-Boc protected to form **101**. The commercially available methyl 4-(aminomethyl)benzoate **102** is

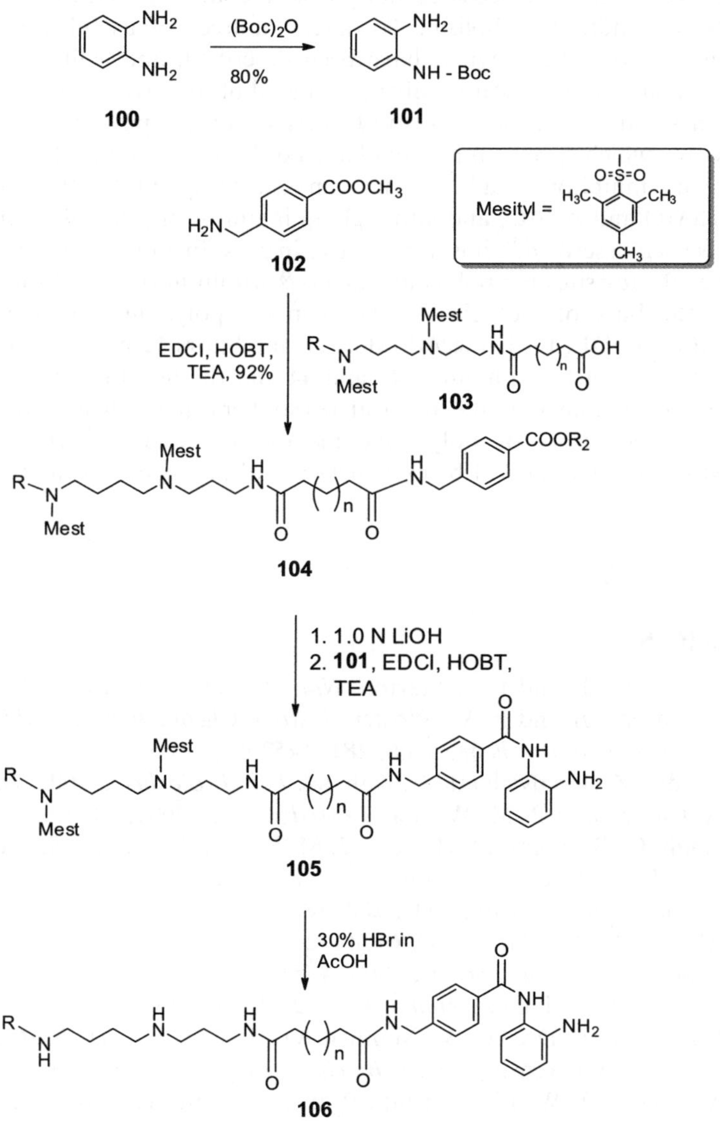

Scheme 1.19   Synthesis of polyaminobenzamide HDAC inhibitors.

then coupled to the backbone intermediate **103**[64] (EDCl, HoBT, TEA) to yield **104**, followed by hydrolysis of the ester (LiOH) and coupling to the previously synthesized intermediate **101** to produce **105**. Removal of the mesityl protection groups using 30% HBr in acetic acid then affords the desired PABA analogs **106**.

1.4   Conclusion

It is clear that the reaction sequences described in this chapter can be used to access a diverse set of nucleosides and polyamine analogs that can be used to modulate polyamine metabolism. To date, hundreds of useful inhibitors of polyamine biosynthetic enzymes have been described, and a like number of polyamine analogs that disrupt polyamine metabolism have been produced.[2,5] Despite these advances, there are still targets within the polyamine metabolic pathways for which specific inhibitors have not been discovered. For example, there are no inhibitors available that can selectively inhibit the polyamine oxidases SMO and APAO, and although an inhibitor of purified recombinant SSAT has been described,[65] it is not effective in cells. In addition, little is known about the effector sites for polyamine analogs within normal and tumor cells. To date, the bulk of biological research in the polyamine arena has been conducted using DFMO **1**, the AdoMet-DC inhibitors **12** and **13**, the nucleoside-based AdoMet-DC inhibitors **41** and **44** and the bis(ethyl)spermine derivatives **59**. Assuming that biochemical researchers in the field expand their arsenal of modulators of polyamine metabolism, analogs synthesized as described in this chapter will be extremely useful in advancing the polyamine field.

References

1. R. A. Casero, Jr and L. J. Marton, *Nat. Rev. Drug Discov.*, 2007, **6**, 373.
2. R. A. Casero, Jr and P. M. Woster, *J. Med. Chem.*, 2009, **52**, 4551.
3. A. E. Pegg, *J. Biol. Chem.*, 2006, **281**, 14529.
4. S. Gandre, Z. Bercovich and C. Kahana, *Eur. J. Biochem.*, 2002, **269**, 1316.
5. R. A. Casero and P. M. Woster, *J. Med. Chem.*, 2001, **44**, 1.
6. V. Singh, G. B. Evans, D. H. Lenz, J. M. Mason, K. Clinch, S. Mee, G. F. Painter, P. C. Tyler, R. H. Furneaux, J. E. Lee, P. L. Howell and V. L. Schramm, *J. Biol. Chem.*, 2005, **280**, 18265.
7. E. Albers, *IUBMB Life*, 2009, **61**, 1132.
8. N. Seiler, *Curr. Drug Targets*, 2003, **4**, 537.
9. S. Bale and S. E. Ealick, *Amino Acids*, 2010, **38**, 451.
10. P. Marchant, S. Dredar, V. Manneh, O. Alshabanah, H. Matthews, D. Fries and J. Blankenship, *Arch. Biochem. Biophys.*, 1989, **273**, 128.
11. S. A. Dredar, J. W. Blankenship, P. E. Marchant, V. Manneh and D. S. Fries, *J. Med. Chem.*, 1989, **32**, 984.
12. R. A. Casero, Jr and A. E. Pegg, *Faseb J.*, 1993, **7**, 653.

13. Y. Wang, A. Hacker, T. Murray-Stewart, B. Frydman, A. Valasinas, A. V. Fraser, P. M. Woster and R. A. Casero, Jr, *Cancer Chemother. Pharmacol.*, 2005, **56**, 83.
14. T. Uemura, H. F. Yerushalmi, G. Tsaprailis, D. E. Stringer, K. E. Pastorian, L. Hawel, 3rd, C. V. Byus and E. W. Gerner, *J. Biol. Chem.*, 2008, **283**, 26428.
15. Y. Wang, W. Devereux, P. M. Woster, T. M. Stewart, A. Hacker and R. A. Casero, Jr, *Cancer Res.*, 2001, **61**, 5370.
16. Y. Wang, A. Hacker, T. Murray-Stewart, J. G. Fleischer, P. M. Woster and R. A. Casero, Jr, *Biochem. J.*, 2005, **386**, 543.
17. K. Igarashi and K. Kashiwagi, *Plant Physiol. Biochem*, 2010, **48**, 506.
18. B. W. Metcalf, P. Bey, C. Danzin, M. J. Jung, P. Casara and J. P. Vevert, *J. Am. Chem. Soc.*, 1978, **100**, 2551.
19. R. Poulin, L. Lu, B. Ackermann, P. Bey and A. E. Pegg, *J. Biol. Chem.*, 1992, **267**, 150.
20. C. J. Bacchi, J. Garofalo, M. Ciminelli, D. Rattendi, B. Goldberg, P. P. McCann and N. Yarlett, *Biochem. Pharmacol.*, 1993, **46**, 471.
21. C. J. Bacchi, H. C. Nathan, T. Livingston, G. Valladares, M. Saric, P. D. Sayer, A. R. Njogu and A. B. Clarkson, Jr, *Antimicrob. Agents Chemother.*, 1990, **34**, 1183.
22. A. J. Bitonti, T. L. Byers, T. L. Bush, P. J. Casara, C. J. Bacchi, A. B. Clarkson, Jr, P. P. McCann and A. Sjoerdsma, *Antimicrob. Agents Chemother.*, 1990, **34**, 1485.
23. P. P. McCann, C. J. Bacchi, A. B. Clarkson, Jr, P. Bey, A. Sjoerdsma, P. J. Schecter, P. D. Walzer and J. L. Barlow, *Am. J. Trop. Med. Hyg.*, 1986, **35**, 1153.
24. J. Pepin, N. Khonde, F. Maiso, F. Doua, S. Jaffar, S. Ngampo, B. Mpia, D. Mbulamberi and F. Kuzoe, *Bull. World Health Organ.*, 2000, **78**, 1284.
25. M. A. Cornbleet, A. Kingsnorth, G. P. Tell, K. D. Haegele, A. M. Joder-Ohlenbusch and J. F. Smyth, *Cancer Chemother. Pharmacol.*, 1989, **23**, 348.
26. D. A. Kendrick, C. Danzin and M. Kolb, *J. Med. Chem.*, 1989, **32**, 170.
27. F. Wu, D. Grossenbacher and H. Gehring, *Mol. Cancer Ther.*, 2007, **6**, 1831.
28. A. E. Pegg and P. P. McCann, *Pharmacol. Ther.*, 1992, **56**, 359.
29. J. Stanek, G. Caravatti, H. G. Capraro, P. Furet, H. Mett, P. Schneider and U. Regenass, *J. Med. Chem.*, 1993, **36**, 46.
30. J. Stanek, G. Caravatti, J. Frei, P. Furet, H. Mett, P. Schneider and U. Regenass, *J. Med. Chem.*, 1993, **36**, 2168.
31. U. Regenass, H. Mett, J. Stanek, M. Mueller, D. Kramer and C. W. Porter, *Cancer Res.*, 1994, **54**, 3210.
32. P. Bey, C. Danzin, V. Van Dorsselaer, P. Mamont, M. Jung and C. Tardif, *J. Med. Chem.*, 1978, **21**, 50.
33. D. L. Kramer, R. M. Khomutov, Y. V. Bukin, A. R. Khomutov and C. W. Porter, *Biochem. J.*, 1989, **259**, 325.
34. Y. Wu and P. M. Woster, *J. Med. Chem.*, 1992, **35**, 3196.

35. Y. Q. Wu, T. Lawrence, J. Q. Guo and P. M. Woster, *Bioorg. Med. Chem. Lett.*, 1993, **3**, 2811.
36. Y. Q. Wu and P. M. Woster, *Bioorg. Med. Chem.*, 1993, **1**, 349.
37. J. Guo, Y. Q. Wu, D. Rattendi, C. J. Bacchi and P. M. Woster, *J. Med. Chem.*, 1995, **38**, 1770.
38. Y. Q. Wu and P. M. Woster, *Biochem. Pharmacol.*, 1995, **49**, 1125.
39. D. E. McCloskey, S. Bale, J. A. Secrist, 3rd, A. Tiwari, T. H. Moss, 3rd, J. Valiyaveettil, W. H. Brooks, W. C. Guida, A. E. Pegg and S. E. Ealick, *J. Med. Chem.*, 2009, **52**, 1388.
40. P. Casara, J. Marchal, J. Wagner and C. Danzin, *J. Am. Chem. Soc.*, 1989, **111**, 9111.
41. L. M. Shantz, B. A. Stanley, J. A. Secrist, III and A. E. Pegg, *Biochemistry*, 1992, **31**, 6848.
42. C. J. Bacchi, R. H. Barker, Jr, A. Rodriguez, B. Hirth, D. Rattendi, N. Yarlett, C. L. Hendrick and E. Sybertz, *Antimicrob. Agents Chemother.*, 2009, **53**, 3269.
43. B. Hirth, R. H. Barker, Jr, C. A. Celatka, J. D. Klinger, H. Liu, B. Nare, A. Nijjar, M. A. Phillips, E. Sybertz, E. K. Willert and Y. Xiang, *Bioorg. Med. Chem. Lett.*, 2009, **19**, 2916.
44. R. H. Barker, Jr, H. Liu, B. Hirth, C. A. Celatka, R. Fitzpatrick, Y. Xiang, E. K. Willert, M. A. Phillips, M. Kaiser, C. J. Bacchi, A. Rodriguez, N. Yarlett, J. D. Klinger and E. Sybertz, *Antimicrob. Agents Chemother.*, 2009, **53**, 2052.
45. K. C. Tang, R. Mariuza and J. K. Coward, *J. Med. Chem.*, 1981, **24**, 1277.
46. P. M. Woster, A. Y. Black, K. J. Duff, J. K. Coward and A. E. Pegg, *J. Med. Chem.*, 1989, **32**, 1300.
47. A. E. Pegg, R. Wechter, R. Poulin, P. M. Woster and J. K. Coward, *Biochemistry*, 1989, **28**, 8446.
48. C. W. Porter, B. Ganis, P. R. Libby and R. J. Bergeron, *Cancer Res.*, 1991, **51**, 3715.
49. C. W. Porter, J. McManis, R. A. Casero and R. J. Bergeron, *Cancer Res.*, 1987, **47**, 2821.
50. R. J. Bergeron, A. H. Neims, J. S. McManis, T. R. Hawthorne, J. R. Vinson, R. Bortell and M. J. Ingeno, *J. Med. Chem.*, 1988, **31**, 1183.
51. R. J. Bergeron, Y. Feng, W. R. Weimar, J. S. McManis, H. Dimova, C. Porter, B. Raisler and O. Phanstiel, *J. Med. Chem.*, 1997, **40**, 1475.
52. R. J. Bergeron, T. R. Hawthorne, J. R. Vinson, D. E. Beck, Jr and M. J. Ingeno, *Cancer Res.*, 1989, **49**, 2959.
53. R. J. Bergeron, J. S. McManis, C. Z. Liu, Y. Feng, W. R. Weimar, G. R. Luchetta, Q. Wu, J. Ortiz-Ocasio, J. R. Vinson, D. Kramer and C. Porter, *J. Med. Chem.*, 1994, **37**, 3464.
54. R. J. Bergeron, J. Wiegand, J. S. McManis, W. R. Weimar, R. E. Smith, S. E. Algee, T. L. Fannin, M. A. Slusher and P. S. Snyder, *J. Med. Chem.*, 2001, **44**, 232.
55. N. H. Saab, E. E. West, N. C. Bieszk, C. V. Preuss, A. R. Mank, R. A. Casero, Jr and P. M. Woster, *J. Med. Chem.*, 1993, **36**, 2998.

56. F. H. Bellevue III, M. Boahbedason, R. Wu, P. M. Woster, J. R. A. Casero, D. Rattendi, S. Lane and C. J. Bacchi, *Bioorg. Med. Chem. Lett.*, 1996, **6**, 2765.
57. H. Yajima, M. Takeyama, J. Kanaki, O. Nishimura and M. Fujino, *Chem. Pharm. Bull.*, 1978, **26**, 3752.
58. X. Bi, C. Lopez, C. J. Bacchi, D. Rattendi and P. M. Woster, *Bioorg. Med. Chem. Lett*, 2006, **16**, 3229.
59. Y. Huang, E. Greene, T. Murray Stewart, A. C. Goodwin, S. B. Baylin, P. M. Woster and R. A. Casero, Jr, *Proc. Natl. Acad. Sci. U. S. A.*, 2007, **104**, 8023.
60. Y. Huang, L. J. Marton, P. M. Woster and R. A. Casero, *Essays Biochem.*, 2009, **46**, 95.
61. Y. Huang, T. Murray-Stewart, Y. Wu, S. B. Baylin, L. J. Marton, P. M. Woster and R. A. Casero, Jr, *Cancer Chemo. Pharmacol.*, 2009, **15**, 7217.
62. S. K. Sharma, Y. Wu, N. Steinbergs, M. L. Crowley, A. S. Hanson, R. A. Casero and P. M. Woster, *J. Med. Chem.*, 2010, **53**, 5197.
63. S. Varghese, D. Gupta, T. Baran, A. Jiemjit, S. D. Gore, R. A. Casero, Jr and P. M. Woster, *J. Med. Chem.*, 2005, **48**, 6350.
64. S. Varghese, T. Senanayake, T. Murray-Stewart, K. Doering, A. Fraser, R. A. Casero and P. M. Woster, *J. Med. Chem.*, 2008, **51**, 2447.
65. R. Wu, N. H. Saab, H. Huang, L. Wiest, A. E. Pegg, R. A. Casero, Jr and P. M. Woster, *Bioorg. Med. Chem.*, 1996, **4**, 825.

CHAPTER 2

Structural Biology in Polyamine Drug Discovery

SHRIDHAR BALE AND STEVEN E. EALICK*

Department of Chemistry and Chemical Biology, 120 Baker Lab, Cornell University, Ithaca, NY 14853-1301, USA

2.1 Structural Biology and Drug Design

Structural biology is a branch of science that examines biological processes using the three-dimensional coordinates of atoms involved in the process. Methods such as X-ray crystallography, nuclear magnetic resonance spectroscopy (NMR), electron microscopy, laser spectroscopy and circular dichrosim provide vital structural details of the molecule of interest to varying levels of accuracy. The structural information of biologically relevant molecules aids in understanding phenomena such as catalysis, binding, signaling and transport. X-ray crystallography has been applied to a large range of molecules varying from inorganic salts and small organic molecules to macromolecules, such as proteins and nucleic acids, and large macromolecular assemblies, such as virus particles and ribosomes. X-ray crystallography is the most widely used method for obtaining the atomic positions of macromolecules.

X-ray crystallography requires the availability of crystals of the molecule of interest to be arranged in an ordered three-dimensional lattice. Crystals of molecules are usually obtained by incubating a homogenous sample of the macromolecule with suitable precipitating agents. The crystals are exposed to a focused X-ray beam to obtain a set of diffraction images created by rotating the crystal. The electron density of the macromolecule is a Fourier transform

RSC Drug Discovery Series No. 17
Polyamine Drug Discovery
Edited by Patrick M. Woster and Robert A. Casero, Jr.
© Royal Society of Chemistry 2012
Published by the Royal Society of Chemistry, www.rsc.org

in which the amplitudes are obtained from the intensities of spots in the diffraction pattern and a phase angle, which is not directly measured in the experiment. This is known as the phase problem in X-ray crystallography. Instead, the phase angle is determined using a variety of techniques including multiple isomorphous replacement (MIR), multi-wavelength anomalous diffraction (MAD), single-wavelength anomalous diffraction (SAD), molecular replacement and direct methods.[1–3] The resulting electron density aids in building an initial model of the crystallized macromolecule, which is further refined to obtain the best fit to the diffraction data.

The determination of the three-dimensional structure of macromolecules of therapeutic interest has always been a major area for research in industry as well as academia. In most of the cases pertaining to inhibitor design, the structure would reveal the ligand binding site, which could be used for rational design of molecules with better potency.[4,5] Virtual screening programs screen libraries of millions of compounds against the active site to obtain a lead compound for further development.[6–8] The virtual screening process is an effective way to obtain novel lead compounds and reduce drug-discovery costs. The development of HIV protease inhibitors Nelfinavir and Amprenavir and the influenza drug Relenza has been aided by structural studies.[5] The drugs Captopril, Dorzolamide and Zanamivir are a few examples that have been rationally developed from a structural viewpoint.[9,10]

Structure-based drug design requires that targets have an experimentally determined structure to a high accuracy. However, there are many targets that do not have an experimentally determined structure. In such cases, homology modeling provides reasonably accurate structural information on macromolecules that are difficult to crystallize.[11,12] In many cases, the structural information from homology modeling has been vital in making key changes to the targets for successful crystallization. Currently, more than 40 drugs designed through structural means have entered clinical trials. In addition, the number of algorithms used for virtual screening for the discovery of lead compounds and associated compound libraries are growing exponentially. As a result, the use of structure in drug discovery has recently been gaining in importance.[13,14]

2.2 Structural Biology of Polyamine-Related Enzymes

The polyamines are ubiquitous aliphatic cations implicated in cellular growth and differentiation, and elevated levels of polyamines in rapidly proliferating cells are of therapeutic interest in various forms of cancer and in the development of antiparasitic agents.[15–18] Polyamine biosynthesis across various species (including micro-organisms) involves six different enzymes, and the associated catabolic pathway involves three different enzymes. The crystal structures of all the biosynthetic enzymes and two of the catabolic enzymes (belonging to various species) have been determined (see Table 2.1 and Figure 2.1 for further details).[19–26] In addition, crystal structures of many enzymes with key inhibitors have also been determined. The structural

Table 2.1 Representative enzymes with determined structures in the polyamine pathway.

Enzyme	PDB codes	Species	Oligomeric states
Arginine decarboxylase	1N13, 1MT1, 1N2M, 3NZP, 3NZQ, 3N2O	*Methanococcus jannaschii, Campylobacter jejuni, Vibrio vulnificus, Escherichia coli*	Dimer, Trimer
Agmatine ureohydrolase	1WOG, 1WOH, 1WOI, 3LHL	*Deinococcus radiodurans, Clostridium difficile*	Hexamer, Trimer
Ornithine decarboxylase	1ORD, 7ODC, 1QU4, 1D7K, 2PLK, 2ON3	*Lactobacillus, Mus musculus, Trypanosoma brucei, Homo sapiens, Vibrio vulnificus, Leishmania donovani*	Dimer
S-Adeno-sylmethionine decarboxylase	1JEN, 1I7B, 2III, 1TLU, 1MHM, 3IWB	*Homo sapiens, Solanum tuberosum, Thermotoga maritima, Aquifex aeolicus*	Dimer, Monomer
Spermidine synthase	3O4F, 2ZSU, 2PSS, 3BWB, 2CMG, 2O05, 2B2C, 1XJ5, 1INL, 1IY9	*Escherichia coli, Pyrococcus horikoshii, Plasmodium falciparum, Trypanosoma cruzi, Helicobacter pylori, Homo sapiens, Caenorhabditis elegans, Arabidopsis thaliana, Thermotoga maritima*	Dimer, Tetramer
Spermine synthase	3C6K, 3C6M	*Homo sapiens*	Dimer
Spermine/ spermidine N^1-acetyltransferase	3BJ7, 2JEV, 2G3T, 2B5G, 2FXF	*Homo sapiens, Mus musculus*	Dimer
Polyamine oxidase	1B37, 1H81, 1XPQ, 1Z6L	*Zea mays, Saccharomyces cerevisiae*	Monomer, Dimer
Antizyme	1ZO0	*Rattus norvegicus*	–

characterization of the enzymes in the pathway provided details on the biosynthesis of the polyamines at a molecular level, thus making viable the
structure-based drug design of enzymes at various stages along the pathway. In
addition to the biosynthetic and catabolic enzymes, antizyme, an enzyme-
inhibitor of ornithine decarboxylase (ODC), plays a critical role in regulating
ODC and hence affects putrescine levels in cells.[27,28] The crystal structure of
antizyme is yet to be determined; however, the NMR structure of an antizyme
isoform from rats has been determined.[29]

The biosynthetic enzymes ODC and S-adenosylmethionine decarboxylase
(AdoMetDC) are of special interest for intervention, as the enzymes catalyze
rate-limiting steps at an early stage in the pathway. Early inhibitor design to
ODC involving various substrate analogs resulted in the synthesis of
2-difluoromethylornithine (DFMO), a potent enzyme-activated irreversible
inhibitor of mammalian ODC.[30,31] DFMO has been a subject of multiple
clinical trials over the last few decades as an anti-cancer agent, but has yet to be
approved by the Food and Drug Administration (FDA) for treatment. However, the success of DFMO for the treatment of African trypanosomiasis
renewed the interest in developing DFMO as a chemotherapeutic agent.[32–36]

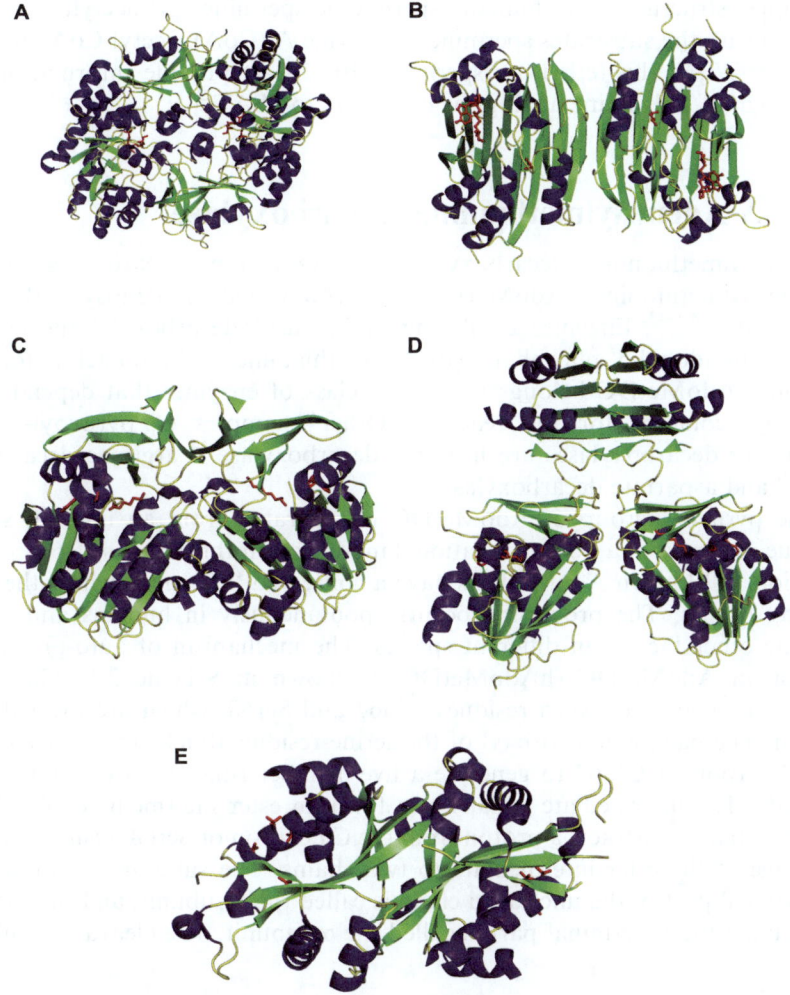

Figure 2.1 Cartoon representations of enzymes. (A) Ornithine decarboxylase, (B) *S*-adenosylmethionine decarboxylase, (C) spermidine synthase, (D) spermine synthase and (E) spermine/spermidine N^1-acetyltransferase from *Homo sapiens* in the polyamine biosynthetic pathway. Bound ligands are shown in stick representation and colored red.

In addition, the crystal structure of *Trypanosoma brucei* ODC bound to DFMO and pyridoxal-5'-phosplate (PLP) has been determined.[37] To our knowledge, this is the only structure of ODC with an inhibitor bound and provides vital information for the further development of new inhibitors from a structural viewpoint. The structural biology of AdoMetDC is discussed in the next sections. The crystal structures of the aminopropyl transferase spermidine synthase with inhibitors *S*-adenosyl-1,8-diamino-3-thio-octane (AdoDATO) and trans-4-methylcyclohexylamine (4MCHA) have also been reported.[38,39]

Multiple structures of human spermidine/spermine N^1-acetyltransferase (SSAT) with the substrates spermine, coenzyme A (CoA), acetyl-CoA, and the inhibitor N^1,N^{11}-bis-(ethyl)-norspermine (BE-3-3-3) provide information for the design of polyamine analogs as inhibitors to SSAT.[25]

2.3 *S*-Adenosylmethionine Decarboxylase

S-adenosylmethionine decarboxylase catalyzes the decarboxylation of *S*-adenosylmethionine (AdoMet) to decarboxylated *S*-adenosylmethionine (dcAdoMet).[40–43] Enzymes catalyzing amino acid decarboxylations usually require pyridoxal-5'-phosphate (PLP) or thiamine as a cofactor for the reaction. AdoMetDC belongs to a small class of enzymes that depend on a pyruvoyl cofactor for the reaction. Other examples of pyruvoyl-group-dependent decarboxylases are histidine decarboxylase,[44] arginine decarboxylase,[19] and aspartate decarboxylase.[45]

The pyruvoyl group of AdoMetDC is generated from an internal serine residue as a part of a post-translational modification through a self-processing reaction. All known AdoMetDCs have a conserved 'ES' motif where the processing occurs. The processing occurs spontaneously in humans and might include other factors in different species. The mechanism of auto-processing in human AdoMetDC (hAdoMetDC) is shown in Scheme 2.1. The auto-processing occurs between residues Glu67 and Ser68, which are located in a β-turn. The backbone hydroxyl of the serine residue attacks the adjacent carbonyl carbon of Glu67 to generate a five-member ring oxyoxazolidine intermediate. The intermediate collapses to form an ester intermediate. The basic residue His243 abstracts a proton from the Cα carbon of Ser68, resulting in the cleavage of the ester intermediate to two chains. The subunit containing the N-terminal part of the uncleaved chain is called the β subunit, and the subunit containing the C-terminal part is called the α subunit. The cleavage results in

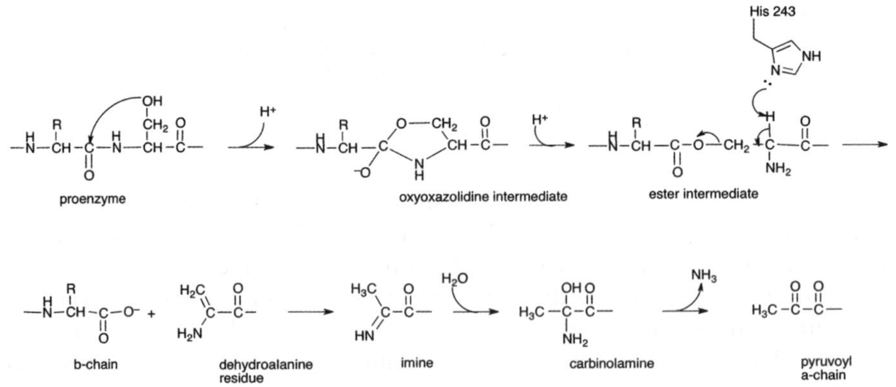

Scheme 2.1   Autoprocessing mechanism in AdoMetDC.

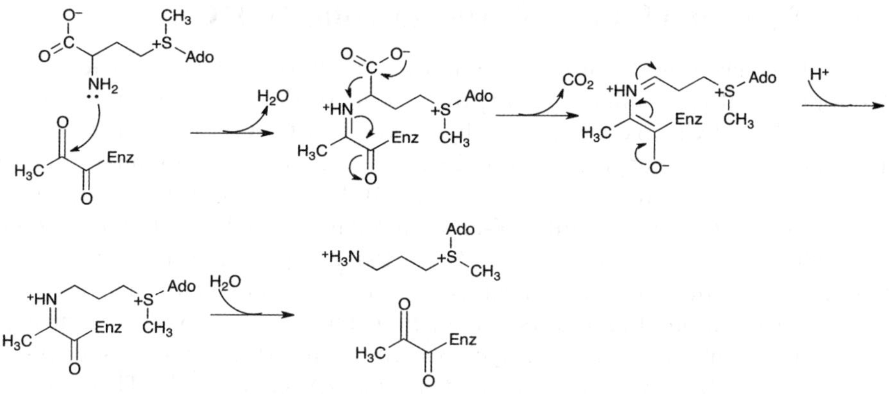

Scheme 2.2 Mechanism of decarboxylation catalyzed by AdoMetDC.

the formation of a dehydroalanine residue at the N-terminus of the α subunit. The dehydrolanine residue is tautomerized to form an imine that is further hydrolyzed to form the active pyruvoyl group. Activity assays on hAdoMetDC have revealed that the enzyme is completely processed to α and β subunits within 30 min of incubation at 37 °C.[46]

The mechanism of decarboxylation in AdoMetDC is shown in Scheme 2.2. The substrate AdoMet binds to the enzyme by making a Schiff base to the active site pyruvoyl group. The decarboxylation reaction proceeds with the pair of electrons from the leaving group (CO_2) delocalized into the pyruvoyl group. The acidic residue Cys82 protonates the Cα carbon of the substrate to generate the imine intermediate. The imine is further hydrolyzed to release the product dcAdoMet and regenerate the pyruvoyl group for further rounds of catalysis. It has also been observed that the incorrect protonation (on the pyruvoyl group) of the intermediate would result in the transamination of the pyruvoyl group resulting in a dead enzyme. Activity assays of hAdoMetDC have indicated that on average, the enzyme performs ~15 000 decarboxylation reactions before being transaminated.[46] In addition, it has been observed that the C82A mutant is much less active than the wild-type enzyme, thus suggesting a role for Cys82 in processing and decarboxylation.

An interesting aspect of AdoMetDCs is the role of the polyamine putrescine in regulating the activity of AdoMetDC by affecting the rates of autoprocessing and decarboxylation. In effect, high concentrations of putrescine in cells drive the production of more dcAdoMet, hence leading to production of more spermidine and spermine (as dcAdoMet is committed to polyamine biosynthesis). The effects of putrescine on AdoMetDC are, however, species-specific. In humans, putrescine activates the rate of autoprocessing by four-fold. Putrescine accelerates the decarboxylation rates in humans, *T. brucei*, *Trypanosoma cruzi* and *Caenorhabditis elegans*. Putrescine is essential for the decarboxylation process in *Neurospora crassa*, but there is no effect on the processing rate.[47-50]

2.4   Early Inhibitors of Human AdoMetDC

AdoMetDC has been the subject of inhibitor design for over three decades. The development of early inhibitors (Figure 2.2) of AdoMetDC was based on the compound methylglyoxal bis(guanylhydrazone) (MGBG), a potent competitive inhibitor of the enzyme.[51] MGBG caused acute toxicity in humans due to anti-mitochondrial effects, and variants of MGBG were pursued for development. One of the variants, 4-amidinoindan-1-one-2'-amidinohydrazone (CGP48664A), a Ciba-Geigy (now Novartis) compound, was a subject of several Phase I and Phase II clinical trials as a cancer therapeutic agent.[17,52–54] The next generation of inhibitors developed included S-(5'-deoxy-5'-adenosyl)-1-ammonio-4-(methylsulfonio)-2-cyclopentene (AdoMac) and 5'-deoxy-5'-[(Z)-4-amino-2-butenyl]methylaminoadenosine (AbeAdo).[55,56] These compounds were designed to act as enzyme-activated irreversible inhibitors but inactivated the enzyme by transamination of the active site pyruvoyl group.

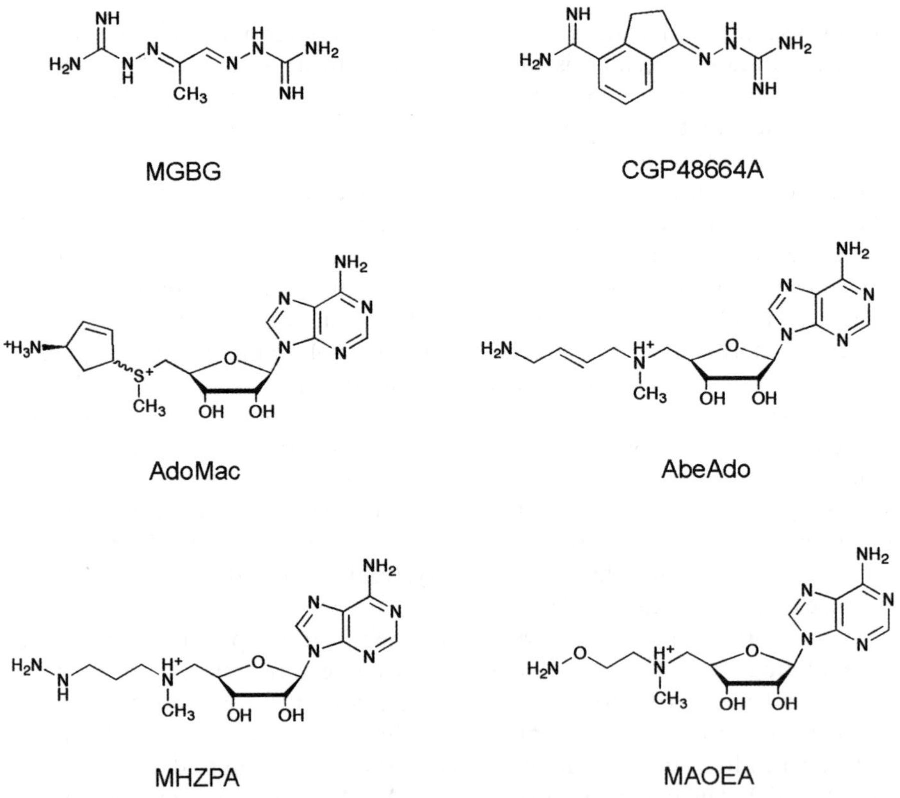

Figure 2.2   Early inhibitors of AdoMetDC. **MGBG**, methylglyoxal bis(guanylhydrazone); **CGP48664A**, 4-amidinoindan-1-one-2'-amidinohydrazone; **AdoMac**, S-(5'-deoxy-5'-adenosyl)-1-ammonio-4-(methylsulfonio)-2-cyclopentene; **AbeAdo**, 5'-deoxy-5'-[(Z)-4-amino-2-butenyl]methylaminoadenosine; **MHZPA**, 5'-deoxy-5'-[N-methyl-N-(3-hydrazinopropyl)amino]adenosine; **MAOEA**, 5'-deoxy-5'-[N-methyl-N-[(2-aminooxy)ethyl]amino]adenosine.

In parallel, mechanism-based inhibitors, such as 5′-deoxy-5′-[*N*-methyl-*N*-(3-hydrazinopropyl)amino]adenosine (MHZPA) and 5′-deoxy-5′-[*N*-methyl-*N*-[(2-aminooxy)ethyl]amino]adenosine (MAOEA), that make a Schiff base to the pyruvoyl group to inactivate AdoMetDC were developed.[57–59]

These early inhibitors demonstrated that blocking AdoMetDC activity was a promising therapeutic strategy. With the increase in the application of structure-based drug design in pharmaceutical industries and academia, our laboratory has been involved in more than a decade of research related to the structure, mechanism, evolution and drug design of AdoMetDC.[60] The following sections summarize the key structural features, inhibitors designed and evolutionary features of AdoMetDC.

2.5 Crystal Structure of Human AdoMetDC and Mutants

The crystal structure of hAdoMetDC and the non-processing mutants H243A and S68A provided the first glimpse into the overall structure of the enzyme and details of autoprocessing.[22,49,61] HAdoMetDC crystallizes as a dimer with the protomers related to each other by a twofold symmetry (Figure 2.1(B)). The overall secondary structure of the protomer is a four-layer αββα sandwich. The central core of the protomer comprises two anti-parallel β-sheets of eight β-strands each that are flanked on either side by amphipathic α-helices and 3_{10}-helices (Figure 2.3). The pyruvoyl group is formed at the N-terminus of the α chain and is located at the edge of the β-sheets. The active site is located between

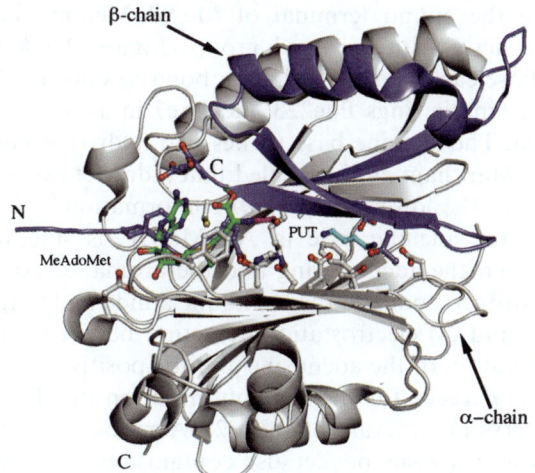

Figure 2.3 Cartoon representation of the protomer of hAdoMetDC (from PDB code 1I7B). The α chain is colored light gray, and the β chain is colored blue. Putrescine, MeAdoMet and critical residues for putrescine and substrate binding and processing are shown as a ball-and-stick representation. The carbon atoms of MeAdoMet are colored green and of putrescine are colored light blue.

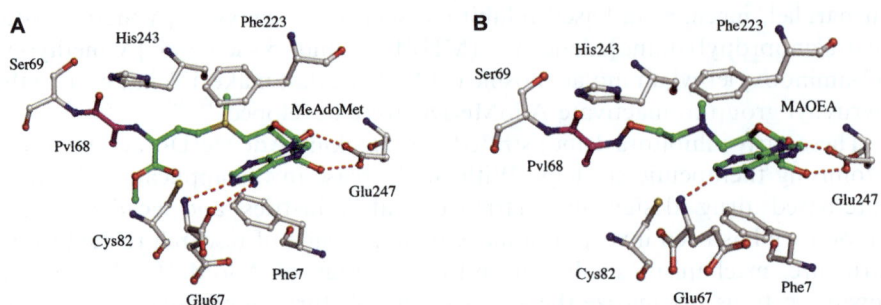

Figure 2.4 Interactions of (A) MeAdoMet and (B) MAOEA in the active site of
hAdoMetDC. The carbon atoms of the pyruvoyl group are colored
magenta, and those of MeAdoMet and MAOEA are colored green.
Hydrogen bonds are shown as dashed red lines.

the central β-sheets away from the dimer interface and comprises residues from
both the β-sheets. The H243A and S68A mutants trapped the enzyme in the ester
and proenzyme form, respectively, providing the first structural insights into the
mechanism of autoprocessing. In addition, the high-resolution crystal structure
of the H243A mutant revealed a molecule of putrescine bound between the
central β-sheets. The β-sheets are linked together through the putrescine molecule by numerous hydrogen bonds and hydrophobic interactions.

To gain further insights into the substrate binding of AdoMetDC, the crystal
structure of the hAdoMetDC complex with the methyl ester of AdoMet
(MeAdoMet) and putrescine was determined.[62] MeAdoMet binds covalently to
the enzyme with the amino terminal of MeAdoMet making a Schiff base
linkage with the active site pyruvoyl group (Figure 2.4(A)). The hydroxyl
groups of the ribose each make a hydrogen bond to Glu247. The adenine ring
stacks between aromatic rings Phe223 and Phe7 in an unusual higher-energy
syn conformation. The adenine base makes two hydrogen bonds with Glu67.
NMR studies by Markham *et al.* revealed that AdoMet exists in solution with
the adenine base in the lower-energy *anti* conformation.[63] The higher-energy
conformation of the adenine base in AdoMetDC is stabilized by: (1) π–π
interactions between the adenine ring and the aromatic residues Phe223 and
Phe7; (2) two hydrogen bonds to the backbone amide and C-terminal carboxyl
group of Glu67; and (3) electrostatic interaction between the partially negatively charged N3 atom of the adenosine and the positively charged sulfonium
group.[64,65] The preference for the *syn* conformation of MeAdoMet was also
observed in the crystal structure of the F223A mutant of hAdoMetDC with
MeAdoMet.[64] The active site pocket also contains residues Cys82, His243 and
Ser229, which are implicated in the processing and decarboxylation reactions.

Multiple crystal structures reveal that putrescine binds between the central β
sheets 16–20 Å from the active site (Figure 2.3) of the enzyme. Putrescine makes
extensive hydrophobic and hydrogen-bonding interactions with the enzyme.
Putrescine is hydrogen-bonded directly and through water molecules to residues Asp174, Glu15, Thr176, Glu178, Glu256 and Ser113. The positively

charged terminus of putrescine partly neutralizes the negatively charged acidic residues pointing towards each other between the β sheets. The aliphatic carbon chain of putrescine stacks against the aromatic rings of Phe285 and Phe111. Biochemical studies reveal positive cooperativity in putrescine and substrate binding to the dimeric form of the enzyme.[66] The mechanistic and structural details of hAdoMetDC have revealed the active site, proenzyme and the putrescine binding site as three potential targets of inhibition of the enzyme.

2.6 Inhibitor Design for Active Site of AdoMetDC

The crystal structure of hAdoMetDC with the substrate analog MeAdoMet provides a very good case for rational drug design. In addition, the crystal structures of hAdoMetDC with early inhibitors MAOEA, MHZPA, MGBG and CGP48664 have also been determined.[62] The inhibitors MAOEA (Figure 2.4(B)) and MHZPA bind covalently to the pyruvoyl group and make similar hydrogen-bonding and stacking interactions to that of MeAdo-Met (Figure 2.4(A)). The inhibitors CGP48664 and MGBG stack between the aromatic rings of Phe223 and Phe7 and make hydrogen-bonding interactions with Glu247, Leu65 and Ser229.

The inhibitor–enzyme complexes reveal the key residues in the active site of the enzyme. One of our approaches was to design new inhibitors to the active site that are substrate analogs based on the interactions of MAOEA/MHZPA with the enzyme. Key modifications to MAOEA included changes to: (1) the terminal group for formation of a Schiff base; (2) the linker length between the central nitrogen atom and terminal group; (3) substitutions to the adenine ring; (4) the central nitrogen atom; and (5) the substituent on the central nitrogen atom (see Figure 2.5). A series of new compounds have been chemically synthesized and tested for inhibitory activity against hAdoMetDC (see Table 2.2 and Figure 2.6).[64]

2.6.1 Role of the Central Nitrogen/Sulfonium Atom

Third-generation inhibitors all contained a central nitrogen/sulfonium atom that is positively charged. Complexes of hAdoMetDC with 5′-deoxy-5′-dimethyl thioadenosine (MMTA) and 5′-deoxy-5′-(N-dimethyl)amino-8-methyl

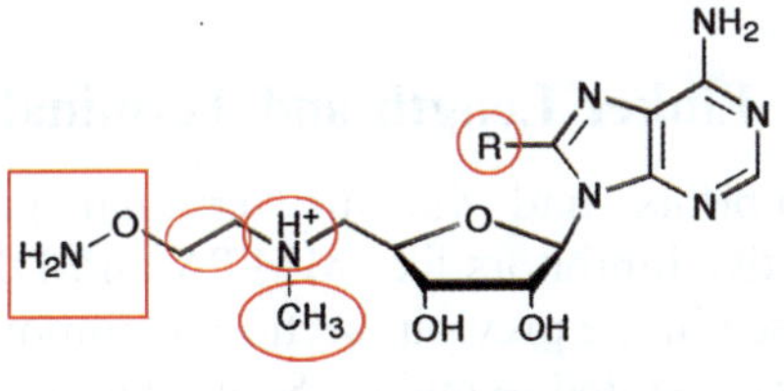

MAOEA

Figure 2.5 MAOEA is a model template for generation of new inhibitors. Sites on MAOEA for various modifications are marked in red.

Table 2.2 Inhibitory activities of substrate analogue inhibitors.

Compound	IC_{50} (μM)	IC_{50} of unsubstituted analogue (if applicable)	Schiff base to pyruvoyl group[a]	PDB code
1	0.8	12	No	3H0W
2	3	17	No	3H0V
3	14	18	No	3DZ6
4	7	55	Yes	3DZ5
5	0.35	6	No	3DZ4
6	70	450	No	3DZ2
7	240	NA	Yes	–
8	0.17	1.9	Yes	–
9	12.4	NA	No	–
10	17.3	NA	Yes	–

[a]As seen in the crystal structures.

adenosine (DMAMA) suggest that the ribose, adenine ring and positive charge of the ligand are the minimal features of a ligand that are required to bind to the enzyme. Biochemical analysis, quantum calculations and crystallization trials have revealed that a positive charge at that position on ligands is critical for binding to AdoMetDC and that the neutral ligands 5′-deoxy-5′-methyl thioadenosine (MTA) and *S*-adenosylhomocysteine (SAH) do not bind to the enzyme. In addition, *in vitro* assays of inhibition of AdoMetDC by substrate analogs by Pankaskie *et al.* revealed that a positive charge at the sulfonium position is critical for inhibitory activity.[67] The complexes of hAdoMetDC with MMTA and DMAMA revealed that the positive charge is involved in favorable cation–π interactions with aromatic rings Phe223 and Phe7.[68] In addition, stopped-flow kinetic experiments showed that the rate of substrate binding depended on residues Phe223 and Phe7, supporting the importance of the cation–π interactions. The positive charge plays an additional role in stabilizing the *syn* conformation of the ligands bound through an electrostatic interaction with the partially negatively charged N3 atom of the adenine base.[65]

Substitution of the sulfonium atom with a nitrogen atom retains the positive charge at that position at physiological pH. Most of the compounds in this family of inhibitors have a central tertiary nitrogen with a methyl substituent on the nitrogen atom. An ethyl substituent on the central nitrogen is also well tolerated.

2.6.2 Role of the Linker Length and Terminal Group

Third-generation inhibitors had various terminal groups at the 5′ end (Figure 2.6). Competitive inhibitors like MMTA and DMAMA with no terminal groups are discussed in the previous section. Compounds with an aminooxy or hydrazino terminus are suited to make a Schiff base to the active site pyruvoyl group. However, the linker length between the terminal group and the central nitrogen/sulfonium atom is critical. A linker length of three to four atoms makes the formation of the Schiff base linkage geometrically feasible. The role of the

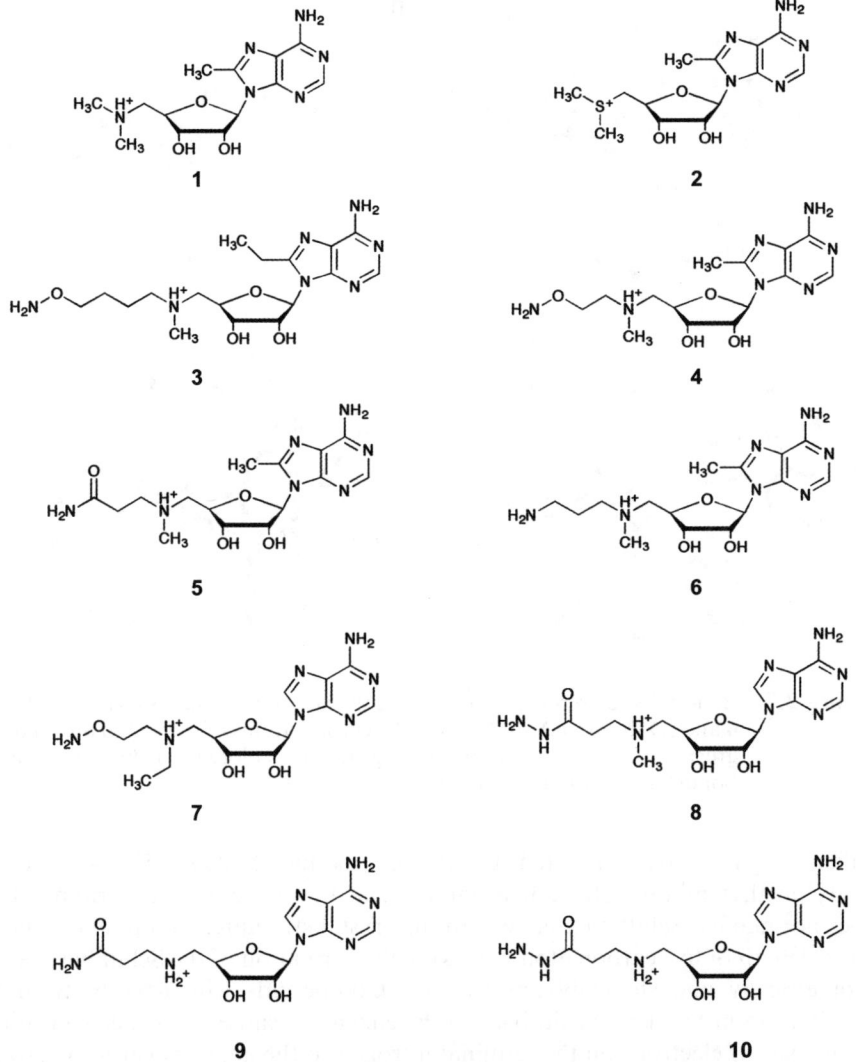

Figure 2.6 Representative newly designed inhibitors to hAdoMetDC. The crystal structures of all the inhibitors bound to hAdoMetDC have been determined (includes unpublished data). **1**, 5'-deoxy-5'-(*N,N*-dimethylamino)-8-methyladenosine; **2**, 5'-deoxy-5'-dimethylsulfonio-8-methyladenosine; **3**, 5'-deoxy-5'-[(4-aminooxybutyl)methylamino]-8-ethyladenosine; **4**, 5'-deoxy-5'-[(2-aminooxyethyl)methylamino]-8-methyladenosine; **5**, 5'-deoxy-5'-[(2-carboxamidoethyl)methylamino]-8-methyladenosine; **6**, 5'-deoxy-5'-[(3-aminopropyl)methylamino]-8-methyladenosine; **7**, 5'-deoxy-5'-[(2-aminooxyethyl)ethylamino]adenosine; **8**, 5'-deoxy-5'-[(2-hydrazino-carbonyl ethyl)methylamino]adenosine; **9**, 5'-deoxy-5'-[2-aminooxoethyl) amino]adenosine; **10**, 5'-deoxy-5'-[(2-hydrazinocarbonylethyl) amino] adenosine.

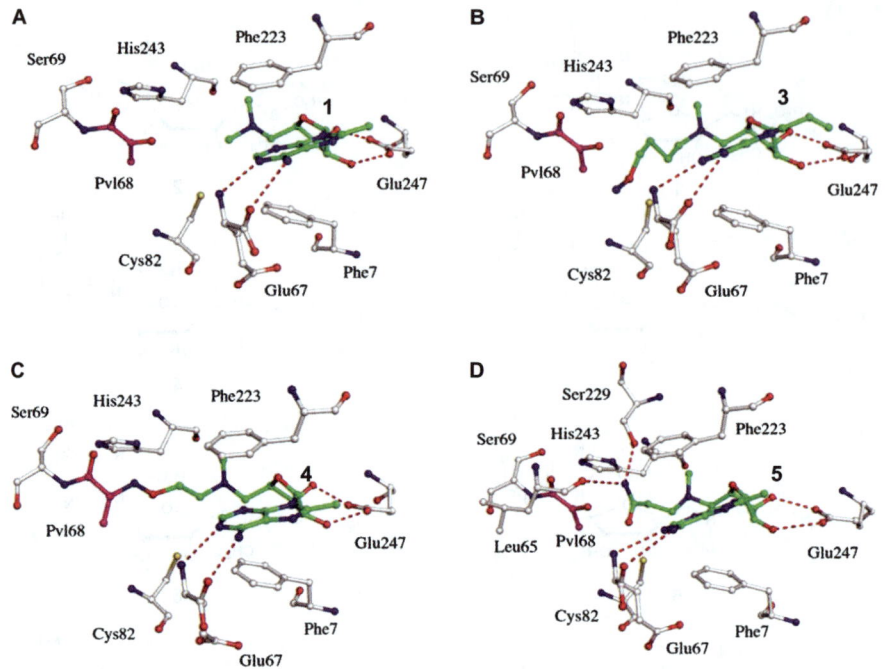

Figure 2.7 Ball-and-stick representation of complexes of compounds (A) **1**, (B) **3**, (C) **4** and (D) **5** with hAdoMetDC. The carbon atoms of the ligand are colored green, and those of the pyruvoyl group are colored magenta. Hydrogen bonds are shown as dashed red lines.

linker length is demonstrated by the compound 5′-deoxy-5′-[(4-aminoox-ybutyl)methylamino]-8-ethyladenosine (**3**) that has an aminooxy terminus but does not make a Schiff base as seen in the crystal structure. Compound **3** has a linker length of five carbon atoms, making the formation of the Schiff base bond geometrically and sterically not feasible. Compounds with a carboxyamido terminus do not make a Schiff base to the enzyme because of the delocalization of the pair of electrons on the terminal nitrogen to the adjacent carbonyl group and are hence competitive inhibitors. The compound 5′-deoxy-5′-[(3-amino-propyl)methylamino]-8-methyladenosine (**6**) with an aminoalkyl terminus does not make a Schiff base to the pyruvoyl group. This unexpected finding is probably due to the lack of enhancement of the nucleophilicity of the terminal nitrogen atom when attached to an alkyl chain. Complexes of several of these inhibitors to hAdoMetDC are shown in Figure 2.7.

2.6.3   Effect of 8-Substitution

The preference of MeAdoMet, MAOEA and MHZPA for binding in the *syn* conformation was exploited to synthesize ligands with a substitution on the

8-position of the adenine ring. Ligands with the C^8 substitution prefer a *syn* conformation in solution (owing to clashes between the substituent group and the ribose ring in the *anti* conformation) and are expected to have enhanced binding to AdoMetDC. A series of compounds with various C^8 substitutions were synthesized, and the inhibitory activities were tested against hAdoMetDC.[64] The results reveal that in general, a C^8-methyl substitution improved the inhibition by eight- to 18-fold (see Table 2.2). A C^8-ethyl or methylamino substitution provided little or no improved inhibition over the unsubstituted compound. Larger groups, such as phenyl group substitutions, abolish the inhibitory activity. The effect is largely due to the energy penalty of exposing a large hydrophobic group to the external solvent to maintain a *syn* conformation of the ligands to bind to AdoMetDC.

2.7 Search for New Inhibitors by Virtual Screening

First-generation (MGBG) and second-generation (CGP48664A) inhibitors of AdoMetDC are not substrate analogs but nevertheless are competitive inhibitors. High-throughput structure-based virtual screening coupled with experimental testing was employed to discover inhibitors to hAdoMetDC that had novel binding cores and function as lead compounds for further drug design.[69] A combination of various modeling/docking programs, such as MAESTRO, LIGPREP, MACROMODEL and GLIDE (grid-based ligand docking with energetics) (developed by Schrödinger, Inc., New York), were used for virtual screening. The National Cancer Institute (NCI) diversity set containing 1990 drug-like compounds was virtually screened against the active site of hAdoMetDC and scored using the docking scoring protocol in GLIDE. Top scoring hits (133 out of 1990) were experimentally tested for inhibition against hAdoMetDC.[69] A representative selection of various top-scoring compounds is shown in Figure 2.8. The compounds obtained by screening invariably had a π-ring system and probably inhibited AdoMetDC by stacking against Phe223 and Phe7 while making additional favorable contacts. Of special interest is the compound (*E*)-9-amino-6-[(2,6-diaminopyridin-3-yl)diazenyl]-2-ethoxyacridin-10-ium (NSC 354961; I) that inhibited hAdoMetDC with an IC_{50} in the low micromolar range.

2.8 Inhibitor Design for the Putrescine-Binding Site and Proenzyme

The polyamine putrescine is known to activate the processing and decarboxylation rates in hAdoMetDC. To gain structural insights into the mechanism of activation by putrescine, the crystal structures of putrescine free hAdoMetDC, D174N, E256Q and E178Q mutants were determined.[66] A comparison of the putrescine free and putrescine bound structures revealed that the binding of putrescine leads to the rearrangement of the side chains of aromatic residues Phe285, Phe315, Tyr318 and Phe320. In the absence of putrescine, the site is filled with four water molecules, and there is no significant change in the overall

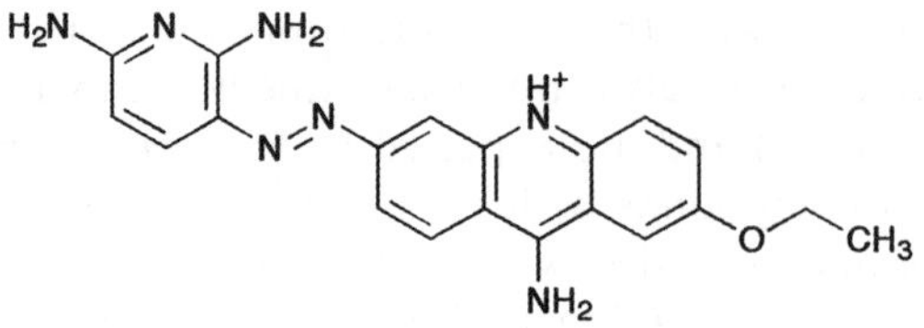

I

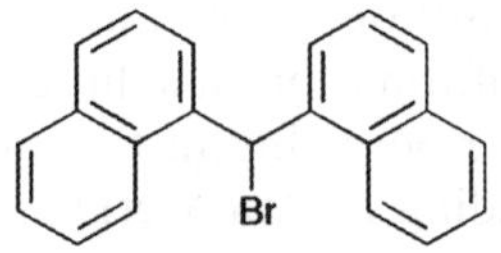

II

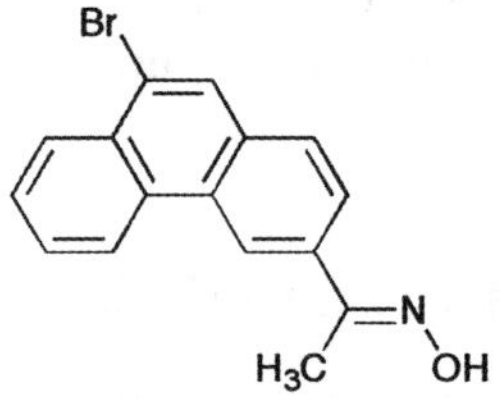

III

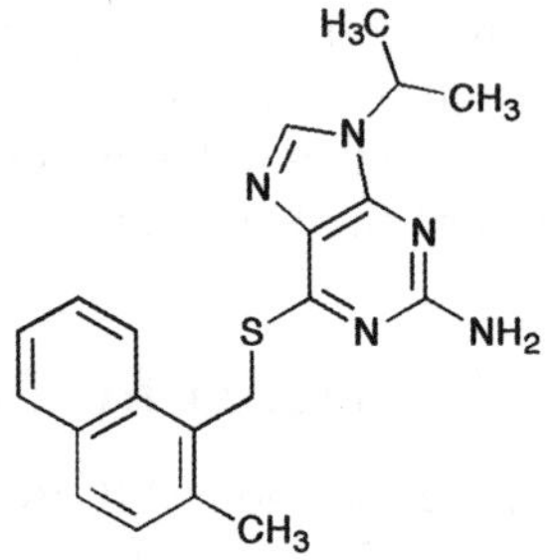

IV

Figure 2.8   Selected inhibitors from the virtual screening of the NCI diversity set.
I, (*E*)-9-amino-6-((2,6-diaminopyridin-3-yl)diazenyl)-2-ethoxyacridin-10-
ium; **II**, 1,1′-(bromomethylene)dinaphthalene, **III**, (*E*)-1-(9-bromophe-
nanthren-3-yl)ethanone oxime; **IV**, 9-isopropyl-6-(((2-methylnaphthalen-
1-yl)methyl)thio)-9*H*-purin-2-amine.

structure of the enzyme suggesting that the structural stabilization by putrescine is minimal. However, the rearrangement of the residues mentioned above
is associated with the closing of loop 312–320, thereby shielding putrescine
from external solvent and enhancing its electrostatic effects. Putrescine is
involved in a hydrogen-bonding network (through residues Glu178, Glu256
and Lys80) to the active site residues Glu11, Ser229 and His243 orienting them
in the proper conformation for processing and catalysis.[66] Lys80 is positioned
to play a critical role in mediating the hydrogen-bonding effects between the
putrescine binding site and the active site. Loop 312–320 undergoing the conformational change upon putrescine binding is located at the dimer interface.
The position of the loop at the dimer interface suggests a role of the interface in
the cooperativity in putrescine binding to the dimeric enzyme. Compounds that
deregulate putrescine effects or mimic the effects of the putrescine-related
mutants are promising candidates to evaluate for inactivation of AdoMetDC.

The proenzyme also presents an additional target for inhibition in the life-
cycle of AdoMetDC. As the proenzyme lacks the active site pyruvoyl group for
binding Schiff-base-forming inhibitors, complexes of the proenzyme with
MMTA and DMAMA were obtained (unpublished data). These competitive
inhibitors bind at the partially formed 'active site' of the proenzyme and alter
the conformation of the loop 65–70 where processing occurs. The ribose

hydroxyl oxygen atoms make two hydrogen bonds to Glu247, and the adenine base stacks in the *syn* conformation between Phe7 and Phe223. The altering of the processing loop from the native state has an effect on the rate of processing. The binding of this class of inhibitors would affect the activity of AdoMetDC by altering the processing rate and blocking the active site for substrate binding. Further development of compounds with dual modes of inhibitory action would be of special interest.

2.9 Interspecies Correlations

2.9.1 Classification of AdoMetDC

The classification of AdoMetDC proteins into various classes based on species, quaternary structure and activation factor has been discussed in detail previously.[60,70] Briefly, AdoMetDCs from bacteria and archaea fall in Group 1, and those from eukaryotes fall in Group 2. The crystal structure of hAdo-MetDC was the first to be determined among all the classes. The crystal structures of AdoMetDC from *Solanum tuberosum* (potato), *T. maritima* and *A. aeolicus* (PDB ID 2III) have revealed evolutionary links between Group 1 and 2 enzymes.[70,71] A comparison of the protomer of hAdoMetDC with dimeric *T. maritima* and *A. aeolicus* AdoMetDCs shows a strong correlation of secondary structure in spite of the low sequence identity between them. The similarity in the C-terminal and N-terminal halves of hAdoMetDC with the protomers of prokaryotic AdoMetDCs makes a strong case of evolution of the eukaryotic enzyme from the prokaryotes by gene duplication, fusion and mutational drift (Figure 2.9).

The activation of AdoMetDC also varies between various groups. Certain Group 1 AdoMetDCs (*e.g.*, *Escherichia coli*) depend on metal ion for activation.[72] Activation factors (if any) in other Group 1 AdoMetDCs are yet to be identified. Group 2 AdoMetDCs present an interesting case of oligomerization and activation effects of putrescine. HAdoMetDC exists as a dimer and needs putrescine for its full activity. Further, binding of putrescine in hAdoMetDC is cooperative, suggesting a functional role for dimerization. However, potato AdoMetDC (pAdoMedDC) exists as a monomer and is fully active in the absence of putrescine. Dimerization in pAdoMetDC is hindered by mutations in the proposed dimer interface. In addition, an insertion of one amino acid residue in the β-strand at the interface is observed that would probably lead to steric clashes upon dimerization. The crystal structure of AdoMetDC from potato revealed that the critical residues in the buried charge site are mostly conserved in humans with the exception of Phe285, Asp174, Phe111 and Leu13 (which are His294, Val181, Arg114 and Arg18 respectively in potato). The side chains of positively charged arginine residues Arg114 and Arg18 are positioned to mimic the effect of putrescine for full activity of pAdoMetDC. However, in Group 1 enzymes, the buried charge site is present in the dimer interface. In *T. maritima*, residues His68, Glu11, Arg112, His64, His7, His110 and Asp79

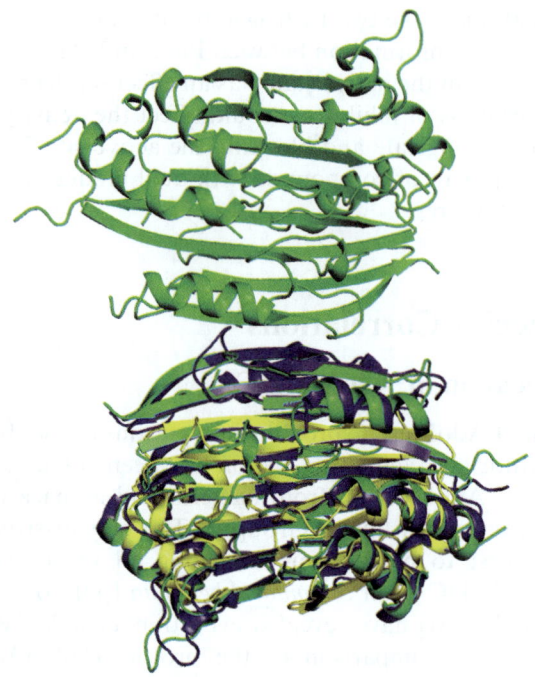

Figure 2.9 Superposition of human (green), potato (blue), *T. maritima* (yellow) AdoMetDC shown in cartoon representation.

from one protomer are connected through hydrogen bonding to the two-fold related residues in the other protomer. The extensive hydrogen-bonding network suggests that both active sites are connected to each other, and the activation occurs through charge relay. A superposition of buried charged sites among AdoMetDCs of known structures is shown in Figure 2.10.

2.9.2 Ligand Binding in Prokaryotic AdoMetDC

The crystal structure of the processed form of *T. maritima* AdoMetDC (TmAdoMetDC) provided the first insights into the active site of Group 1 AdoMetDCs.[73] Processing in TmAdoMetDC was induced by heating the enzyme for 1 h at 80 °C; probably a physiologically relevant step as thermophilic bacteria like *T. maritima* have an optimum growth temperature at 80 °C. TmAdoMetDC exists as a dimer, and processing occurs in both protomers. The active site contains residues from both protomers, suggesting that dimerization is essential for activity. In addition, the superposition of the TmAdoMetDC dimer with a protomer of hAdoMetDC (Figure 2.9) reveals that the prokaryotes lack strands β_7, β_8, β_{15} and β_{16} involved in dimerization in hAdoMetDC. The loss of the strands probably hinders further oligomerization of dimeric TmAdoMetDC.

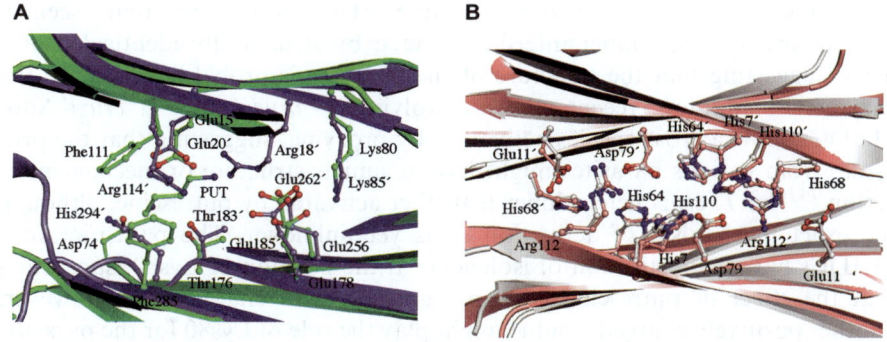

Figure 2.10 Superposition of buried charged sites in (A) Group 2 and (B) Group 1 AdoMetDCs. Hydrogen bonds and water molecules are omitted for clarity. For (A), human AdoMetDC carbon atoms are shown in green, and potato AdoMetDC carbon atoms are shown in violet (prime denotes residues from potato AdoMetDC). For (B), *T. maritima* AdoMetDC carbon atoms are shown in light gray, and *A. aeolicus* AdoMetDC carbon atoms are shown in pink. Labels of residues of *A. aeolicus* AdoMetDC are omitted for clarity (prime denotes residues from the twofold-related protomer in *T. maritima* AdoMetDC).

The complexes of TmAdoMetDC with MeAdoMet and MMTA reveal a conservation in ligand-binding mode as seen in humans. All the active-site residues in TmAdoMetDC are conserved in humans except for Trp70, that is Phe7 in humans. The substrate analog MeAdoMet binds to TmAdoMetDC by making a Schiff base link with the active site pyruvoyl group. The adenine base stacks in the *syn* conformation between Phe49′ and Trp70′ (prime represents residues from the twofold related protomer). The ribose makes two hydrogen bonds to Glu72′, and the adenine ring makes two hydrogen bonds to Glu62. A similar binding mode is also seen for MMTA binding to TmAdoMetDC. The conservation of the roles of active site residues and the ligand-binding mode reveal common themes of substrate recognition and binding across Group 1 and Group 2 AdoMetDCs.

2.9.3 Implications of the Prozyme

The discovery of the prozyme in parasites presents a novel feature of allosteric activation and regulation in AdoMetDCs.[50] In *T. brucei*, AdoMetDC is fully active as a heterodimer formed between the enzyme and a regulatory dead homolog called the prozyme (a 1000-fold activation upon heterodimer formation). A sequence comparison suggests that the prozyme evolved from the enzyme by a gene duplication event followed by a significant mutational drift of the prozyme from the enzyme. The prozyme lacks the 'ES' motif and residues critical for processing, indicating a complete loss of decarboxylation activity. The activation of parasitic AdoMetDC upon heterodimer formation is similar to the cooperativity seen in dimeric AdoMetDC. The cooperativity seen in humans involves the dimer interface formed by structurally identical protomers, suggesting that the structure of the prozyme is probably similar to the enzyme. In addition, recent studies involving *T. brucei* and *T. cruzi* AdoMetDCs show a cross-species activation by prozyme, suggesting that the prozymes from various parasites might have a similar structure or mechanism of action.[74] The *T. cruzi* heterodimer is further activated by putrescine, although the exact mechanism of activation is as yet unknown. The parasitic AdoMetDCs have a substitution of isoleucine at the position of Lys80, suggesting that the effect of putrescine is more allosteric than charge relay; however, another positively charged residue might play the role of Lys80 for the parasitic enzyme.

The structure of the heterodimer of parasitic AdoMetDCs would provide newer insights into inhibitor design for parasitic infections. The structure would also reveal the mechanism of activation brought about by the prozyme. The residues involved in substrate binding in hAdoMetDC are strictly conserved in the parasitic AdoMetDCs (Glu266, Tyr242, Phe28 and Glu85 in *T. brucei*), suggesting that ligands bind to AdoMetDC in a similar mode across various species.[73] In addition, the preference of ligands binding in the *syn* conformation to hAdoMetDC has also been used recently to design potent inhibitors to the parasitic AdoMetDC heterodimer.[75,76] The conservation of active site residues and inhibitory activity of C^8 substituted inhibitors suggest that the library of inhibitors developed for chemotherapeutic activity in humans also merits investigation as anti-trypanosomal compounds.

References

1. W. A. Hendrickson, *Science*, 1991, **254**, 51–58.
2. I. Uson and G. M. Sheldrick, *Curr. Opin. Struct. Biol.*, 1999, **9**, 643–648.
3. G. Taylor, *Acta Crystallogr. D: Biol. Crystallogr.*, 2003, **59**, 1881–1890.
4. P. J. Whittle and T. L. Blundell, *Annu. Rev. Biophys. Biomol. Struct.*, 1994, **23**, 349–375.
5. T. L. Blundell, B. L. Sibanda, R. W. Montalvao, S. Brewerton, V. Chelliah, C. L. Worth, N. J. Harmer, O. Davies and D. Burke, *Philos. Trans. R. Soc. Lond. B Biol. Sci.*, 2006, **361**, 413–423.
6. H. Koppen, *Curr. Opin. Drug Discov. Devel.*, 2009, **12**, 397–407.
7. G. Klebe, *Drug Discov. Today*, 2006, **11**, 580–594.
8. G. Schneider, *Nat. Rev. Drug Discov.*, 2010, **9**, 273–276.
9. R. E. Babine and S. S. Abdel–Meguid, *Protein Crystallography in Drug Discovery*, Wiley-VCH, Weinheim, 2004.
10. D. R. Flower, ed., *Drug Design: Cutting Edge Approaches*, The Royal Society of Chemistry, London, 2002.
11. N. Eswar, M. A. Marti–Renom, B. Webb, M. S. Madhusudhan, D. Eramian, M. Shen, U. Pieper and A. Sali, Current Protocols in Protein Science, Wiley, New York, 2007, pp. 2.9.1–2.9.31.

12. A. Sali and T. L. Blundell, *J. Mol. Biol.*, 1993, **234**, 779–815.
13. B. O. Villoutreix, R. Eudes and M. A. Miteva, *Comb. Chem. High Throughput Screen.*, 2009, **12**, 1000–1016.
14. K. J. Simmons, I. Chopra and C. W. Fishwick, *Nat. Rev. Microbiol.*, 2010, **8**, 501–510.
15. H. M. Wallace, A. V. Fraser and A. Hughes, *Biochem. J.*, 2003, **376**, 1–14.
16. E. W. Gerner and F. L. Meyskens, Jr, *Nat. Rev. Cancer*, 2004, **4**, 781–792.
17. R. A. Casero, Jr and L. J. Marton, *Nat. Rev. Drug Discov.*, 2007, **6**, 373–390.
18. R. A. Casero, Jr, B. Frydman, T. M. Stewart and P. M. Woster, *Proc. West. Pharmacol. Soc.*, 2005, **48**, 24–30.
19. W. D. Tolbert, D. E. Graham, R. H. White and S. E. Ealick, *Structure*, 2003, **11**, 285–294.
20. H. J. Ahn, K. H. Kim, J. Lee, J. Y. Ha, H. H. Lee, D. Kim, H. J. Yoon, A. R. Kwon and S. W. Suh, *J. Biol. Chem.*, 2004, **279**, 50505–50513.
21. J. J. Almrud, M. A. Oliveira, A. D. Kern, N. V. Grishin, M. A. Phillips and M. L. Hackert, *J. Mol. Biol.*, 2000, **295**, 7–16.
22. J. E. Ekstrom, I. I. Matthews, B. A. Stanley, A. E. Pegg and S. E. Ealick, *Structure*, 1999, **7**, 583–595.
23. H. Wu, J. Min, Y. Ikeguchi, H. Zeng, A. Dong, P. Loppnau, A. E. Pegg and A. N. Plotnikov, *Biochemistry*, 2007, **46**, 8331–8339.
24. H. Wu, J. Min, H. Zeng, D. E. McCloskey, Y. Ikeguchi, P. Loppnau, A. J. Michael, A. E. Pegg and A. N. Plotnikov, *J. Biol. Chem.*, 2008, **283**, 16135–16146.
25. M. C. Bewley, V. Graziano, J. Jiang, E. Matz, F. W. Studier, A. E. Pegg, C. S. Coleman and J. M. Flanagan, *Proc. Natl. Acad. Sci. U. S. A.*, 2006, **103**, 2063–2068.
26. C. Binda, A. Coda, R. Angelini, R. Federico, P. Ascenzi and A. Mattevi, *Structure*, 1999, **7**, 265–276.
27. S. Hayashi, *Ornithine Decarboxylase: Biology, Enzymology and Molecular Genetics*, Pergamon Press, New York, 1989.
28. A. E. Pegg, *J. Biol. Chem.*, 2006, **281**, 14529–14532.
29. D. W. Hoffman, D. Carroll, N. Martinez and M. L. Hackert, *Biochemistry*, 2005, **44**, 11777–11785.
30. P. Bey, C. Danzin, V. Van Dorsselaer, P. Mamont, M. Jung and C. Tardif, *J. Med. Chem.*, 1978, **21**, 50–55.
31. B. W. Metcalf, P. Bey, C. Danzin, M. J. Jung, P. Casara and J. P. Vevert, *J. Am. Chem. Soc.*, 1978, **100**, 2551–2553.
32. F. Milord, J. Pépin, L. Loko, L. Ethier and B. Mpia, *Lancet*, 1992, **340**, 652–655.
33. R. R. Love, R. Jacoby, M. A. Newton, K. D. Tutsch, K. Simon, M. Pomplum and A. K. Verma, *Cancer Epidemiol. Biomark. Prev.*, 1998, **7**, 989–992.
34. R. R. Love, P. P. Carbone, A. J. Verma, D. Gilmore, P. Carey, K. D. Tusch, M. Pomplun and G. Wilding, *J. Natl. Cancer Inst.*, 1993, **85**, 732–737.
35. M. D. Prados, W. M. Wara, P. K. Sneed, M. McDermott, S. M. Chang, J. Rabbit, M. Page, M. Malec, R. L. Davis, P. H. Gutin, K. Lamborn, C.

B. Wilson, T. L. Phillips and D. A. Larson, *Int. J. Radiation Oncol. Biol. Phys.*, 2001, **49**, 71–77.

36. V. A. Levin, J. H. Uhm, K. A. Jaeckle, A. Choucair, P. J. Flynn, W. K. A. Yung, M. D. Prados, J. M. Bruner, S. M. Chang, A. P. Kyritsis, M. J. Gleason and K. R. Hess, *Clin. Cancer Res.*, 2000, **6**, 3878–3884.

37. N. V. Grishin, A. L. Osterman, H. B. Brooks, M. A. Phillips and E. J. Goldsmith, *Biochemistry*, 1999, **38**, 15174–15184.

38. V. T. Dufe, W. Qiu, I. B. Muller, R. Hui, R. D. Walter and S. Al-Karadaghi, *J. Mol. Biol.*, 2007, **373**, 167–177.

39. S. Korolev, Y. Ikeguchi, T. Skarina, S. Beasley, C. Arrowsmith, A. Edwards, A. Joachimiak, A. E. Pegg and A. Savchenko, *Nat. Struct. Biol.*, 2002, **9**, 27–31.

40. M. L. Hackert and A. E. Pegg, in *Comprehensive Biological Catalysis*, ed. M. L. Sinnott, Academic Press, London, 1997, pp. 201–216.

41. A. E. Pegg, H. Xiong, D. Feith and L. M. Shantz, Biochem. Soc. Trans., 1998, 26, 580–586 526.

42. C. W. Tabor and H. Tabor, *Advan. Enzymol. Related Areas Mol. Biol.*, 1984, **56**, 251–282.

43. P. D. van Poelje and E. E. Snell, *Ann. Rev. Biochem.*, 1990, **59**, 29–59.

44. T. Gallagher, D. A. Rozwarski, S. R. Ernst and M. L. Hackert, *J. Mol. Biol.*, 1993, **230**, 516–528.

45. A. Albert, V. Dhanaraj, U. Genschel, G. Khan, M. K. Ramjee, R. Pulido, B. L. Sibanda, F. von Delft, M. Witty, T. L. Blundell, A. G. Smith and C. Abell, *Nat. Struct. Biol.*, 1998, **5**, 289–293.

46. H. Xiong, B. A. Stanley and A. E. Pegg, *Biochemistry*, 1999, **38**, 2462–2470.

47. T. C. Beswick, E. K. Willert and M. A. Phillips, *Biochemistry*, 2006, **45**, 7797–7807.

48. D. Ndjonka, A. Da'dara, R. D. Walter and K. Luersen, *Biol. Chem.*, 2003, **384**, 83–91.

49. J. L. Ekstrom, W. D. Tolbert, H. Xiong, A. E. Pegg and S. E. Ealick, *Biochemistry*, 2001, **40**, 9495–9504.

50. E. K. Willert, R. Fitzpatrick and M. A. Phillips, *Proc. Natl. Acad. Sci. U. S. A.*, 2007, **104**, 8275–8280.

51. H. G. Williams–Ashman and A. Schenone, *Biochem. Biophys. Res. Commun.*, 1972, **46**, 288–295.

52. U. Regenass, H. Mett, J. Stanek, M. Mueller, D. Kramer and C. W. Porter, *Cancer Res.*, 1994, **54**, 3210–3217.

53. M. J. Millward, A. Joshua, R. Kefford, S. Aamdal, D. Thomson, P. Hersey, G. Toner and K. Lynch, *Invest. New Drugs*, 2005, **23**, 253–256.

54. L. L. Siu, E. K. Rowinsky, L. A. Hammond, G. R. Weiss, M. Hidalgo, G. M. Clark, J. Moczygemba, L. Choi, R. Linnartz, N. C. Barbet, I. T. Sklenar, R. Capdeville, G. Gan, C. W. Porter, D. D. Von Hoff and S. G. Eckhardt, *Clin. Cancer Res.*, 2002, **8**, 2157–2166.

55. C. Danzin, P. Marchal and P. Casara, *Biochem. Pharmacol.*, 1990, **40**, 1499–1503.

56. Y. Wu and P. M. Woster, *J. Med. Chem.*, 1992, **35**, 3196–3201.

57. J. A. Secrist III, *Nucleosides & Nucleotides*, 1987, **6**, 73–84.
58. E. Y. Artamonova, L. L. Zavalova, R. M. Khomutov and A. R. Khomutov, *Biorg. Khim.*, 1986, **12**, 206–212.
59. L. M. Shantz, B. A. Stanley, J. A. Secrist and A. E. Pegg, *Biochemistry*, 1992, **31**, 6848–6855.
60. S. Bale and S. E. Ealick, *Amino Acids*, 2010, **38**, 451–460.
61. W. D. Tolbert, Y. Zhang, S. E. Cottet, E. M. Bennett, J. L. Ekstrom, A. E. Pegg and S. E. Ealick, *Biochemistry*, 2003, **42**, 2386–2395.
62. W. D. Tolbert, J. L. Ekstrom, I. I. Mathews, J. A. I. Secrist, P. Kapoor, A. E. Pegg and S. E. Ealick, *Biochemistry*, 2001, **40**, 9484–9494.
63. G. D. Markham, P. O. Norrby and C. W. Bock, *Biochemistry*, 2002, **41**, 7636–7646.
64. D. E. McCloskey, S. Bale, J. A. Secrist, A. Tiwari, T. H. Moss, J. Valiyaveettil, W. H. Brooks, W. C. Guida, A. E. Pegg and S. E. Ealick, *J. Med. Chem.*, 2009, **52**, 1388–1407.
65. S. Bale, W. Brooks, J. W. Hanes, A. M. Mahesan, W. C. Guida and S. E. Ealick, *Biochemistry*, 2009, **48**, 6423–6430.
66. S. Bale, M. M. Lopez, G. I. Makhatadze, Q. Fang, A. E. Pegg and S. E. Ealick, *Biochemistry*, 2008, **47**, 13404–13417.
67. M. Pankaskie and M. M. Abdel–Monem, *J. Med. Chem.*, 1980, **23**, 121–127.
68. J. C. Ma and D. A. Dougherty, *Chem. Rev.*, 1997, **97**, 1303–1324.
69. W. H. Brooks, D. E. McCloskey, K. G. Daniel, S. E. Ealick, J. A. Secrist, 3rd, W. R. Waud, A. E. Pegg and W. C. Guida, *J. Chem. Inf. Model.*, 2007, **47**, 1897–1905.
70. A. V. Toms, C. Kinsland, D. E. McCloskey, A. E. Pegg and S. E. Ealick, *J. Biol. Chem.*, 2004, **279**, 33837–33846.
71. E. M. Bennett, J. E. Ekstrom, A. E. Pegg and S. E. Ealick, *Biochemistry*, 2002, **41**, 14509–14517.
72. Z. J. Lu and G. D. Markham, *Biochemistry*, 2007, **46**, 8172–8180.
73. S. Bale, K. Baba, D. E. McCloskey, A. E. Pegg and S. E. Ealick, *Acta Crystallogr. D Biol Crystallogr*, 2010, **66**, 181–189.
74. E. K. Willert and M. A. Phillips, *Mol. Biochem. Parasitol.*, 2009, **168**, 1–6.
75. B. Hirth, R. H. Barker, Jr, C. A. Celatka, J. D. Klinger, H. Liu, B. Nare, A. Nijjar, M. A. Phillips, E. Sybertz, E. K. Willert and Y. Xiang, *Bioorg. Med. Chem. Lett*, 2009, **19**, 2916–2919.
76. R. H. Barker, Jr, H. Liu, B. Hirth, C. A. Celatka, R. Fitzpatrick, Y. Xiang, E. K. Willert, M. A. Phillips, M. Kaiser, C. J. Bacchi, A. Rodriguez, N. Yarlett, J. D. Klinger and E. Sybertz, *Antimicrob. Agents Chemother.*, 2009, **53**, 2052–2058.

CHAPTER 3

Antiparasitic Drug Discovery for the Polyamine Pathway

NIGEL YARLETT* AND MARY MORADA

Haskins Laboratories and Department of Chemistry and Physical Sciences, Pace University, New York, NY 10038, USA

This review details the advances made in the design and use of antimicrobials targeting the polyamine biosynthetic pathways of parasites in the past 5 years. Readers wanting to have a historical perspective are encouraged to read the earlier review by Bacchi and Yarlett.[1]

It is evident that polyamine metabolism in parasites is as diverse as the parasites themselves. It has evolved to serve the needs of the parasite in its specific niche and often demonstrates similar characteristics for groups parasitizing the same anatomical site (bloodstream, intracellular, gastrointestinal, vaginal). That polyamines are indispensable for the growth and development of all living cells, coupled with the significant differences that are evident between mammalian and parasite polyamine metabolism, makes this a valid druggable target. The review is organized in sections that cover the major polyamine biosynthetic enzymes present in parasites. There is no common pathway for all parasites, so representative pathways are presented.

3.1 Ornithine Decarboxylase

Ornithine decarboxylase (ODC EC 4.1.17) is the first and rate-limiting enzyme in polyamine biosynthesis, catalyzing the decarboxylation of L-ornithine to

RSC Drug Discovery Series No. 17
Polyamine Drug Discovery
Edited by Patrick M. Woster and Robert A. Casero, Jr.
© Royal Society of Chemistry 2012
Published by the Royal Society of Chemistry, www.rsc.org

putrescine. As the polyamine biosynthetic pathway is essential for the growth and survival of parasites, the inhibition of this enzyme is an excellent strategy for antiparasitic therapy. ODC has several important factors. In mammalian cells, ODC is highly regulated, and its turnover is extremely rapid with half-lives of less than 1 h.[2] In contrast with mammalian cells, ODC has been demonstrated to be a stable enzyme in *Trypanosoma brucei*[3] and *Leishmania donovani*.[4] Like other short-lived proteins, the mammalian ODC is degraded by the 26 S proteasome.[5] However, ODC degradation is not triggered by ubiquitination which otherwise is the usual mechanism for targeting proteins for degradation by the 26 S proteasome. Instead, the degradation of ODC by the 26 S proteasome is induced by the binding of an inducible protein called antizyme to the enzyme.[6] The polyamines induce the degradation of ODC by affecting the synthesis of antizyme. The synthesis of antizyme is dependent on a unique mechanism involving ribosomal frame shifting.[7] Antizyme binds to the monomer of ODC with high affinity, which results in a shift from enzymatically active dimers to enzymatically inactive monomers. ODC also requires pyridoxal phosphate (PLP) as a cofactor and thiol group reducing agents, which are necessary for enzyme activity. ODC expression also is regulated by transcription, stability and the efficiency of translation of the mRNA.

The inhibition of ODC has been well studied in various systems including humans, plants and protozoa. The classical ODC inhibitor, DL-α-difluoromethylornithine (DFMO) functions as an irreversible suicide inhibitor of the enzyme and is to date the most effective inhibitor tested. DFMO had shown remarkable therapeutic efficacy in treating *Trypanosoma brucei brucei*,[8] *Trypanosoma brucei gambeinse* (African sleeping sickness),[9–11] has met with limited success in treating *Leishmania* sp.[12] *Plasmodium falciparum* and *Plasmodium berghei*.[13]

Because of the limited success of DFMO, new ODC inhibitors have been tested: 3-aminooxy-1-aminopropane (APA), and its derivatives, CGP 54169A and CGP 52622A. APA, is a structural analog of putrescine, and the aminooxy group in compound forms an oxime with the pyridoxal phosphate cofactor in the active site of ODC. APA has been shown to inhibit the growth of *Leishmani donovani* promastigotes in a dose-dependent manner over 72 h with an IC_{50} of 42 µM in promatigotes and an IC_{50} of 5 µM in intracellular amastigotes.[14] This effect can be compared to DFMO with an IC_{50} of > 10 000 µM in promatigotes and 50 µM in amastigotes (Table 3.1).[14] Treatment with 50 µM APA for 48 h inhibited ODC *L. donovani* activity by 76%, resulting in a 44% reduction in putrescine and a 70% reduction in spermidine levels when compared to control cells (Table 3.1).[14] At 50 µM of APA, trypanothione levels decreased by 79%. However, at concentrations up to 200 µM, APA had no effect on macrophages, therefore establishing the usefulness of *L. donovani* ODC as a target for the treatment of leishmaniasis.

In *Plasmodium falciparum*, the ODC and *S*-adenosylmethionine decarboxylase (AdoMetDC, see below) enyzmes are both located on a single open reading frame that encodes a bifunctional ODC-AdoMetDC protein, a property unique to *Plasmodia*.[15] This bifunctional enzyme is considered to be a druggable

Table 3.1 *In vitro* activity of ODC inhibitors.

Name	Structure	P. faciparum		L. donovani promastigote		L. donovani amastigote		Reference
		K_i (μM)	IC_{50} (μM)	IC_{50} (μM)	K_i (μM)	IC_{50} (μM)	K_i (μM)	
DFMO		87.6	1250	>10,000	NA	50	NA	4, 17, 18
APA		0.0027	1	42	NA	5	NA	18, 14
CGP 54169A		0.0079	2	NA	NA	NA	NA	18
CGP 52622A		0.002	2.7	NA	NA	NA	NA	18
PT3			14					23

IC_{50}, is the concentration of compound that causes a 50% reduction in product formation; K_i, is the concentration of inhibitor causing 50% inhibition of the enzyme; NA, data not available.

protein with an index of 0.8 out of 1.[16] Despite this unusual organization, both domains act independently and exhibit specific regulatory features that are distinct from those of the monofunctional host enzymes.[17] Studies have demonstrated that APA causes a decrease in polyamine concentrations in *Plasmodium falciparum* trophozoite stages with an IC_{50} of 1.0 µM.[18] However, several putrescine analogs (CGP 54169A and CGP 52622A) have K_i values in the low nM range because of the parasite-specific interactions with ODC,[19] and are more than 600 times more effective than DFMO *in vitro* (Table 3.1).[18] All of these inhibitors decreased the levels of putrescine and spermine but resulted in an increase in intracellular spermine, which could be antagonized with the addition of exogenous putrescine of 0.5 mM.[20]

One of the first AdoMetDC inhibitors, methylglyoxal bis(guanlhydrazone) (MGBG), prevented *P. falciparum* growth *in vitro*[21] and an inhibitor that resembles AdoMet; MDL-73811, with a K_i of 1.6 µM and an *in vitro* IC_{50} of 1–3 µM, was 1000-fold more effective than DFMO.[22] However, MDL-73811 when used alone did not cure mice that were infected with *P. berghei*.[22] This is attributed to a rapid clearance of the inhibitor, which has a plasma half-life of about 10 min. In contrast to MDL 73811, CGP-40215A did not perturb intracellular polyamine concentrations, suggesting a mechanism of action against *P. falciparum* independent of polyamine synthesis.[18] With a K_i of 3 µM, CGP 48664A (an AdoMetDC inhibitor) was least effective against the plasmodium parasite (Table 3.2).[18]

Müller and coworkers[23] presented a new approach to target the malaria parasite by designing novel pyridoxal-adducts as antimalarials. These compounds are trapped upon phosphorylation; in this state, they mimic PLP, and together with the substrate moiety, they bind and inhibit PLP-dependent enzymes, which eventually kill the parasite. PLP is a crucial cofactor during erythrocytic schizogony in *P. falciparum*. One specific novel compound, PT3, which is a cyclic pyridoxyl-tryptophan methyl ester, inhibits the proliferation of the parasite very effectively with an IC_{50} of 14 µM and is not toxic to human cells (Table 3.1).[23] Using *in silico* modeling, PT3 was shown to fit well into the PLP-binding site of the plasmodial ODC and abolish enzyme activity *in vitro*.

3.2 *S*-Adenosylmethionine Decarboxylase

S-Adenosylmethionine decarboxylase (AdoMetDC EC 4.1.1.50) generates decarboxylated *S*-adenosylmethionine (dcAdoMet), which is required as the aminopropyl group donor for spermidine and spermine synthase to form spermidine and spermine, respectively.[24] AdoMetDC is a major regulatory enzyme in polyamine biosynthesis and is considered a potentially important drug target for the chemotherapy of proliferative and parasitic diseases.

Polyamine metabolism in trypanosomes has been shown to be a valid chemotheraputic target for inhibitors aimed at critical points in the pathway such as AdoMetDC. AdoMetDC is activated by a novel mechanism unique to the trypanosomatid parasites; the functional form of the enzyme is a heterodimer

Table 3.2 AdoMetDC inhibitors activites against *T.b. brucei* LAB 110 EATRO, *T.b. rhodesiense* KETRI 243, *T.b. rhodesiense* KETRI 2538 and *P. falicparum*.

	Structure	*T. b. brucei Lab 110 EATRO* IC_{50} *(µM)*	*T. b. rhodesiense KETRI 243* IC_{50} *(µM)*	*T.b. rhodesiense KETRI 2538* IC_{50} *(µM)*	*P. falciparum* IC_{50} *(µM)*	*Reference*
CGP 40215A					1.8	18
CGP 48664A					8.8	18
MDL-73811		50	20	21	3	29
Genz-644131		0.58	0.375	0.3		29
Genz-643990-1				0.172		30
Genz-644043-1				0.319		30
Genz 644053-1				0.259		30

IC_{50}, is the concentration of compound that causes a 50% reduction in product formation.

between the active subunit and a prozyme that arose from gene duplication,[25] which are both essential for growth of the blood form *Trypanosoma brucei* and *Trypanosoma cruzi*.[26] Gene knockouts and small interfering RNA experiments have shown that the suppression of this enzyme by knockdown of either AdoMetDC or its regulatory subunit prozyme is lethal to the parasite.[27] Prozyme is an essential activator of AdoMetDC activity and in its absence only low activity of the AdoMetDC homodimer occurs which is insufficient to maintain cell growth. Therefore, inhibitors that block heterodimerization, or lock the AdoMetDC structure into an inactive confirmation, would provide a parasite-specific mechanism to inhibit this essential enzyme.

The compound MDL-73811 (5'-{[(Z)-4-aminobut-2-enyl]-(methylamino)}-adenosine) was the lead compound for Merrell Dow's development of AdoMetDC inhibitors in the 1980s. MDL-73811 is an irreversible inhibitor of AdoMetDC. The compound has a favorable level of selectivity for killing trypanosomes compared to its toxicity for mammalian cells.[28] The IC_{50} for MDL-73811 in trypanosomes ranges from 0.004 µg/ml to 0.05 µg/ml depending on the strain (Table 3.2) [29,30] and was found to cure *T. brucei brucei* and murine laboratory model infections.[31] Even though MDL-73811 has trypanocidal activity, it is not an attractive drug candidate. The compound has excellent activity versus acute model infections, but when used alone it is incapable of curing a model CNS infection,[32] and serum drug levels in rat-model infections appeared to indicate that the compound is cleared with a $t_{1/2}$ of 10 to 20 min.[33]

New and novel *S*-adenosylmethionine decarboxylase inhibitors have been synthesized for the treatment of human African trypanosomiasis. According to Barker *et al.*,[30] MDL 73811 and four other analogs when tested for trypanocidal activity *in vitro* and for the ability to inhibit purified AdoMetDC prozyme–enzyme complex were found to be active. Other analogs (Genz-644390, Genz 644043, Genz 644053) showed a > 100-fold loss of activity towards the parasite (Table 3.2). The Genzyme analog, 8-methyl-5'-{[(Z)-4-aminobut-2-enyl]-(methylamino)}adenosine (Genz-644131), was 10-fold more effective in blocking parasite growth and fivefold-enhanced inhibition of the purified *T.b. rhodesiense* strain STIB 900 AdoMetDC heterodimer. This analog differs from the parent in having an 8-methyl group on the purine ring that has favorable pharmacokinetic, biochemical and trypanocidal activities. Genz-644131 showed a five- to 85-fold greater potency against the trypanosome strains *T.b. brucei Lab 110 EATRO*, *T.b. rhodesiense KETRI 243* and *T.b. rhodesiense KETRI 2538 in vitro* and a fivefold higher activity against MDL-73811, a threefold longer blood half-life, an eightfold higher concentration in the brain and 41% serum protein binding.[29,30]

3.3   Spermidine Synthase

Spermidine synthase (SpdSyn) EC 2.5.1.16) catalyzes the transfer of an aminopropyl group from *S*-adenosylmethionamine to putrescine forming spermidine.[34] SpdSyn in bloodstream form (BSF) *T. brucei* has been shown to represent a valid drug target for chemotherapy. A tetracyline-inducible RNA interference (RNAi) system was used to determine the functional importance of *T. brucei* SpdSyn in BSF parasites.[35] In the absence of tetracycline, the RNAi cell lines grew at the same rate as the parental cultures, and the addition of tetracycline showed a dramatic effect. Within 72 h, there was a significant decrease in the growth rate, with a reduction in the endogenous TbSpdSyn mRNA and a 60% decline in spermidine levels.[35] Xiao *et al.*[36] studied the effects of the RNAi-induced silencing of SpdSyn and ODC genes in *T. brucei*. The knockdown of either gene product led to the depletion of polyamines and trypanothione pools resulting in cell death. dcAdoMet levels were elevated, while AdoMet was not affected. There was also no significant effect on the protein levels of other polyamine pathway enzymes. This cellular response to the loss of AdoMetDC activity is distinctive, suggesting that AdoMetDC activity controls the expression levels of the other spermidine biosynthetic enzymes.

The regulation of the intracellular polyamine pools in many eukaryotes is accomplished by a series of regulatory and degradation pathways with disruption systems promoting unwanted side-effects such as uncontrolled cell proliferation or apoptosis.[37] The downregulation of the corresponding mRNA correlated with a cessation in growth, and a decrease in intracellular spermidine, but there was no significant build up of putrescine. It was also shown that the activity of TbSpdSyn could be dramatically reduced by dicyclohexylamine and cyclohexylamine (CHA), compounds that act as competitive inhibitors with respect to putrescine. The K_m for human SpdSyn (HsSpdsyn) is 80 µM, but the TbSpdSyn appears to have a high affinity for dcSAM compared to the human enzyme, with HsSpdSyn exhibiting a K_m for dcSAM of 7 µM.[38] This is 70-fold higher than its trypanosomal counterpart, and this difference may represent a property for inhibitor design. TbSpdSyn is essential in BSF *T. brucei*, and its inability to scavenge enough spermidine from the environment is validation that TbSpdSyn is an excellent potential target for future drug development.

Targeting the *P. falciparum* form of SpdSyn as a drug target has not been studied extensively. The analog CHA has been studied as an inhibitor of the aminopropyltransferase activity of SpdSyn by competitively binding to the putrescine binding pocket.[16] Several derivatives of CHA has been investigated: 4-MCHA (trans-1,4-methylcyclohexlamine), dicyclohexylamine, as well as APA (an inhibitor of ODC) and its derivative, 5-amino-1-pentene (APE). The inhibitor that exhibited the best *in vitro* growth IC_{50} was 4-MCHA at 34 µM (Table 3.3). 4-MCHA was also the most potent inhibitor of SpdSyn activity at a K_i of 0.18 µM (Table 3.3). APA only inhibited SpdSyn at an IC_{50} of 84 µM, less effective than its activity towards ODC (Table 3.3). When 4-MCHA was administered at 25 mg/kg in mice infected with *P. berghei*, there was no effect on parasite proliferation or survival of the mice.[39] This may be a cause of metabolism of 4-MCHA in the host organism. The inhibition of SpdSyn by CHA showed a significant decrease in both spermidine and spermine levels, associated with a fourfold increase in putrescine levels.[40] *S*-adenosyl-1,8-diamino-3thiooctane (AdoDATO) was reported to have an inhibitory activity towards SpdSyn with an IC_{50} of 8.5 µM (Table 3.3).[41]

Table 3.3 Effects of SpdSyn inhibitors in *P. falciparum.*

	Structure	*K$_i$ (μM)* *IC$_{50}$*	In vitro *(μM) IC$_{50}$*	*Reference*
		P. falciparum		
CHA		19.7	198	107
4-MCHA		0.18	34.2	107
APA		84	1	107
AdoDATO		8.5	8.5	41
Dicyclohexylamine		> 1000	342	107
APE		6.5	83.3	107

IC_{50}, is the concentration of compound that causes a 50% reduction in product formation; K_i, is the concentration of inhibitor causing 50% inhibition of the enzyme activity.

A theoretical low-resolution 3D model of the cytosolic *P. falciparum* spermidine synthase (PfSpdSyn) has been made based on the crystal structure of *Arabidopsis thaliana* (Figure 3.1).[42] The comparison between the active sites of PSpdSyn and human SpdSyn revealed key differences that could be useful for the design of new selective inhibitors of PfSpdSyn.

3.4 Trypanothione Synthase

Trypanothione is a low-molecular-mass thiol found in members of the order Kinetoplastida (African trypanosomes, S. American trypanosomes and *Leishmania* sp.), which is responsible for maintaining the intracellular reducing environment during aerobic metabolism. The presence of trypanothione has also been reported in *Entamoeba histolytica*,[43] an anaerobic parasite that lacks glutathione and glutathione-metabolizing enzymes,[44] and *Naegleria fowleri*, an aerobic parasite that possess both glutathione and trypanothione-metabolizing enzymes.[45] The identification of trypanothione and trypanothione-metabolizing enzymes in non-kinetoplastid species (Figure 3.2) is interesting from an evolutionary standpoint, but it is not unequivocal and has been questioned.[46] Trypanothione synthase (T$_{SH}$S) is the sole enzyme responsible for the formation of trypanothione [N^1,N^8-bis-(glutathionyl)spermidine] from glutathione and spermidine. The protein contains an ATP-dependent synthetase domain that generates the intermediate glutathionylspermidine and the final product trypanothione, and also an amidase domain that hydrolyzes these thiols back

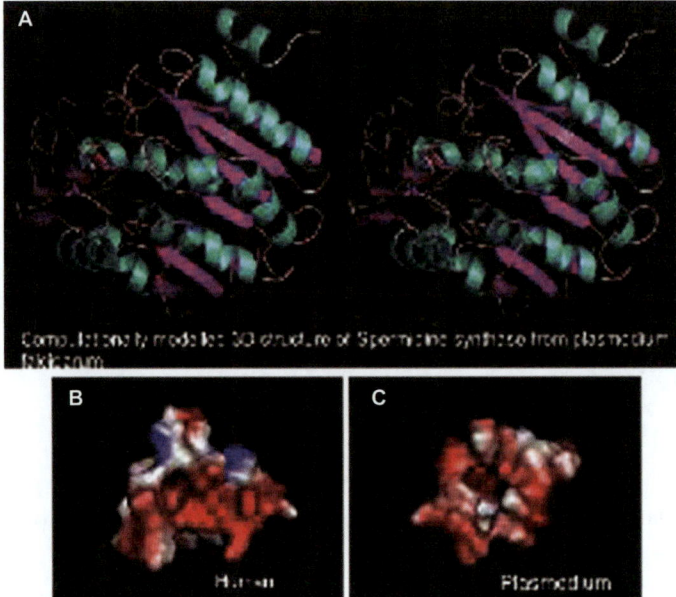

Figure 3.1 (A) Predicted model of *P. falciparum* spermidine synthase in sterio view. (B & C) Electrostatic potential differences mapped on the active site residues within the 5 Å distance.

to glutathione and spermidine.[47] Trypanothione together with trypanothione reductase ($T_{SH}R$) replaces the function of glutathione, glutathione reductase ($G_{SH}R$) and thioredoxin/thioredoxin reductase in mammalian cells. In conjunction with type I and II tryparedoxin peroxidases, trypanothione is essential in combating host-cell-derived oxidative stress mechanisms and redox cycling of oxygen-reactive drugs such as nifurtimox.[48] The essential function of trypanothione in these parasites is supported by the discovery of trypanothione *S*-transferase and glyoxylase I and II, thiol-dependent enzymes that are involved in defense against chemical stress. These findings have resulted in several seminal papers exploring $T_{SH}S$ as a target for rational drug design. Trypanothione is synthesized from glutathione and spermidine by an ATP-dependent C:N ligase ($T_{SH}S$) with N^1 and N^8 glutathionyl spermidine as intermediates. Phosphonates and phosphinates are known inhibitors of C:N ligases and have been explored as inhibitors of glutathione and trypanothione synthases (Table 3.4). A series of phosphinopeptides structurally related to glutathione have shown promise as antiproliferative agents against the hemoflagellated protozoan *Trypanosoma cruzi*; two of these, butyl-{[(L-γ-glutamyl-L-leucyl) amino]methyl} phosphinate and pentyl-{[L-γ-glutamyl-L-leucyl) amino]methyl} phosphinate, were potent growth inhibitors of *T. cruzi* amastigotes grown in myoblasts.[49] The presence of the phosphinic acid moiety on these compounds mimics the tetrahedral transition state of trypanothione synthase ($T_{SH}S$), a typical C:N ligase, and demonstrates slow-tight binding kinetics.[49] Phosphonic

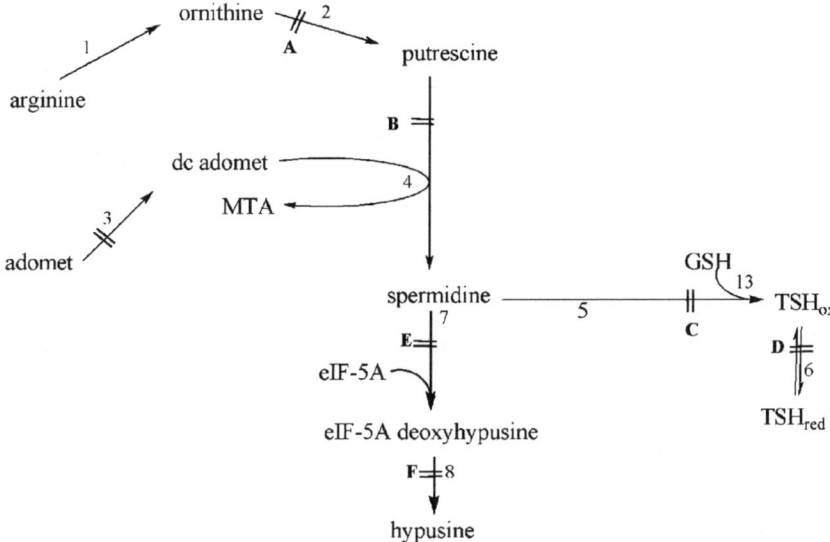

Figure 3.2   **Polyamine biosynthesis by Trypanosomatids.** Enzymes involved in poly-amine biosynthesis from arginine: 1. arginase, 2. ornithine decarboxylase, 3. adomet decarboxylase, 4. spermidine synthase, 5. trypanothione syn-thase, 6. trypanothione reductase, 7. deoxyhypusine synthase, 8. deoxy-hypusine hydrolase. Inhibitors that block specific parts of the pathway: A. DL-α-difluoromethylornithine, B. MDL73811, C. 2-[3-{(3-fluoro-phenyl)-1H-Indazol-1-yl}-3, 3-Dimethylbutan-2one, D. 1,1-(benzo[γ]thio-phen-2-yl-cyclohexyl)-piperidine. E. N^1-guanyl-1,7-diaminoheptane, F. ciclopirox. Abbreviations: adomet, adenosylmethionine; dcadomet, dec-arboxylated adenosylmethionine; MTA, methylthioadenosine; GSH, glutathione; TSH, trypanothione; eIF-5A, elongation factor 5A.

and phosphinic acid derivatives of glutathione have been shown to inhibit purified $T_{SH}S$ from *Crithidia fasciculata, Trypanosoma brucei, T. cruzi* and *Leishmania major* (Table 3.1). The tetrahedral phosphonate acts as a simple ground-state analog of glutathione with a K_i 156 μM for *C. fasciculata* trypa-nothione synthase, whereas the phosphinate behaves as a stable mimic of the postulated unstable tetrahedral intermediate.[48] Kinetic studies indicate that the phosphinate acts as a slow binding bisubstrate competitive inhibitor (with respect to glutathione and spermidine) with a dissociation constant K_i of 18.6 nM for *C. fasciculata* glutathione synthase ($G_{SH}S$). However, the inhibitor was 20–40 fold less effective towards $T_{SH}S$ from *T. brucei, T. cruzi* and *L. major* which may result from the substitution of Asp610 in the active site of $G_{SH}S$ for Pro610 in the active site of $T_{SH}S$[47]. It has also been suggested that the polyamine binding site of $T_{SH}S$ has to accommodate N^1-glutathionylspermidine or N^8-glutathionyl-spermidine in addition to spermidine which may be a critical difference in sen-sitivity to phosphinate inhibition by $G_{SH}S$ and $T_{SH}S$.[47] The phosphinate inhibitor showed no growth inhibition at 100 μM over 72 h towards *T. brucei* procyclics, *T. cruzi* epimastigotes or *L. major* promastigotes, which is proposed to

Table 3.4 Inhibitors of trypanothione synthase. The concentration of inhibitor required to cause 50% inhibition of growth is shown for *T. b. rhodesiense*, *T. cruzi* and *L. major*.

Trypanothione synthase

Name	Structure	T. brucei rhodesiense EC_{50}	T. cruzi EC_{50}	L. major EC_{50}	Reference
Phosphinate		1.3 mM	0.53 mM	0.65 mM	48
Butyl-[{(L-γ-glutamyl-L-leucyl)amino}methyl]phosphinate			9.8 μM		49
N-(3-{Dimethylamino}propyl)-2-(3{3-fluorophenyl}-1*H*-indazol-1-yl)}acetamide		6.9 μM			50
2-[3-{(3-Fluorophenyl)-1*H*-indazol-1-yl]-3,3-dimethylbutan-2one		0.095 μM			51

EC_{50} is the concentration of inhibitor causing 50% inhibition of growth.

be a penetration problem. Interestingly, boronic acid derivatives of glutathione were shown to have low K_i and IC_{50} values for $G_{SH}S$ (81 µM and 17 µM, respectively) suggesting the utility of boron substituted compounds for inhibition of $T_{SH}S$.[49] Other potent inhibitors of trypanothione synthase have recently been synthesized, including *N*-[3-(dimethylamino)propyl]-2-[3-(3-fluorophenyl)-1*H*-indazol-1-yl] acetamide which has an IC_{50} of 45nM for *T. b. brucei* $T_{SH}S$.[50] *N*-[3-(dimethylamino)propyl]-2-[3-(3-fluorophenyl)-1*H*-indazol-1-yl] acetamide causes a dose-dependent reduction in *T. b. brucei* trypanothione and an EC_{50} against the parasite in the low micromolar range (Table 3.4).[50] Using a high-throughput screen, a novel series of $T_{SH}S$ inhibitors were identified, which had mixed, uncompetitive and allosteric-type inhibition with respect to spermidine, ATP and glutathione, respectively. Several of these inhibitors had low EC_{50} values towards *T. b. brucei*, e.g. 2-{3-[(3-fluorophenyl)-1*H*-indazol-1-yl]-3,3-dimethylbutan-2one} which had an IC_{50} for the enzyme of 95 nM.[51]

3.5 Trypanothione Reductase

The reduction in trypanothione disulfide is catalyzed by trypanothione reductase ($T_{SH}R$), a homologous reaction to the reduction in glutathione disulfide by glutathione reductase ($G_{SH}R$; EC 1.8.1.7) in mammals. $T_{SH}R$ and $G_{SH}R$ share 40% sequence identity and have the same three amino acids at the catalytic site: Cys53, Cys58 and His4610 of the second subunit, in *T. cruzi*.[52] *T. cruzi* contains a single gene for $T_{SH}S/T_{SH}R$, which codes for a single bifunctional protein.[53,54] Due to its critical function in recycling oxidized trypanothione, it is an attractive drug target. Selectivity of inhibitors towards $T_{SH}R$ versus the host $G_{SH}R$ is possible because of the presence of an extra protonated amino group in trypanothione; hence, $T_{SH}R$ possesses a negatively charged active site, whereas that of human $G_{SH}R$ carries an overall positive charge.[52] $T_{SH}R$ also has a larger substrate-binding pocket to fit its bulkier endogenous ligand, trypanothione. A number of different $T_{SH}R$ inhibitors have been reported in the literature that can be broadly classified into five groups:

3.5.1 Tricyclics

Tricyclic neuroleptics have been used as a backbone for the synthesis of novel trypanothione reductase inhibitors. The 2-, 4- and 5-chlorophenyl analogs prepared by quaternization of the nitrogen atom of chlorpromazine resulted in the synthesis of a series of *N*-acyl-2-amino-(4-chlorophenyl) and (2-chlorophenyl) sulfides which demonstrated mixed inhibition kinetics for the *T. cruzi* $T_{SH}R$ (K_i, $K_i' = 11.3$–42.8 mM). The quaternized analogs of the 2-chlorophenyl phenyl sulfides had strong antitrypanosomal and antileishmanial activity *in vitro* against *T. b. rhodesiense*, *T. cruzi* and *L. donovani*, whereas the *N*-acyl-2-amino-4-chlorophenyl sulfides were active against *P. falciparum*.[55] A second series based on the structure of the known tricyclic (neuroleptic) inhibitor of $T_{SH}R$, prochlorperazine (PCP (2,1-(1-phenyl-cyclohexyl)-piperidine, phenylcyclidine)

(Table 3.5) was tested for the ability to inhibit *T. cruzi* trypanothione reductase.[56] Kinetic testing of this series revealed a fivefold improvement over PCP. One of these, arylcyclohexylamine (BTCP; 1,1-(1-benzo[γ]thiophen-2-yl-cyclohexyl)-piperidine), was considered to be a promising screening hit for further development due to its low molecular weight, low micromolar potency against *T. cruzi* $T_{SH}R$ (IC_{50} = 3.7 μM), a promising ligand efficiency (0.35 kcal/mol 1 L), lack of activity against the human homolog of $T_{SH}R$, glutathione reductase ($G_{SH}R$), and the fact that phencyclidines are known to cross the blood–brain barrier, an essential property for the successful treatment of stage 2 human African trypanosomasis (HAT). BTCP also has the advantage of being a drug-like molecule, in contrast to some of the more potent reported $T_{SH}R$ inhibitors, many of which are polyamine analogs designed to mimic the spermidine moiety of the enzyme substrate trypanothione.[57]

3.5.2   Polyamine Analogs

The most promising compound in this area is the macrocyclic spermidine alkaloid lunarine isolated from the biennial *Lunaria biennis* (money plant or honesty). Various analogs related to the alkaloid have been synthesized in racemic form and evaluated for ability to inhibit $T_{SH}R$. Kinetic data are consistent with an inactivation mechanism involving a conjugate addition of an active site cysteine residue onto the C24–C25 double bond of the tricyclic nucleus of the alkaloid.[58] The 3,5-disubstituted benzofuran mimetics of the naturally occurring spermidine-bridged macrocyclic alkaloid lunarine behaved as time-dependent inhibitors of $T_{SH}R$. In this series of compounds, the bis-polyaminoacrylamide derivatives were all shown to be competitive inhibitors, but only the bis-4-methyl-piperazin-1-yl-propylacrylamide derivative 4 displayed time-dependent activity (Table 3.5).[59] Analogs based on the structure of norspermidine peptidic inhibitors have also been evaluated for inhibition of $T_{SH}R$ with some success.[60] Kinetic analysis of these analogs indicates that subtle structural changes altered their mechanism of action, switching between competitive and non-competitive (allosteric) inhibition.

3.5.3   Redox Inhibitors

The ability of naphtho- and anthraquinones to inhibit redox cycling of the $T_{SH}R$ in trypanosomes has been known for some time.[61] More recently, a series of 1,4-naphthoquinones starting from menadione, plumbagin and juglone were synthesized as potential inhibitors of *T. cruzi* $T_{SH}R$. The 3,3'-[poly-aminobis(carbonylalkyl)]bis(1,4-naphthoquinone) series, constructed by the addition of various polyamine chains to carbons 2 and/or 3 of the three parent molecules and optimized for specificity of $T_{SH}R$ inhibition versus human disulfide reductases, were identified.[61] In these analogs, an optimum chain length was determined for inhibition of trypanothione disulfide reduction. Kinetically,

Table 3.5 Inhibitors of trypanothione reductase. The concentration of inhibitor required to cause 50% inhibition of growth is shown for *T. b. rhodesiense*, *T. cruzi* and *L. major*.

Trypanothione reductase	Structure	*T. brucei rhodesiense* EC_{50}	*T. cruzi* EC_{50}	*L. major* EC_{50}	Reference
Tricyclic: 1,1-(benzo[γ]thiophen-2-yl-cyclohexyl)-piperidine		10 μM	3.7 μM		57
Polyamine analog: bis-4-methyl-piperazin-1-yl-propylacrylamide					58
Redox inhibitor: 3,3'-[polyaminobis(carbonylalkyl)]bis(1,4-naphthoquinone)					61
Substrate analog: dimethyaminopropylamide analogue trypanothione			30 μM		64
Compound library screen: quinazoline			3.68 μM	2.76 μM	67

members of this series behaved as potent subversive substrates and effective uncompetitive inhibitors (versus trypanothione disulfide and NADPH) of $T_{SH}R$. The dimensions and accessibility of the cavity at the dimer interface, as a putative naphthoquinone binding site, for *T. cruzi* $T_{SH}R$ and $hG_{SH}R$ were compared using published X-ray analysis. *T. cruzi* lipoamide dehydrogenase reduced the plumbagin derivatives by an order of magnitude faster than the corresponding menadione derivatives.[61] The most efficient and specific naphthoqinone analog exhibited potent antitrypanosomal activity *in vitro* against *T. brucei* and *T. cruzi* cultures (Table 3.5).[61] Palladium (II) complexes with bioactive nitrofuran-containing thiosemicarbazones having the formulas [$PdCl_2(HL)$] and [$Pd(L)_2$] have also shown promise as *in vitro* growth inhibitors towards *T. cruzi*. The presence of the ligand was maintained or even increased as a result of palladium complexation. Although the presence of Pd will result in DNA binding, the data presented by these authors suggest that this group of compounds act by irreversible inhibiton of $T_{SH}R$ and therefore are trypanocidal due to the production of oxidative stress as a result of their bioreduction and extensive redox cycling.[62]

3.5.4   Substrate Analogs

Nonreducible substrate inhibitors of $T_{SH}R$ have been prepared in which the cystine moeity is replaced successively by djenkolic acid, lanthionine and cystathionine. These analogs retain sulfur atoms in the bridge connecting the peptidic halves of the molecules.[63] Similar analogs in which the bridging group is composed exclusively of carbon atoms have also been made with the intent of ultimately using the olefin as a means of introducing an epoxide moiety, to provide a potential irreversible $T_{SH}R$ inhibitor. These inhibitors contain either 3-dimethylaminopropylamide (DMAPA, Table 3.5) or two 3-propylamino-propylamide (PAPA) groups in place of trypanothione's spermidine moiety.[64] The acyclic trypanothione analog with two PAPA chains in the place of the spermidine moiety displayed both a lower K_m and higher k_{cat}/K_m than its analog with two DMAPA chains. The inhibition kinetics for the recombinant *T. cruzi* $T_{SH}R$ revealed modest competitive inhibition. $T_{SH}R$ had a slight preference for the DMAPA chain over the PAPA chain and tolerates the different structural geometries of the saturated versus the unsaturated inhibitors, displaying approximately a twofold preference for the saturated inhibitors.[64] Czechowicz and coworkers[65] synthesized the substrate analog dethio-trypanothione and related trypanothione analogs featuring ring-closing olefin metathesis (RCM). Dethiotrypanothione however proved to be a very poor inhibitor of *T. cruzi* $T_{SH}R$, despite its close structural similarity to the natural substrate. However, the more hydrophobic analogs which replaced the charged γ-glutamyl moieties of with Cbz groups resulted in a nearly sixfold stronger binding, while inclusion of the larger Fmoc groups in place of the γ-glutamyl moieties, resulted in the most potent inhibitor. These studies underscore the enzyme's general affinity for hydrophobic ligands.[65]

3.5.5 Compounds Identified Through Screening of a Library of Drug-Like Compounds

One of the limiting factors of a high-throughput screen is the availability of a reproducible and validated assay. Recent advances in this area include the development of a *T. cruzi* line expressing firefly luciferase or the tandem tomato fluorescent protein (tdTomato) for *in vitro* screens.[66] This breakthrough has the advantage of requiring relatively low numbers of parasites, and it can be adapted to *in vivo* studies using mouse footpad models infected with parasites containing fluorescent or bioluminescent parasites that have signal intensities sufficiently high to allow determination of changes following drug treatment.[66] This innovative method of high-throughput screening lends itself to the rapid screening of compound libraries.

By the addition of various polyamine chains to carbons 2 and/or 3 of 1,4-naphthoquinones, a series of 3,3′-[polyaminobis(carbonylalkyl)]bis(1,4-naptho-quinones) inhibitors were developed with specificity for trypanothione reductase versus human disulfide reductases (Table 3.5). These inhibitors were shown to be potent subversive substrates and effective uncompetitive inhibitors versus trypanothione disulfide and NADPH.[61] Molecular-modeling studies based on the known X-ray structures of *T. cruzi* trypanothione reductase and human equivalent enzyme glutathione reductase indicated distinct structural features, dimensions and accessibility of the cavity at the dimer interface of *T. cruzi* $T_{SH}R$ compared with that of $hG_{SH}R$ to the putative naphthaquinone binding sites. The most efficient and specific subversive substrates of *T. cruzi* trypanothione reductase and *T. cruzi* lipoamide dehydrogenase exhibited potent *in vitro* antitrypanosomal activity to *T. brucei* and *T. cruzi* cultures. The validation of this class of compound toward trypanothione reductase is supported by recent studies in which a series of quinazoline-like analogs were designed based on a scaffold protonated at a physiological pH to selectively bind trypanothione reductase versus glutathione reductase. A series of 2-piperazin-1-yl-quinazolin-4-ylamine derivatives (Table 3.5) were designed and docked with trypanothione reductase demonstrating that such a scaffold was able to interact with the trypanothione reductase active site.[67] Different moieties, such as methoxy and dimethylaminoethoxy groups in R1 and indole, quinone and tetrahydroquinoline moieties in R2, were attached to the quinazoline core to provide additional interactions with the enzyme. The designed compounds were effective inhibitors of *T. cruzi* $T_{SH}R$, demonstrated selectivity for $T_{SH}R$ over human $G_{SH}R$ and inhibited parasite growth *in vitro*.[67]

High-throughput screens have the advantage of featuring unique classes of compounds. A classical example is the results of a screen published by Holloway and colleagues[68] which identified five unique compound classes that act as *T. cruzi* $T_{SH}R$ designated as aryl/alkyl piperidines, basic benzyhydryls, nitrogenous heterocycles, conjugated indoles and iminobenzimidazoles. Further investigation of some of these compounds revealed that many of these compounds displayed greater antiparasitic activity towards *T. b. brucei* or *P. falciparum* than *T. cruzi*, but this method succeeded in the identification of novel $T_{SH}R$ inhibitor chemotypes that are drug-like and display antiparasitic activity. Based upon gene-knockout studies at the University of Dundee[69] it would be necessary to cause $>90\%$ inhibition of $T_{SH}R$ activity for cell death to occur. Therefore, an inhibitor effective in the low-nanomolar range or irreversible inhibition kinetics would be needed for therapeutic utility. Furthermore, the $T_{SH}R$ active site is large and relatively solvent-exposed, which may make the discovery of low-molecular-weight, highly potent, competitive inhibitors problematic, especially in the absence of structural information from inhibitors bound into the active site.[70]

3.6 Hypusine

Spermidine can be used for post-translational hypusination of the precursor of eukaryotic translation initiation factor 5A (eIF-5A). Hypusine modification of eIF-5A by spermidine is the most specific post-translational modification known. Only the second lysine residue of the eIF-5A peptide chain is modified, and the sequence around the hypusine residue is conserved in all domains of life from elongation factor P (EF-P) in bacteria, the archea initiation factor 5A (aIF-5A) to the eukaryotic eIF-5A, suggesting that post-translational modification by hypusination is highly conserved.[71] This hypusine modification is catalyzed by transfer of the 4-aminobutyl moiety to the ε-amino group of lysine forming a deoxyhypusine residue in the eIF-5A precursor by deoxyhypusine synthase (DHS). This is followed by hydroxylation of eIF-5A by deoxyhypusine hydroxylase (DOHH), forming hypusinated eIF-5A. The significant biological role of hypunsination in parasite growth and development deserves evaluation as a druggable target. To date, however, eIF-5A and the associated enzymes leading to hypusination have only been demonstrated in *Plasmodium falciparum*,[72] *P. vivax*,[73] *Leishmania donovani*[74] and *Trichomonas vaginalis* (Figures 3.2 and 3.3).[75]

The eIF-5A gene has been cloned and expressed from *P. falciparum* and *T. vaginalis*.[72,75] The *P. falciparum* eIF-5A gene is present as a single copy which codes for a 20 kDa cytoplasmic protein, whereas in *T. vaginalis* two copies were identified (*tveif-5a1* and *tveif-5a2*) coding for 19 kDa cytoplasmic proteins;[72,75] the presence of multiple gene copies has previously been reported for other genes identified in the *T. vaginalis* genome.[76] Phylogenetic analysis of the eIF-5A amino acid sequences of *T. vaginalis* and *P. falciparum* with those from mammals, yeast, protozoa and prokaryotes indicated that there is a high degree of conservation among the different lineages, TveIF-5A grouping with the EF-P from *E. coli* and *Marimonas* sp., supporting lateral gene transfer from a prokaryotic ancestor,[75] and *P. falciparum* grouping with other apicomplexa and having a higher homology to plant proteins.[72,75]

The DHS gene has been cloned and sequenced from *P. falciparum*, *P. vivax* and *L. donovani*,[73,74,77] but its presence in other parasites has been inferred based upon genome analysis and is often cited as present in all eukaryotes. The human and yeast recombinant protein is a single gene product of 33–48 kDa which exists as a tetramer of 160 kDa.[78] The plasmodium DHS is a bifunctional protein with

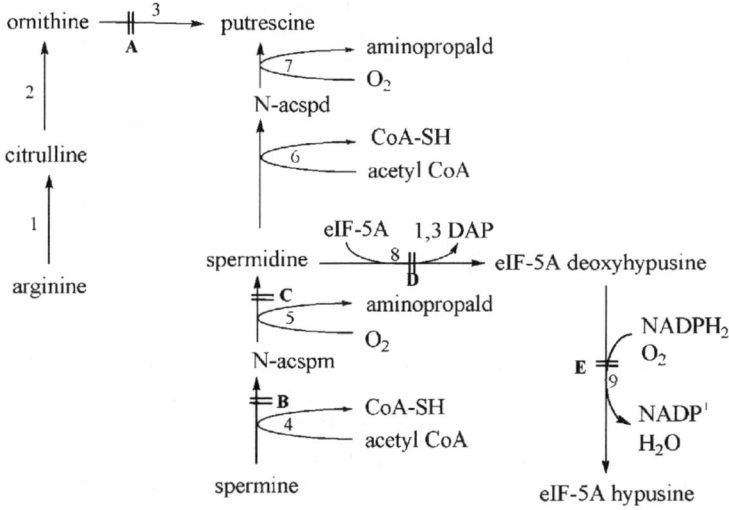

Figure 3.3 **Polyamine biosynthesis by *T. vaginalis*.** Enzymes involved in polyamine biosynthesis from arginine and spermine are: 1. arginine deiminase, 2. ornithine carbamyltransferase, 3. ornithine decarboxylase, 4. spermine: spermidine N^1-acetyltransferase, 5. polyamine oxidase, 6. spermidine acetyltransferase, 7. polyamine oxidase, 8. deoxyhypusine synthase, 9. deoxyhypusine hydrolase. Inhibitor that block specific parts of the pathway: A. DL-α-difluoromethylornithine, B. SL11158, C. BW-1, D. N^1-guanyl-1,7-diaminoheptane, E. ciclopirox. Abbreviations: eIF-5A, elongation factor 5A; 1,3-DAP, 1,3-diaminopropane; aminopropionald, aminopropionaldehyde; acspm, N^1-acetylspermine.

dual enzymatic specificities as DHS and homospermidine synthase (HSS).[77] A bi-functional DHS/HSS has been demonstrated in members of the Crotalarieae that has been proposed to result from recruitment of HSS from deoxyhypusine synthase (DHS) by independent gene duplication in several different angiosperm lineages during evolution;[79] the origin of the bifunctional protein in *Plasmodium* sp. has yet to be inferred. *L. donovani* has two DHS sequences in their genome which occur on two separate chromosomes, termed *DHSL20* and *DHSL34*.[74] The recombinant *DHSL20* protein is approximately 42 kDa and has a non-polar Leu[334] in place of the human DHS Lys[329]. This amino acid is present in the catalytic site and is indispensable for enzyme–substrate intermediate formation; hence the *L. donovani DHSL20* recombinant enzyme is inactive.[74] However, the recombinant *DHSL34* codes for a protein with a molecular mass of approximately 66 kDa (it exists as a dimer of 122 kDa) and has Lys[535] which is equivalent to the human Lys[329] catalytic center, and Lys[287], which is important for covalent intermediate formation.[80] Also conserved are the NAD^+-binding motif (equivalent to human DHS Asn[106], Asp[238], His[288] and Asp[313]) and the spermidine-binding motif (equivalent to human DHS His[288], Trp[327], Lys[329], Asp[316] and Glu[323]).[74] The *P. vivax* DHS gene (*PvDHS*) had significant conservation of gene synteny with other plasmodium sp. and coded for a 50 kDa recombinant protein

Table 3.6   Inhibitors of deoxyhypusine synthase. The concentration of inhibitor
causing 50% inhibition (K_i) of the enzyme from *P. falciparum*.

Deoxyhypusine synthase		*P. falciparum*	
N^1-guanyl-1,7- diaminoheptane		0.1 µM	77
Deoxyhypusine hydrolase Ciclopirox		5 µM	85

which had conserved spermidine and NAD^+-binding motifs.[73] The structural
analog of spermidine, N^1-guanyl-1,7-diaminoheptane (GC-7), is an effective
competitive inhibitor of the plasmodium DHS (K_i 0.1 µM for *P. falciparum*,
Table 3.6).[77] The presence of Asp^{316} in the spermidine-binding site is critical to
the binding of GC-7,[73] which is present in both *L. donovani* and *P. vivax* enzymes.
However, GC-7 was found to be a weak inhibitor of the *L. donovani* DHS, and
parasite growth was unaffected by this compound.[74] To date, only the guanylhydrazone CNI-1493 has shown chemotherapeutic potential towards this
target with a 72 h IC_{50} of 136 µM.[81]

As of now, the only parasitic DOHH gene that has been cloned and expressed is
from *P. falciparum*,[82] making this the only complete parasite pathway that has
been characterized. The *Pfdohh* gene is present as a single copy that is expressesd
as a 46.5 kDa protein which contains five E-Z-type HEAT-like repeat proteins
(Huntington protein elongation factor 3) similar to those present in the phyco-
cyanin lyase subunits of cyanobacteria, leading to the suggestion that the plasmodial DOHH may have arisen from an E/F-type phycobilin lyase that gained a
new role in hydroxylation.[82] The human DOHH has four E-Z-type HEAT-like
repeat proteins.[83] In common with the mammalian DOHH, the parasite active
site contains an Fe^{2+} as a cofactor, which in mammals can be blocked by two
catecholpeptides containing a 3,4,-dihydroxybenzoyl- and a 2,3-dihydroxy-
benzoyl moiety (Table 3.6).[84] Modeling of the parasite enzyme has only recently
been proposed[85] but holds great promise. The α-hydroxypyridones target the
active metalloenzyme and are effective inhibitors of DOHH with a decreasing
effect from: ciclopirox, deferoxamine, 2,2'-dipyridyl, deferiprone and mimosine
(IC_{50} 5–200 µM). However, many of these are toxic when used in experimental
models of rodent malaria infection.[85] The potential of alkyl-4-oxo-piperidone 3-
carboxylates which are structurally related to dihydropyrimidines are the most
potent inhibitors of DOHH so far examined.[85,86]

3.7   Polyamine Retroconversion Pathways: SSAT/PAO/SMO

The presence of a polyamine retro-conversion pathway has been described in
several protists including *Cryptosporidium parvum*[87] (Figure 3.4), *Encephalitozoon*

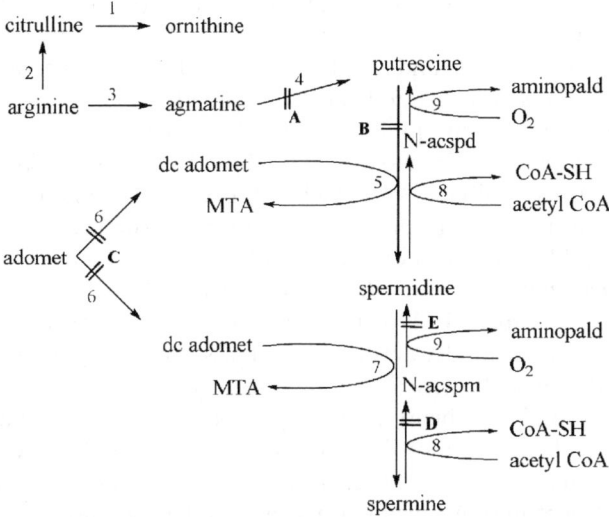

Figure 3.4 **Polyamine biosynthesis by *C. parvum*.** Enzymes involved in polyamine biosynthesis from arginine and spermine are: 1. ornithine carbamyltransferase, 2. arginine deiminase, 3. agmatine imminohydrolase, 4. arginine decarboxylase, 5. spermidine synthase, 6. adomet decarboxylase, 7. spermine synthase, 8. spermine:spermidine N^1-acetyltransferase, 9. polyamine oxidase. Inhibitors that block specific parts of the pathway: A. difluoromethylarginine, B. 4-MCHA, C. BW-1, D. SL11158, E. Bis benzyl polyamine. Abbreviations: eIF-5A, elongation factor 5A; 1,3-DAP, 1,3-diaminopropane; aminopropionald, aminopropionaldehyde; acspm, N^1-acetylspermine. Abbreviations: aminopropionald, aminopropionaldehyde; acspm, N^1-acetylspermine; adomet, adenosylmethionine; dcadomet, decarboxylated adenosylmethionine; MTA, methylthioadenosine.

cuniculi,[88] *Toxoplasma gondii*,[89] *Pneumocystis carinii*[90] and *Trichomonas vaginalis*[91] (Figure 3.3), and its utility as a target for chemotherapy has been explored in some of these parasites. Targeting polyamine retroconversion in the parasite has proved challenging due to side effects on host cells. The resulting upregulation of host-cell SSAT produces polyamine depletion due to excessive acetylation and excretion of host intracellular amines, resulting in apoptosis and cell death.[92] A series of (bis)alkyl-polyamines which have variable repeats have been synthesized and tested for anti-parasitic activity. The (bis)ethyl 3-3-3 ((bis)ethylnorspermine) and (bis)ethyl 3-4-3 ((bis)ethylhomospermine), although effective inhibitors of microsporidia growth, were found to have short half-lives in serum and have unacceptable toxicity.[93] The (bis)ethyloligoamines based on repeating aminobutyl groups with sterically hindered central carbons were found to be effective inhibitors of *Encephalitozoon cuniculi* with IC_{50} of 8.2 µM.[94] The chain length and substitution pattern appear to play an important role in activity of these compounds as inhibitors of parasite polyamine oxidase, such that the tetra- and pentamines were not active PAO inhibitors (42.5–9000 µM), whereas the octamines (1.67–8.2 µM) and decamines exhibited increasing inhibitory potential

(0.62–0.63 µM), and the dodec- and tetradecamines (0.40–0.54 µM) exhibited similar inhibitory activity.[95] The octamine analog SL11158 was a competitive inhibitor of *E. cuniculi* PAO and acted as a weak substrate.[88] These analogs have extended polyamine features, and hence SL11158 (Table 3.7) was also examined kinetically for ability to inhibit spermidine:spermine-N^1-acetyltransferase (SSAT). The octamine 11158 had mixed inhibition kinetics for SSAT consistent with the competitive inhibition observed with PAO.[88] The microsporidian poly-amine oxidase has not been purified, thus limiting structure–activity modeling which would enable refinement of these inhibitors. Compounds demonstrating *in vitro* and *in vivo* activity towards *E. cuniculi*, SL11158 (PG 11158, CGC-11158) along with the *cis* isomer SL11157 (PG 11157) and PG11302 were tested *in vivo* using an SCID mouse model developed at Tufts University. It was found that SL11157 was the most effective at clearing the infection at 75 mg/kg body weight per day for 7 days.[96] A (bis)phenylbenzyl-substituted 3-7-3 compound, 1,15-bis{N-[o-(phenyl)benzylamino}-4,12-diazapenta]-decane (BW-1, Table 3.7)) was found to be a better substrate of *E. cunniculi* PAO (K_m 2 µM compared to 30 µM with N^1-acetylspermine) producing a hydrated aldehyde and 2(phenyl)benzyla-mine as products.[97]

Cryptosporidium parvum has also been shown to have an active SSAT/PAO retroconversion pathway (Figure 3.4). The *C. parvum* SSAT is the only parasite retroconversion enzyme to date that has been cloned and expressed.[98] The *C. parvum* SSAT has distinct biochemical characteristics that distinguish it from the host SSAT enabling designer polyamine analogs to be performed. Two polyamine analogs based on the *cis* 3-4-3 backbone [N^1,N^{12}]bis(ethyl)-*cis*-6,7-dehydrospermine (SL11047) and the *cis* 4-4-4 backbone SL11157 (CGC-11157, PG-11157) were compared for *in vivo* activity using a T-cell receptor α-deficient mouse model. In all cases, the *cis* analogs were significantly more active than the *trans* isomer.[98,99] Kinetically the *cis* analogs consistently had lower inactivation half-lives and affinity for the enzyme (K_i).[98] Structure modeling of these analogs using the B88-PW91 alogarithm for the solution of the density functional equations revealed that the *cis* isomers have a nucleo-philic center that is directed as a point on the bend of the molecule resulting in a focused charge that is held tightly in the enzyme pocket giving rise to long residence times; this compares to the results obtained for the *trans* isomers which have a nucleophilic center that is dispersed at right angles across the center of the molecule resulting in a weaker attraction to the enzyme pocket giving rise to shorter residence times.[98] *T. gondii*, like *C. parvum*, is dependent upon polyamine uptake and back-conversion to satisfy its polyamine require-ment,[89] and *T. gondii* SSAT is inhibited by polyamine analogs,[89] but anti-microbial therapy using polyamine analogs has not been explored in this parasite. Similarly, despite the presence of an active ornithine decarboxylase (ODC) that is inhibited by DL-α-difluoromethylornithine (DFMO),[100] *Tricho-monas vaginalis* due to the lack of AdoMetdc is also dependent upon uptake and retroconversion of spermine to satisfy its polyamine requirements.[91] The trichomonad SSAT is competitively inhibited by di(eithyl)norspermine (K_i 28 µM), and the PAO had the highest activity for N^1-acetylspermine as

Table 3.7 Inhibitors of SSAT and polyamine oxidase. The concentration of inhibitor causing 50% inhibition of SSAT or polyamine oxidase (K_i) in *E. cuniculi*, *C. parvum* and *T. vaginalis*.

SSAT/polyamine oxidase	*E. cuniculi*	*C. parvum*	*T. vaginalis*
SL11158	240 μM	100 μM	88
BW-1	2 μM		97
Di(ethyl)-norspermine		140 μM	91
Bis benzyl polyamine			90

substrate.[91] Like *T. gondii*, polyamine analogs have not been explored in detail as potential anti-trichomonad agents. Inhibition of putrescine synthesis by *T. vaginalis* results in increased VEC binding but paradoxically did not result in contact dependent host cell death.[101] The association between polyamine metabolism, particularly putrescine synthesis and secretion, and less efficient VEC binding, which, paradoxically, allows for contact-dependent host cell killing[102] for nutrient acquisition.[103] The greatly increased levels of adherence achieved by abolishing putrescine synthesis would clearly not be favorable to survival of the parasite *in vivo*, since this might produce a more vigorous host response that would limit infection. Spermine acquisition requires putrescine synthesis and secretion into vaginal fluids in order to have intact polyamine metabolism.[91] In this way, the ability to lyse host cells, albeit with less-than-optimal adherence, is complemented, and among the many nutrients available through host cell lysis may be spermine. Indeed, spermine concentrations in women undergo distinct cyclic changes during the menstrual cycle,[104] necessitating a mechanism for spermine acquisition that relies on the cytotoxic ability at the expense of better adherence. Altogether, this complex, orchestrated interaction between polyamines, host proteins and macromolecules, and the parasite surface permits successful host parasitism that in turn ensures the non-self-limiting nature of infection.[101]

In common with *C. parvum*, *E. biennusi* and *T. gondii*, *Pneumocystis carinii* is a major cause of morbidity and mortality among AIDS patients and, although possessing an active ODC, relies upon transport and interconversion of host derived polyamines.[90,105] Bis benzylpolyamine (MDL-27695, Table 3.7) was shown to be transported into *P. carinii* where it displaces the natural polyamines resulting in growth inhibition.[90]

The presence of a putative spermine oxidase (SMO) has been reported in *Amoeba proteus* crude extracts.[106] Spermine was the preferred substrate, and N^1-acetylspermine resulted in 64% of the activity, while spermidine, its acetyl derivatives and putrescine were oxidized very poorly by this enzyme (<20%). These authors found that spermine or N^1-acetylspermine when added to the culture medium was more toxic than emetine, an amoebicidal reference drug. The *A. proteus* SMO activity was inhibited by phenylhydrazine or isoniazid, but unaffected by mepacrine.

References

1. C. J. Bacchi and N. Yarlett, *Mini Rev. Med. Chem.*, 2002, **2**, 553.
2. P. Coffino, *Nat. Rev. Mol. Cell Biol.*, 2001, **2**, 188.
3. M. A. Phillips, P. Coffino and C. C. Wang, *J. Biol. Chem.*, 1987, **262**, 8721.
4. S. Hanson, J. Adelman and B. Ullman, *J. Biol. Chem.*, 1992, **267**, 2350.
5. Y. Murankami, S. Matsufuji, T. Kameji, S. Hayashi, K. Igarashi, T. Tamura, K. Tanaka and A. Ichihara, *Nature*, 1992, **360**, 597.

6. Y. Rosenburg Hasson, Z. Bercovich, A. Ciechanover and C. Kahana, *Eur. J. Biochem.*, 1989, **185**, 469.
7. S. Matsufuji, T. Matsufuji, Y. Miyazaki, J. F. Atkins, R. F. Gesteland and S. Hayashi, *Cell*, 1995, **80**, 51.
8. C. J. Bacchi, H. C. Nathan, S. H. Hutner, P. P. McCann, A. J. Bitonti and A. Sjoerdsma, *Science*, 1980, **210**, 332.
9. C. J. Bacchi and P. P. McCann, in *Parasitic Protozoa and Polyamines*, ed. P. P. McCann, A. E. Pegg and A. Sjoerdsma, Academic Press, Orlando, FL, 1987, p. 317.
10. C. Burri and Brun, *Parasitol.*, 2003, **90**, S39.
11. P. J. Schechter, J. L. R. Barlow and A. Sjoerdsma, in *Inhibition of Polyamine Metabolism: Biological Significance and Basis for New Therapies*, ed., P. P. McCann, A. E. Pegg and A. Sjoerdsma, Academic Press, Orlando, FL, 1987, p. 345.
12. J. S. Keithly and A. H. Fairlamb, in *Leishmaniasis: Current Status and New Strategies for Control*, ed., D. T. Hart, Plenum Press, New York, 1987, p. 749.
13. A. J. Bitonti, P. P McCann and A. Sjoerdsma, *Exp. Parasitol.*, 1987, **64**, 237.
14. S. Singh, A. Mukherjee, A. R. Khoutov, L. Persson, O. Heby, M. Chatterjee and R. Madhubala, *Antimicrob. Agents Chemo.*, 2007, **51**, 528.
15. S. Müller, A. Da'dara, K. Lüersen, C. Wrenger, R. Das Gupta, R. Madhubala and R. D. Walter, *J. Biol. Chem.*, 2000, **275**, 8097.
16. K. Clark, J. Niemand, S. Resksting, S. Smit, A. C. van Brummelen, M. Williams, A. I. Louw and L. Birkholtz, *Amino Acids*, 2010, **38**, 633.
17. T. Krause, K. Lüersen, C. Wrenger, T. W. Gillberger, S. Müller and R. D. Walter, *Biochem. J.*, 2000, **352**, 287.
18. R. Das Gupta, T. Krause-Ihle, B. Bergmann, I. B. Muller, A. R. Khomutov, S. Muller, R. Walter and K. Luersen, *Antimicrob. Agents Chemo.*, 2005, **49**, 2857.
19. L. Birkholtz, F. Joubert, A. W. Neitz and A. I. Louw, *Proteins*, 2003, **50**, 464.
20. Y. G. Assaraf, L. Abu-Elheiga, D. T. Spira, H. Desser and U. Bachrach, *Biochem. J.*, 1987, **242**, 221.
21. J. E. Heidrich, L. A. Hunsaker and D. L. Vadar Jagt, *IRCS Med Sci.*, 1983, **11**, 929.
22. P. S. Wright, T. L. Byers, D. E. Cross-doersen, P. P. McCann and A. J. Bitonti, *Biochem. Pharmacol.*, 1991, **41**, 1713.
23. I. B. Müller, F. Wu, B. Bergmann, J. Knöckel, R. D. Walter, H. Gehring and C. Wrenger, *Plos One*, 2009, **4**(2), e4406.
24. L. J. Martin and A. E. Pegg, *Annu. Rev. Pharmacol. Toxicol.*, 1995, **35**, 55.
25. E. K. Willert, R. Fitzpatrick and M. A. Phillips, *Proc. Natl. Acad Sci.*, 2007, **104**, 8275.
26. E. K. Willert and M. A. Phillips, *Mol. Biochem Parasitol.*, 2009, **168**, 1.
27. E. K. Willert and M. A. Phillips, *PLoS Pathog.*, 2008, **4**, e1000183.
28. C. J. Marasco, D. L. Kramer, J. Miller, C. W. Porter, C. J. Bacchi, D. Rattendi, L. Kucera, N. Iyer, R. Bernacki, P. Pera and J. R. Sufrin, *J. Med. Chem.*, 2002, **45**, 5112.

29. C. J. Bacchi, R. H. Barker Jr, A. Rodriguez, B. Hirth, D. Rattendi, N. Yarlett, C. L. Hendrick and E. Sybertz, *Antimicrob. Agents Chemother.*, 2009, **53**, 3269.
30. R. H. Barker, H. Liu, B. Hirth, C. A. Celatka, R. Fitzpatrick, Y. Xiang, E. K. Willert, M. A. Phillips, M. Kaiser, C. J. Bacchi, A. Rodriguez, N. Yarlett, J. D. Klinger and E. Sybertz, *Antimicrob. Agents Chemother.*, 2009, **53**, 2052.
31. A. J. Bitonti, T. L. Byers, T. L. Bush, P. J. Casara, C. J. Bacchi, A. B. Clarkson Jr, P. P. McCann and A. Sjoerdsma, *Antimicrob. Agents Chemother.*, 1990, **34**, 1485.
32. C. J. Bacchi, H. C. Nathan, N. Yarlett, B. Goldberg, P. P. McCann, A. J. Bitonti and A. Sjoerdsma, *Antimicrob. Agents Chemother.*, 1992, **36**, 2736.
33. T. L. Byers, T. L. Bush, P. P McCann and A. J. Bitonti, *Biochemistry*, 1991, **274**, 527.
34. A. E. Pegg, *Biochem. J.*, 1986, **234**, 249.
35. M. C. Taylor, H. Kuar, B. Blessington, J. M. Kelly and S. R. Wilkinson, *Biochem. J.*, 2008, **15**, 409.
36. Y. Xiao, D. E. McCloskey and M. A. Phillips, *Eukaryot. Cell*, **8**, 747.
37. H. M. Wallace, A. V. Fraser and A. Hughes, *Biochem. J.*, 2004, **376**, 1.
38. E. O Kajander, L. I. Kaupinen, R. L. Pajula, K. Karkola and T. O. Eloanta, *Biochem. J.*, 1989, **316**, 481.
39. I. B. Müller, R. Das Gupta, K. Luersen, C. Wrenger and R. D. Walter, *Mol. Biochem. Parasitol.*, 2008, **160**, 1.
40. A. E. Kaiser, A. M. Gottwald, C. S. Wiesch, B. Lindenthal, W. A. Maier and H. M. Seitz, *Parasitol. Res.*, 2001, **87**, 963.
41. V. T. Dufe, W. Qiu, I. B. Müller, R. Hui, R. D. Walter and S. Al-Karadaghi, *J. Mol. Biol.*, 2007, **373**, 167.
42. D. M. R. Reddy, *Bioinformation.*, **1**, 310.
43. E. M. Tamayo, A. Iturbe, E. Hernández, G. Hurtado, X. M. de Lourdes Gutiérrez, J. L. Rosales, M. Woolery and R. N. Ondarza, *Biotechnol. Appl. Biochem.*, 2005, **41**, 105.
44. R. C. Fahey, G. L. Newton, B. Arrick, T. Overdank-Bogart and S. B. Aley, *Science*, 1984, **224**, 70.
45. R. N. Ondarza, G. Hurtado, E. Tamayo, A. Iturbe and E. Hernández, *Exp. Parasitol.*, 2006, **114**, 141.
46. M. R. Ariyanayagam and A. H. Fairlamb, *Mol. Biochem. Parasitol.*, 1999, **103**, 61.
47. S. L. Oza, S. Wyllie and A. H. Fairlamb, *Mol. Biochem. Parasitol.*, 2006, **149**, 117.
48. S. L. Oza, S. Chen, S. Wyllie, J. K. Coward and A. H. Fairlamb, *FEBS*, 2008, **275**, 5408.
49. E. L. Ravaschino, R. Docampo and J. B. Rodriguez, *J. Med. Chem.*, 2006, **49**, 426.
50. S. Wyllie, S. L. Oza, S. Patterson, D. Spinks, S. Thompson and A. H. Fairlamb, *Mol. Microbiol.*, 2009, **74**, 529.

51. L. S. Torrie, S. Wyllie, D. Spinks, S. L. Oza, S. Thompson, J. R. Harrison, I. H. Gilbert, A. H. Fairlamb and J. A. Frearson, *J. Biol. Chem.*, 2009, **284**, 36137.
52. R. L. Krauth-Siegel, H. Bauer and R. H. Schirmer, *Angew. Chem. Int. Ed. Engl.*, 2005, **44**, 690.
53. S. L. Oza, E. Tetaud, M. R. Ariyanayagam, S. S. Warnon and A. H. Fairlamb, *J. Biol. Chem.*, 2002, **277**, 35853.
54. C-H. Pai, B-Y. Chiang, T-P. Ko, C-C. Chou, C-M. Chong, F-J. Yen, S. Chen, J. K. Coward, A. H-J. Wang and C-H. Lin, EMBO J., 2006, 25, 5970.
55. S. Parveen, M. O. Khan, S. E. Austin, S. L. Croft, V. Yardley, P. Rock and K. T. Douglas, *J. Med. Chem.*, 2005, **48**, 8087.
56. J. L. Richardson, I. R. E. Nett, D. C. Jones, M. H. Abdille, I. H. Gilbert and A. H. Fairlamb, *Chem. Med. Chem.*, 2009, **4**, 1333.
57. S. Patterson, D. C. Jones, E. J. Shanks, J. A. Frearson, I. H. Gilbert, P. G. Wyatt and A. H. Fairlamb, *Chem. Med. Chem.*, 2009, **4**, 1341.
58. C. J. Hamilton, A. Saravanamuthu, C. Poupat, A. H. Fairlamb and I. M. Eggleston, *Bioorg. Med. Chem.*, 2006, **14**, 2266.
59. J. Hamilton, A. Saravanamuthu, A. H. Fairlamb and I. M. Eggleston, *Bioorg. Med. Chem.*, 2003, **11**, 3683.
60. M. J. Dixon, R. I. Maurer, C. Biggi, J. Oyarzabal, J. W. Essex and M. C. Bradley, *Bioorg. Med. Chem.*, 2005, **13**, 4513.
61. L. Salmon-Chemin, E. Buisine, V. Yardley, S. Kohler, M. A. Debreu, V. Landry, C. Sergheraert, S. L. Croft, R. L. Krauth-Siegel and E. Davioud-Charvet 2009. *J. Med. Chem.*, 2001, **44**, 548.
62. L. Otero, M. Vieites, L. Boiani, A. Denicola, C. Rigol, L. Opazo, C. Olea-Azar, J. D. Maya, A. Morello, R. L. Krauth-Siegel, O. E. Piro, E. Castellano, M. González, D. Gambino and H. Cerecetto, *J. Med. Chem.*, 2006, **49**, 3322.
63. M. A. Comini, N. Dirdjaja, M. Kaschel and R. L. Krauth-Siegel, *Int. J. Parasitol.*, 2009, **39**, 1059.
64. E. A. Garrard, E. C. Borman, B. N. Cook, E. J. Pike and D. G. Alberg, *Org. Lett.*, 2000, **2**, 3639–42.
65. J. A. Czechowicz, A. K. Wilhelm, M. D. Spalding, A. M. Larson, L. K. Engel and D. G. Alberg, *J. Org. Chem.*, 2007, **72**, 3689.
66. A. M. C. Canavaci, J. M. Bustamante, A. M. Padilla, C. M. Perez Brandan, L. J. Simpson, D. Xu, C. L. Boehike and R. L. Tarleton, *Plos Neglected Trop. Diseases*, 2010, **4**, e740.
67. A. Cavalli, F. Lizzi, S. Bongarzone, F. Belluti, L. Piazzi and M. L. Bolognesi, *FEMS Immunol. Med. Microbiol.*, 2009, **58**, 51.
68. G. A. Holloway, W. N. Charman, A. H. Fairlamb, R. Brun, M. Kaiser, E. Kostewicz, P. M. Novello, J. P. Parisot, J. Richardson, I. P. Street, K. G. Watson and J. B. Baell, *Antimicrob. Agents Chemo*, 2009, **53**, 2824.
69. S. Krieger, W. Shwarz, M. R. Ariyanayagam, A. H. Fairlamb, R. L. Krauth Siegel and C. Clayton, *Mol. Microbiol.*, 2000, **35**, 542.

70. D. Spinks, E. J. Shanks, L. A. T. Cleghorn, S. McElroy, D. Jones, D. James, *Chem. Med. Chem.*, 2009, **4**, 2060.
71. Y. A. Joe and M. H. Park, *J. Biol. Chem.*, 1994, **269**, 25916.
72. I. M. Molitor, S. Knöbel, C. Dang, T. Spielmann, A. Alléra and G. M. König, *Mol. Biochem. Parasitol.*, 2004, **137**, 65.
73. J. T. Njuguna, M. Nassar, A. Hoerauf and A. E. Kaiser, *BMC Microbiol.*, 2006, **6**, 91.
74. B. Chawla, A. Jhingran, S. Singh, N. Tyagi, M. H. Park, N. Srinivasan, S. C. Roberts and R. Madhubala, *J. Biol. Chem.*, 2010, **285**, 453.
75. B. Carvajal-Gamez, R. Arroyo, R. Lira, C. López-Camarillo and M. E. Alvarez-Sánchez, *Infect. Genet. Evol.*, 2010, **10**, 284.
76. J. M. Carlton, R. T. Hirt, J. C. Silva, A. L. Delcher, M. Schatz, Q. Zhao, J. R. Wortman, S. L. Bidwell, U. C. Alsmark, S. Besteriro, T. Sicheritz-Ponten, C. J. Noel, J. B. Dacks, P. G. Foster, C. Simillion, Y. Van de Peer, D. Miranda-Saavedra, G. J. Barton, G. D. Westrop, S. Muller, D. Dessi, P. L. Fiori, Q. Ren, I. Paulsen, H. Zhang, F. D. Bastida-Corcuera, A. Simoes-Babosa, M. T. Brown, R. d. Hayes, M. Mukherjee, C. Y. Okumura, R. Schneider, A. J. Smith, S. Vanacova, M. Villalvazo, B. J. Haas, M. Pertea, T. V. Feldblyum, T. R. Utterback, C. L. Shu, K. Osoegawa, P. J. de Jong, I. Hrdy, L. Horvathova, Z. Zubacova, P. Dolezal, S. B. Malik, J. M. Logsdon, K. Henze, A. Gupta, C. C. Wang, R. L. Dunne, J. A. Upcroft, P. Upcroft, O. White, S. L. Salzberg, P. Tang, C. H. Chiu, Y. S. Lee, T. M. Embley, G. H. Coombs, J. C. Mottram, J. Tachezy, C. M. Fraser-Liggett and P. J. Johnson, *Science*, 2007, **315**, 207.
77. A. Kaiser, I. Hammerls, A. Gottwald, M. Nassar, M. S. Zaghloul, B. A. Motaal, J. H. Hauber and A. Hoerauf, *Bioorganic Med. Chem.*, 2007, **15**, 6200.
78. Y. A. Joe, E. C. Wolff and M. H. Park, *J. Biol. Chem.*, 1995, **270**, 22386.
79. D. Nurhayati and D. Ober, *Phytochem.*, 2005, **66**, 1346.
80. E. C. Wolff, J. E. Folk and M. H. Park, *J. Biol. Chem.*, 1997, **272**, 15865.
81. S. Specht, S. R. Sarite, I. Hauber, U. F. Gorbig, C. Meier, D. Bevec, A. Hoerauf and A. Kaiser, *Parasitol. Res.*, 2008, **102**, 1177.
82. D. Frommholz, P. Kush, R. Blavid, H. Scheer, J. M. Tu, K. Marcus, K. H. Zhao, V. Atemnkeng, J. Marciniak and A. E. Kaiser, *FEBS*, 2009, **276**, 5881.
83. J. H. Park, L. Aravind, E. C. Wolff, J. Kaevel, Y. S. Kim and M. H. Park, *Proc. Natl Acad Sci.*, 2006, **103**, 51.
84. A. Abbruzzese, M. H. Park and J. E. Folk, *J. Biol. Chem.*, 1988, **261**, 3085.
85. B. Kerscher, E. Nzukou and A. Kaiser, *Amino Acids*, 2010, **38**, 471.
86. A. Kaiser, D. Ulmer, T. Goebel, U. Holzgrabe, M. Saeftel and A. Hoerauf, *Mini Rev Med. Chem.*, 2006, **6**, 1231.
87. N. Yarlett, G. Wu, W. R. Waters, J. A. Harp, M. J. Wannemuehler, M. Morada, D. Athanasopoulos, M. P. Martinez, S. J. Upton, L. J. Marton and B. J. Frydman, *Mol. Biochem. Parasitol.*, 2007, **152**, 170.

88. C. J. Bacchi, D. Rattendi, E. Faciane, N. Yarlett, L. M. Weiss, B. J. Frydman, P. Woster, B. Wei, L. J. Marton and M. Wittner, *Microbiology*, 2004, **150**, 1215.

89. T. Cook, D. Roos, M. Morada, G. Zhu, J. S. Keithly, J. E. Feagin, G. Wu and N. Yarlett, *Microbiology*, 2007, **153**, 1123.

90. S. Merali, M. Saric, K. Chin and A. B. Clarkson Jr, *Antimic. Ag. Chemother.*, 2000, **44**, 337.

91. N. Yarlett, M. P. Martinez, B. Goldberg, D. L. Kramer and C. W. Porter, *Microbiol.*, 2000, **146**, 2715.

92. R. Hayashi, D. Wang, T. Hara, J. A. Iera, S. R. Durell and D. H. Appella, *Bioorg. Med. Chem.*, 2009, **17**, 7884.

93. C. J. Bacchi, N. Yarlett and L. M. Weiss, *Biochem. Soc. Trans.*, 2003, **31**, 420.

94. C. J. Bacchi, D. Orozco, L. M. Weiss, B. Frydman, A. Valasinas, N. Yarlett, L. J. Marton and M. Wittner, *J. Euk. Microbiol.*, 2001, **48**, 92S.

95. C. J. Bacchi, L. M. Weiss, S. Lane, B. Frydman, A. Valasinas, V. Reddy, J. S. Sun, L. J. Marton, I. A. Khan, M. Moretto, N. Yarlett and M. Wittner, *Antimic. Ag. Chemother.*, 2002, **46**, 55.

96. X. Feng, V. K. Reddy, H. M. Kizza, L. M. Weiss, L. J. Marton and S. Tzipori, *Antimic. Ag. Chemother.*, 2009, **53**, 2417.

97. C. J. Bacchi, N. Yarlett, E. Faciane, X. Bi, D. Rattendi, L. M. Weiss and P. Woster, *Antimic. Ag. Chemother.*, 2009, **53**, 2599.

98. N. Yarlett, W. R. Waters, J. A. Harp, M. J. Wannemuehler, M. Morada, J. Bellcastro, S. J. Upton, L. J. Marton and B. J. Frydman, *Antimic. Ag. Chemother.*, 2007, **51**, 1234.

99. W. R. Waters, B. Frydman, L. J. Marton, A. Valasinas, V. K. Reddy, J. A. Harp, M. J. Wannemuehler and N. Yarlett, *Antimic. Ag. Chemother.*, 2000, **44**, 2891.

100. N. Yarlett, B. Goldberg, M. A. Moharrami and C. J. Bacchi, *Biochem. Pharmacol.*, 1992, **44**, 243.

101. A. F. Garcia, M. Benchimol and J. F. Alderete, *Infect. Immun.*, 2005, **73**, 2602.

102. J. F. Alderete and E. Pearlman, *Br. J. Vener. Dis.*, 1984, **60**, 99.

103. K. M. Peterson and J. F. Alderete, *J. Exp. Med.*, 1984, **160**, 1261.

104. V. H. Gilad, R. Halperin, Z. Chen-Levy and G. M. Gilad, *Life Sci.*, 2002, **72**, 135.

105. C-P. Liao, O. Phanstiel, M. Lasbury, C. Zhang, S. Shao, P. J. Durant, B-H. Cheng and C-H. Lee, *Antimic. Ag. Chemother.*, 2009, **53**, 5259.

106. E. Schenkel, J. G. Dubois, M. Helson-Cambier and M. Hanocq, *Cell Biol. Toxicol.*, 1996, **12**, 1.

107. N. Haider, M. Eschbach, S. de Souza Dias, T. Gilberger, R. D. Walter and K. Lueresn, *Mol. Biochem. Parasitol.*, 2005, **142**, 224.

CHAPTER 4

Inhibitors of Polyamine Biosynthetic Enzymes

ANTHONY E. PEGG

Department of Cellular and Molecular Physiology, Milton S. Hershey Medical Center, Pennsylvania State University College of Medicine, PA, 17033, USA

4.1 Introduction

The polyamine biosynthetic pathway is well established as a target for the production of therapeutically useful drugs. Clinically approved treatments include those for African sleeping sickness and removal of unwanted facial hair, and positive clinical trials have been reported for cancer chemoprevention. Although not all trials were positive, agents interfering with polyamine synthesis have been used successfully for cancer therapy, and many laboratory studies suggest that this is a valid approach provided that issues of toxicity, tumor selectivity and adequate drug potency can be overcome. This chapter provides a brief overview of studies with agents blocking polyamine synthesis that are already in the clinic and describes less well-developed inhibitors that have been the subject of laboratory investigations. There is a long history of studies of the synthesis and effects of such compounds dating back to early work showing the importance of polyamines in cell growth. Details and appropriate citations to such work can be found in numerous earlier reviews.[1–12]

   The general rationale for the use of inhibitors of polyamine production for therapy is based on many experimental studies showing that these amines are essential for growth, but the exact basis for any selective antitumor or

RSC Drug Discovery Series No. 17
Polyamine Drug Discovery
Edited by Patrick M. Woster and Robert A. Casero, Jr.
© Royal Society of Chemistry 2012
Published by the Royal Society of Chemistry, www.rsc.org

antiparasitic actions is less clear. In mammals, polyamines are essential for multiple reasons, including the requirements for hypusine formation in the essential protein eIF5A, multiple roles in protein synthesis and their roles in ion-channel regulation.[13–17] These factors are clearly important to the value of polyamine inhibitors as antiproliferative agents, but are also likely to contribute to toxicity if a high level of reduction in polyamine content is achieved.

Stimulation of polyamine synthesis is a critical downstream event from activation of certain oncogenes such as *myc* and *ras*,[7,8,18–23] and this may be the basis of the selective effects on tumor development and growth. Antiprotozoal effects may relate to a number of factors including the differences in the properties of key target enzymes, differences in purine metabolism, the role of polyamines as precursors of trypanothione [N^1,N^8-bis(L-γ-glutamyl-L-hemicystinyl-glycyl) spermidine] an essential metabolite that is not found in mammalian cells, and interference with rapid growth and alteration of antigenic determinants as described below.

Polyamine synthesis from methionine and arginine is brought about by the action of four enzymes specific to the polyamine pathway.[17,24,25] These are ornithine decarboxylase (ODC), which forms putrescine, *S*-adenosylmethionine (AdoMet) decarboxylase (AdoMetDC), which forms decarboxylated *S*-adenosylmethionine (dcAdoMet) and two discrete aminopropyltransferases that use dcAdoMet as an aminopropyl donor to form the higher polyamines. Spermidine synthase (SpdSyn) uses putrescine exclusively as the amine acceptor forming spermidine and spermine synthase (SpmSyn) uses spermidine exclusively to make spermine. Arginase and AdoMet synthetase are also needed for polyamine production but are not unique to the polyamine pathway and are not suitable drug targets.

4.2 Inhibition of ODC

4.2.1 α-Difluoromethylornithine (DFMO)

By far the most widely studied inhibitors of polyamine biosynthesis are the enzyme-activated irreversible inactivators of ODC. Most of the studies and virtually all the clinical work have been carried out with α-difluoromethylornithine (DFMO; Eflornithine, Figure 4.1), and this drug is approved for use as an treatment for African trypanosomiasis (Ornidyl®) and for the removal of unwanted facial hair (Vaniqa®). As described below, there is also compelling evidence that DFMO may be useful for cancer chemoprevention, and several clinical trials have been carried out to test this hypothesis.

4.2.1.1 Mechanism of Action of DFMO

DFMO is accepted as a substrate by ODC. Its decarboxylation leads to the formation of a reactive species which forms a covalent adduct with the protein predominantly at Cys-360 forming [2-(1-pyrroline))methyl]cysteine. A small

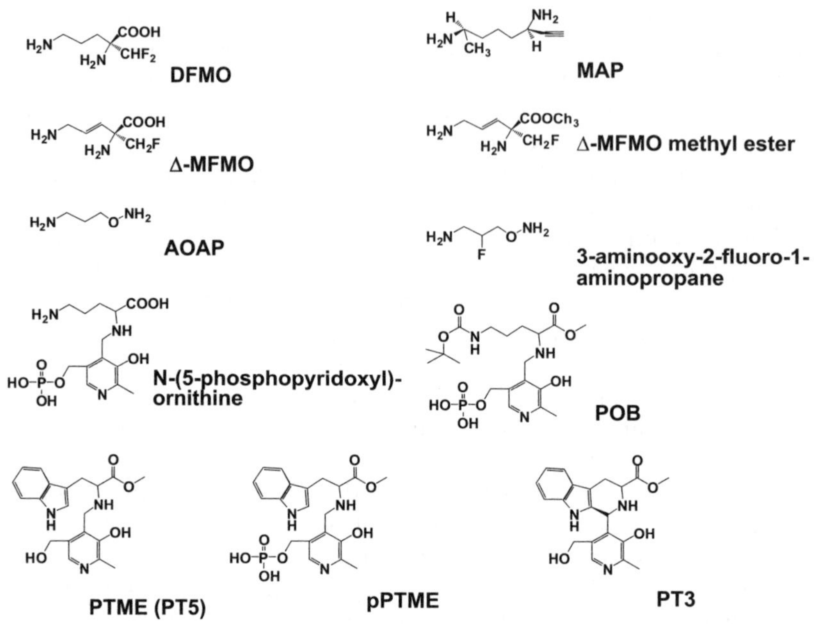

Figure 4.1 Inhibitors of ODC. The structures shown are: α-difluoromethylornithine (DFMO; eflornithine); (E)-α-monofluoromethyldehydroornithine [Δ-MFMO] and (2R,5R)-δ-methylacetylenicputrescine [MAP][3,93]; 3-aminooxy-1-aminopropane (AOAP) and 3-aminooxy-2-fluoro-1-aminopropane[96–100]; N-(5-phosphopyridoxyl)-ornithine[103]; BOC-protected N-(5-phosphopyridoxyl)-ornithine methyl ester derivative (POB)[104]; pyridoxyl-L-tryptophan methyl ester (PTME), its methyl ester (pPTME)[105] and its cyclic derivative PT3.[106]

fraction (*ca.* 10%) of the protein is inactivated via an adduct formed at Lys-69.[26,27] Both of these amino acids are important in the catalytic mechanism of ODC, and it is unlikely that resistance to the drug can arise via alterations in the protein structure. However, amplification of the ODC gene occurs quite readily in response to exposure of mammalian cells to DFMO[28–30] and has been reported in the parasite *Leishmania donovani*.[31]

The half-life of ODC even in the presence of a saturating dose of DFMO is 3.1 min. In mammalian cells, the normal half-life of the enzyme is about 20 min,[20] and therefore, there is always a fraction of the ODC pool that is not irreversibly inactivated. In trypanosomes, such as *T. brucei*, the normal half-life of ODC is much longer and a greater, longer-lasting reduction in activity may be achieved.[32] This may contribute to the value of DFMO for treatment of such organisms.

4.2.1.2 Effect of DFMO Treatment

In mammalian cells, inhibition of ODC by DFMO leads to a major reduction in putrescine and spermidine but, usually, to only a small decline in spermine.

The residual activity of newly synthesized ODC along with a lack of spermine degradation may account for the maintenance of spermine. Growth is significantly reduced when spermidine becomes depleted, but in most cases, the effects of DFMO and other specific ODC inhibitors are cytostatic rather than cytotoxic. The lack of cytotoxicity may be due to the residual spermine in the cells. In trypanosomes (which do not normally contain spermine), both putrescine and spermidine are depleted, and there is a loss of trypanothione. This is likely to be a major contributor to the therapeutic effects, since trypanothione is essential for the parasites to resist oxidative stress.[33,34] DFMO treatment also leads to a major increase in dcAdoMet levels. The extent to which alterations in the ratio of AdoMet:dcAdoMet and in the AdoMet pathway may contribute to the effects of the drug is not well understood. In trypanosomes, which lack purine salvage pathways, this may be a significant factor.[35–38]

Many studies have used DFMO to alter polyamine content and examine downstream pathways affecting polyamines. Such experiments require suitable controls in which putrescine is shown to reverse the effects of DFMO confirming specificity of the inhibitor. Polyamine-derived effects on the levels and phosphorylation status of key regulatory proteins have been reported. These include Mdm2, MEK-1, EGFR, p21Cip1, p27Kip, p53, Src, cdk-4, importin-α1, Akt/protein kinase and GSK-3β[19,39–49] Changes in the activity of these proteins could account for the pharmacological effects of DFMO treatment. DFMO has also been used to show that microtubule assembly requires polyamines.[50,51]

4.2.1.3 Antiparasitic Uses of DFMO

Laboratory studies showing that DFMO could cure acute infections of *Trypanosoma brucei brucei* in mice (reviewed in several previous publications[3,12,52,53]) were validated in clinical trials, and the drug under the trade name Ornidyl® is approved for the treatment of African sleeping sickness caused by *T. brucei gambiense*.[54–56] Major problems are the expense of the drug and the fact that the recommended route of administration involves extended hospitalization for intravenous treatment. The remarkable efficacy of DFMO against *T. brucei* may be related to an increased ability of the immune system to deal with trypanosomes whose growth and development is slowed by the lack of polyamines.[53] Disease caused by many strains of *T. brucei rhodesiense* is significantly less sensitive to DFMO than that caused by *T. brucei gambiense*, and treatment of these strains is less successful and requires combination with other drugs.

Based on laboratory studies, ODC is a validated target, and ODC inhibitors may have utility in treatment of diseases caused by a variety of other parasites including *Acanthamoeba, Leishmania, Giardia, Plasmodia* and *Eimeria*.[3,10,31,57–62] Whether these observations can be exploited therapeutically remains to be seen and may require combination of DFMO with other drugs, but this is certainly worthy of further investigation.

4.2.1.4   DFMO and Tumor Treatment

Despite some success in animal tumor models, in general, DFMO was not effective for cancer therapy, although there are some exceptions described below. A major factor for the lack of response to DFMO in these clinical trials is likely to be the availability of polyamines from other sources including the diet, microbial flora and, via the retroconversion pathway, other cells having significant spermine stores.[63–65] The upregulation of the polyamine transport system after polyamine depletion allows these sources to provide sufficient polyamines for tumor growth. Laboratory and clinical studies in which polyamine deficient diets combined with antibiotics to reduce the contribution of polyamines from intestinal micro-organisms and/or oxidase inhibitors to prevent polyamine-recycling were used have indicated that these factors could be used to improve response,[66,67] but the extension of these experiments to general clinical practice presents many problems.

One possible exception was demonstrated in patients with recurrent glial tumors where treatment with DFMO alone or combined with other agents had a significant effect. Although the combination of DFMO and MGBG (see Section 4.3.1) was discontinued due to unacceptable toxicity,[68] addition of DFMO to the PCV (procarbazine, CCNU, vincristine) regime had a clear benefit.[23,69] The response to such chemotherapy for malignant gliomas correlated well with ODC protein levels measured in formalin-fixed slides, suggesting that outcome could be improved further by screening for patients with tumors having low ODC levels.[70,71]

Recent mechanism based studies suggest that DFMO may be useful for treatment of *Myc*-induced cancers.[72,73] Amplification of *N-Myc* is a frequent event in the development of neurobastoma, and ODC may an intermediary in this effect.[46,74,75] Preclinical studies show that DFMO may be a useful addition to therapeutic regimes for this tumor resulting in downregulation of *myc*-related cell signaling pathways that lead to tumor growth.[44,46,74] Clinical trials of DFMO combined with other agents that may have some efficacy in neuroblastomas are currently underway.

4.2.1.5   DFMO and Cancer Chemoprevention

ODC activity is enhanced by a variety of tumor promoters and oncogenes, and DFMO given well after the initiating agent greatly reduces tumor development in rodents treated with a wide variety of chemical carcinogens and tumor promoters. The production of tumors of the skin, bladder, stomach, intestine, colon, oral cavity and mammary gland has been shown to be blocked by DFMO treatment (reviewed in several previous publications[3,5,7,76]) or by expression of antizyme, a natural regulator of ODC in animal models.[77–80]

Studies with mice also suggest that quite modest reductions in ODC activity may have an effect on the propensity to tumor formation in some tissues.[19,81] These laboratory studies have led to multiple clinical trials with DFMO as a chemopreventive agent, either alone or in combination with other compounds in populations at high risk for the development of neoplasms.[7,22,82]

Trials of oral DFMO showed a reduction in prostate size and polyamine content,[83] and a study to prevent skin cancer with oral DFMO showed some efficacy towards basal-cell cancers.[84] There were no minimal adverse effects in these trials, and it was suggested that further studies should be undertaken. The most promising recent trial showed that daily oral doses of DFMO plus sulindac produced a substantial reduction in colorectal adenoma recurrence.[85,86] Analysis of polyamines and prostaglandin E2 in rectal mucosa showed that polyamines, but not prostaglandin E2, were suppressed by DFMO treatment.[87]

4.2.1.6 DFMO and Facial Hair Removal

Polyamines have a critical role on the growth and development of hair follicles with both excessive and low levels leading to hair loss.[88,89] A preparation (trade name VANIQA®) containing DFMO in a cream allowing dermal uptake is now an approved (and commercially successful) treatment for removal of unwanted facial hair.[90–92]

4.2.2 Other ODC Inhibitors

Many other enzyme-activated irreversible inactivators of ODC have been prepared using the same principles as those underlying DFMO as a lead compound.[3,93] Some of these compounds such as (E)-α-mono-fluoromethyldehydroornithine [Δ-MFMO] and (2R,5R)-δ-methylacetyl-enicputrescine [MAP] (Figure 4.1) have properties that suggest that they may be preferable to DFMO in terms of potency, cellular uptake or pharmacokinetics but, due to perceived lack of commercial potential, have not been developed. In view of the successes with DFMO described above, this is a field worthy of further consideration and modification. For example, synthesis of the methyl ester of Δ-MFMO produces a compound that facilitates entry into the cell where cleavage by intracellular esterases generates Δ-MFMO. This inhibitor is considerably more potent than DFMO both in inactivation of ODC[93] and against laboratory animal models of trypanosomiasis and cancer.[36,94,95]

Other potent ODC inhibitors based on the synthesis of aminooxy derivatives of putrescine including 3-aminooxy-1-aminopropane (AOAP) and 3-ami-nooxy-2-fluoro-1-aminopropane (Figure 4.1) have been shown to be effective in cultured cells and parasites having effects very similar to DFMO but at lower concentrations.[96–100] Recent structural studies have shown that AOAP binds to the active site of *L. donovani* ODC and has a K_i of 1 nM.[101] The long-term stability *in vivo* and suitability for therapeutics of these aminooxy derivatives have not yet been fully evaluated, but they clearly are a promising lead in the search for improved ODC inhibitors, particularly as AOAP and derivatives were effective in blocking growth of *P. falciparum*.[100] A melamine derivative of 2-hydroxy-AOPA, which was designed to take advantage of the ability of trypanosomal nucleoside transporter to take up melamine, has been shown to have improved activity versus *T. brucei*.[102]

Transition-state mimics of the PLP-dependent-ODC are also an interesting potential source of inhibitors. Many years ago, *N*-(5-phosphopyridoxyl)-orni-thine (Figure 4.1) was shown to be a strong *in vitro* inhibitor but could not be used in cells.[103] Recent studies using its BOC-protected methyl esters deriva-tives such as (POB)[104] or pyridoxyl-L-tryptophan methyl ester (PTME, PT5) and its phosphorylated derivative (pPTME) (Figure 4.1)[105] have suggested that suitable transition state analogs or their precursors can be made and are highly antiproliferative, although the specificity of these compounds is not fully established. An interesting use of this strategy to target malaria has been published.[106] Since the *P. falciparum* parasite has a pyridoxine/pyridoxal kinase, exposure to PT3, a cyclic pyridoxyl-tryptophan methyl ester (Fig-ure 4.1), led to synthesis and accumulation of its phospho derivative and subsequent inhibition of ODC and probably other PLP-dependent enzymes.[106] Recently, an enzymatic high-throughput screen of 316 114 unique molecules was used to identify inhibitors of ODC. Several novel families of inhibitors, including some selective for the parasitic enzyme, were identified.[107]

4.3   Inhibition of AdoMetDC

Several potent inhibitors of AdoMetDC have been designed and tested for use as anti-protozoal and cancer chemotherapeutics.[4,108,109] Inhibition of AdoMetDC in cultured mammalian cells leads to a large increase in putrescine and to a decline in spermidine and spermine.[4,110–114] A significant difference in the con-sequences of blockage of AdoMetDC rather than ODC is that putrescine con-tent is increased by the former due to compensatory increases in ODC activity and the lack of its conversion into higher polyamines. This has the potentially useful feature of reducing compensatory exogenous polyamine uptake, but it is also possible that high levels of putrescine may partially compensate for some of the functions of the higher polyamines. This is clearly not the case with regard to hypusine formation for which only spermidine can be used[115], but it is not likely that all of the therapeutic effects of inhibition of polyamine synthesis are exclusively due to reduced activity of eIF5A. Another obvious difference is that AdoMetDC inhibitors cause the disappearance of dcAdoMet, whereas, as described above, ODC inhibitors greatly increase this content.

4.3.1   MGBG, SAM686A and Related Compounds

The first such inhibitor, MGBG (Figure 4.2), predates all of the ODC inhibi-tors. It was first employed (with some success but high toxicity) as an anticancer

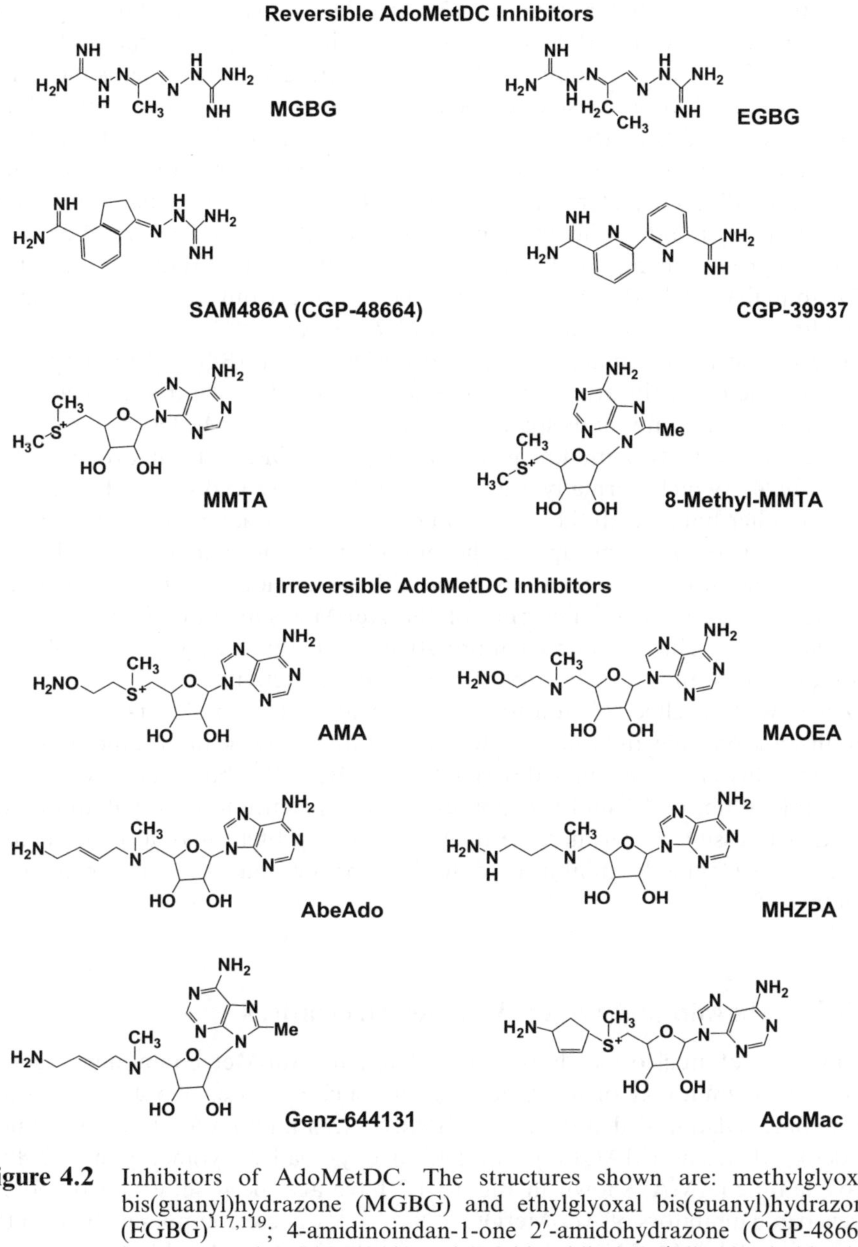

Reversible AdoMetDC Inhibitors

MGBG

EGBG

SAM486A (CGP-48664)

CGP-39937

MMTA

8-Methyl-MMTA

Irreversible AdoMetDC Inhibitors

AMA

MAOEA

AbeAdo

MHZPA

Genz-644131

AdoMac

Figure 4.2 Inhibitors of AdoMetDC. The structures shown are: methylglyoxal bis(guanyl)hydrazone (MGBG) and ethylglyoxal bis(guanyl)hydrazone (EGBG)[117,119]; 4-amidinoindan-1-one 2′-amidohydrazone (CGP-48664, later renamed as SAM486A) and (2,2-bipyridine)-6,6′-dicarboximidamide (CGP-39937)[120–122]; 5′-deoxy-5′-dimethylsulfonioadenosine (MMTA) and 5′-deoxy-5′-dimethylsulfonio-8-methyladenosine (8-methylMMTA)[108]; S-(5′-deoxy-5′-adenosyl)methylthioethylhydroxylamine (AMA)[129], 5′-[(2-aminooxyethyl)methylamino]-5′-deoxyadenosine (MAOEA) and 5′-deoxy-5′-[(3-hydrazinopropyl)methylamino]adenosine (MHZPA)[130]; 5′-{(Z)-4-amino-2-butenyl]methylamino}-5′-deoxyadenosine (AbeAdo)[132] and 8-methyl AbeAdo (Genz644131)[109,140]; S-(5′-deoxy-5′-adenosyl)-1-amino-4-methylthio-2-cyclopentene.[141,142]

drug in the 1960s, but not until 1972 was its ability to inhibit polyamine synthesis at the AdoMetDC step discovered in Williams-Ashman's laboratory.[116,117] Several other compounds with different alkyl substituents including EGBG (Figure 4.2) were also described and tested in animal models but did not show sufficient advantages to enter detailed clinical trials.[111,117–119] However, in the early 1990s, modifications forming 4-amidinoindan-1-one 2′-amidohydrazone (CGP-48664, later renamed as SAM486A) and related diamidines such as (2,2-bipyridine)-6,6′-dicarboximidamide (CGP-39937) (Figure 4.2) were shown to produce compounds that were more potent than MGBG as inhibitors of mammalian AdoMetDC[120–122] and the enzyme from variety of parasites including *T. cruzi*[123], *P. falciparum*[100] and *T. brucei*.[124,125]

The reason for the inhibition of AdoMetDC by MGBG and the improved effect of these modifications was determined in 2001.[126] Structural analysis showed that the compounds span the entrance of the AdoMetDC active site via interactions of the two amidine groups at each end. One forms hydrogen bonds with Glu247 (which normally interacts with the ribose hydroxyls of AdoMet), and the other bonds with the side chain of Ser229 and the main chain carbonyl of Leu65. The spacer groups of the inhibitors are located between the two essential Phe residues at positions 7 and 223, which normally provide stacking interactions with the adenine ring of the AdoMet substrate. SAM486A and derivatives are able to interact more strongly in this way by virtue of their aromatic rings, thus accounting for their stronger activity.[126]

An unwanted effect of non-irreversible AdoMetDC inhibitors is that they stabilize the enzyme and thus lead to a substantial increase in the amount of the protein. This effect was first described with MGBG[127] but occurs with other reversible inhibitors[110] and is probably due to prevention of polyubiquination at the active site.[128] An increase in AdoMetDC protein may also occur also with the irreversible inhibitors described below, but the new protein is inactive.[113]

4.3.2   AbeAdo and Other AdoMet Derivatives

A number of nucleoside derivatives related to AdoMet containing reactive hydrazino-, aminooxy or hydrazido- groups such as 5′-deoxy-5′-[(3-hydrazino-propyl)methylamino]adenosine (MHZPA), 5′-[(2-aminooxyethyl)methylamino]-5′-deoxyadenosine (MAOEA) and S-(5′-deoxy-5′-adenosyl)methylthioethylhydroxylamine (AMA) shown in Figure 4.2 have also been shown to be potent irreversible inhibitors of AdoMetDC.[108,112,129–131] These compounds bind to the active site of AdoMetDC in the same way as the substrate AdoMet adopting the *syn* conformation and with the adenine ring between Phe7 and Phe223.[126] The reactive group then forms a covalent linkage with the pyruvoyl prosthetic group. These compounds are potent inhibitors but seem to be too reactive with other cellular compounds such as pyruvate to be useful drugs. An 8-methyl substitution increases their efficiency by favoring the adoption of the *syn* conformation.[108] Other AdoMet analogs, including some with such 8-methyl substitutions such as

5′-deoxy-5′-dimethylsulfonio-8-methyladenosine (8-methyl-MMTA, Figure 4.2) that cannot be decarboxylated but bind reversibly to the active site and thus act as competitive inhibitors, have also been described.[4,108,130]

A potent enzyme-activated irreversible inhibitor of AdoMetDC was descri- bed in 1989.[132] This compound 5′-{[(Z)-4-amino-2-butenyl]methylamino}-5′- deoxyadenosine (AbeAdo, MDL73811, Figure 4.2) was designed to form a Schiff base with the pyruvate prosthetic group at the active site followed by an enzyme-mediated abstraction of a proton forming a conjugated imine, which could then react with a nucleophilic residue of AdoMetDC. It does indeed lead to a time-dependent loss of mammalian, protozoan and bacterial AdoMetDC activity.[133–135] However, with human AdoMetDC, it was shown that the enzyme was inactivated by a transamination of the pyruvate group,[136] sug- gesting that its primary mechanism of action of AbeAdo was actually to greatly increase the incorrect protonation of the pyruvate during the reaction. Other derivatives of AbeAdo have also been shown to inhibit trypanosomal Ado- MetDC.[137] Recently, its 8-methyl derivative (Genz-644131, Figure 4.2) has been shown to be a significantly improved inhibitor of the trypanosomal AdoMetDC over the parent compound.[109,138,139] This is explained by the greater likelihood of the compound being in the *syn* configuration. Other irreversible inactivators of AdoMetDC such as *S*-(5′-deoxy-5′adenosyl)-1- ammonio-4-(methylsulfonio)-2-cyclopentene (AdoMac, Figure 4.2) have also been produced.[140,141]

4.3.3 AbeAdo and Other AdoMetDC Inhibitors

Attempts to provide additional AdoMetDC inhibitors using virtual screening based on the crystal structure of human AdoMetDC have been described.[142] This may prove a powerful approach to generate species-specific inhibitors, but at present only the mammalian, plant and *T. maritima* structures are avail- able,[143,144] although a model of the malarial AdoMetDC has been published.[145]

4.3.4 Therapeutic Potential of AdoMetDC Inhibitors

4.3.4.1 Antitumor Effects of AdoMetDC Inhibitors

SAM486A inhibited the growth of a range of tumor cell lines and tumor xenografts.[114,146,147] These studies led to preliminary clinical trials against solid tumor and lymphomas either alone or in combination with drugs such as 5- fluorouracil/leucovorin. Some efficacy was seen,[148–154] but no follow-up studies have been reported. More recently, the possibility of using SAM486A for treatment of neuroblastomas has been supported by studies with neuro- blastoma cell cultures where inhibition of polyamine synthesis by this drug alone or combined with DFMO caused apoptosis and tumor death.[72,73,155]

SAM486A accumulated in tumor cells to much lower concentrations than MGBG.[114] This may account for its reduced toxicity. MGBG causes severe mitochondrial damage, and SAM486A had much less effect on mitochondria at doses that affected tumor-cell proliferation. In view of the lack of antitumor activity of AbeAdo described below, it is possible that some of the antitumor activity of the diamidines is related to a target distinct from AdoMetDC. However, there is convincing evidence that for SAM486A, the polyamine biosynthetic pathway is an essential mediator of the response. Inhibition of tumor cell growth by SAM486A was prevented by exogenous spermidine of spermine, and CHO cells derived for resistance to the drug had an amplification of the *AdoMetDC* gene leading to increased enzyme production.[114,146]

Although AbeAdo is highly effective in reducing cellular polyamine content, hypusine production and cell proliferation in tumor-cell cultures[156–158], it had little effect on growth of either L1210 leukemia or Lewis lung carcinoma in mice.[158] One explanation for this is that the very high level of putrescine which accumulates in the cells allows tumor growth in these animal models, but it is not clear why this potent and stable AdoMetDC inhibitor is so ineffective compared to MGBG and SAM486A.

4.3.4.2   *Antiparasitic Effects of AdoMetDC Inhibitors*

Although there are no published reports of clinical efficacy, the AdoMetDC inhibitors have clear promise for therapy of diseases caused by parasitic protozoa. SAM486A reduced polyamine content and hypusine synthesis in *P. falciparum*.[100,159] AdoMac was an effective anti-trypanosomal agent *in vitro*.[160] Both AbeAdo and the more potent 8-methyl derivative Genz-644131 were active against *T. brucei rhodesiense* and *T. brucei brucei*, including strains resistant to DFMO in mouse models.[109,135,137,138,161] These studies provide strong evidence that the AdoMetDC inhibitors may be clinically useful parasitic agents and should be developed further. It is possible that the rapid build-up of AdoMet in treated cells, which correlated with efficacy better than the fall in polyamines or hypusine, is a key factor in the success of this drug.[36] Genz-644131 may be the best candidate for clinical trials based on its biostability, brain penetration and potency.[109]

AbeAdo reduced the infectivity and replication of *T. cruzi*.[162] The Ado-MetDC inhibitors may be particularly useful approaches to therapy for Chagas disease. ODC inhibitors cannot be used, since *T. cruzi* does not contain an ODC but takes up exogenous putrescine. However, the parasite does require the synthesis of spermidine for growth and trypanothione production.[10,163,164]

4.3.4.3   *Antiviral Effects of AdoMetDC Inhibitors*

Laboratory studies have revealed efficacy of some AdoMetDC inhibitors including the acetyl derivative of AbeAdo[137] and SAM486A[165] against human immunodeficiency virus type 1 (HIV-1). This may be related to their ability to reduce the hypusine modification of eIF5A, since there is evidence that this protein is a cellular cofactor of the HIV-1 regulatory protein Rev.[165,166]

4.4 Inhibition of Aminopropyltransferases

Although, as described in the introductory section, mammalian cells contain two highly specific aminopropyltransferases, SpdSyn for spermidine production and SpmSyn for spermine synthesis, some micro-organisms including protozoal parasites contain only a SpdSyn and make little, if any, spermine.[167] There has been some interest in the production of specific inhibitors of aminopropyltransferases both for therapeutic manipulation of the polyamine content and as tools to investigate polyamine function.

4.4.1 Product Inhibition of Aminopropyltransferases

Aminopropyltransferases are quite strongly inhibited by their product 5'-deoxy,5'-methylthioadenosine (MTA, Figures 4.3 and 4.4) with SpmSyn being more sensitive.[167] This inhibition is readily explained by the structures of human SpdSyn and SpmSyn with bound MTA where hydrophobic surfaces in the active site pocket interact with the adenine ring of MTA.[168] MTA is normally rapidly degraded by MTA phosphorylase in mammalian cells[169,170] and by MTA hydrolase[171] in bacteria, so this inhibition is of little practical consequence in mammalian cells. However, very potent inhibitors of these MTA degrading enzymes are now available and undergoing tests for clinical therapeutic value.[172–174] It is highly likely that such inhibitors would also inhibit polyamine synthesis via MTA-mediated inhibition of the aminopropyltransferase steps.

4.4.2 SpdSyn Inhibitors

Several groups have suggested that SpdSyn may be a useful target for therapeutic intervention in Chagas disease, African trypanosomiasis, Leishmaniasis

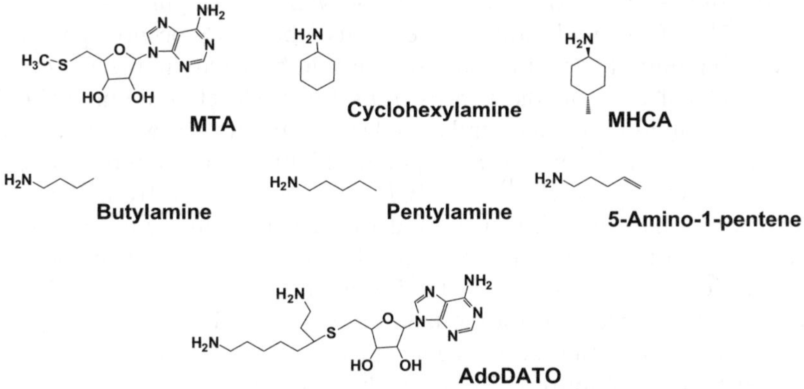

Figure 4.3 Inhibitors of SpdSyn. The inhibitors shown are: 5'-deoxy,5'-methylthioadenosine (MTA)[168]; cyclohexylamine, *trans*-4-methylcyclohexlamine (MCHA), butylamine, 5-amino-1-pentene and pentylamine[181,182]; and *S*-adenosyl-1,8-diamino-3-thiooctane (AdoDATO).[187,188]

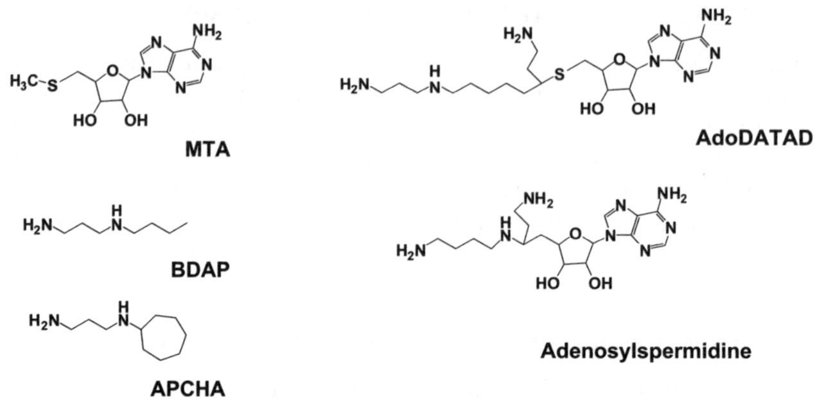

Figure 4.4 Inhibitors of SpmSyn. The inhibitors shown are: 5'-deoxy,5'-methyl-thioadenosine (MTA)[168]; S-adenosyl-1,12-diamino-3-thio-9-azadodecane (AdoDATAD)[196]; adenosylspermidine (3-(RS)-(5'-deoxy-5'-carbaadenos-6'-yl)spermidine)[197]; and N-(butyl-1,3-diaminopropane (BDAP) and N-(3-aminopropyl)cyclohexylamine (APCHA).[181,199]

and malaria.[6,10,60,175,176] Inactivation of trypanosomal SpdSyn using RNA interference (RNAi) has been used to show that this enzyme is a potential target for drug development.[177,178] These studies showed that *T. brucei* cannot scavenge sufficient spermidine from their environment to meet growth requirements when SpdSyn was inactive.

Similar proof of principle is available for the use of SpdSyn inhibitors as potential antitumor agents. Blockage of the pathway at the SpdSyn step has the same effects on polyamine content and hypusine formation as blockage at the AdoMetDC step with the exception that the former increases dcAdoMet content, and the latter reduces dcAdoMet.

Initial studies established that cyclohexylamine (Figure 4.3) inhibited SpdSyn.[179–181] This inhibition is due to its ability to occupy part of the putrescine binding site and an adjacent hydrophobic cavity in the active site.[182] [183] Some related compounds such as *trans*-4-methylcyclohexlamine (MCHA, Figure 4.3), 5-amino-1-pentene and pentylamine (Figure 4.3) were designed and tested, and found to be much more potent inhibitors of mammalian SpdSyn with IC$_{50}$ values *ca.* 2–4 μM.[180,181] MCHA inhibits SpdSyn from *Plasmodium falciparum* and has been shown to bind to the active site by crystallographic studies.[183] Exposure of growing rats to MCHA or 5-amino-1-pentene by oral administration lowered spermidine levels (and increased spermine) without any significant toxicity except for a reduction in body weight.[184,185]

Coward and colleagues subsequently produced much more potent aminopropyltransferase inhibitors. These multisubstrate analog inhibitors were designed using models of the transition state of a direct displacement mechanism.[179,186–188] These studies resulted in the synthesis of S-adenosyl-1,8-diamino-3-thiooctane (AdoDATO, Figure 4.3) which is a very potent and selective inactivator of SpdSyn (IC$_{50}$ *ca.* 50 nM). Crystallographic studies have shown that this compound binds tightly and specifically in the active site.[182,183,189] AdoDATO does alter polyamine levels in cultured cells,[190,191] leading to a fall in spermidine and an increase in spermine, but its poor uptake and complex synthesis have prevented detailed preclinical studies.

Recently, SpdSyn has been shown to be a target of the oncogenic transcription factor *c-Myc*, which is commonly overexpressed in human cancers. *Myc*-induced proliferation was blocked using RNAi directed against SpdSyn.[192] These results support the concept of using SpdSyn inhibitors for chemoprevention, and experiments using MCHA confirmed that this drug delayed the onset of B-cell lymphoma development in λ-*Myc* transgenic mice.[192]

Inhibition of *Plasmodium falciparum* SpdSyn by cyclohexylamine has been examined and confirmed as a potential means for therapy.[176] The use of more potent and specific inhibitors, which are available, might improve this approach. Virtual screening techniques have been used to identify additional compounds interacting with the *Plasmodium falciparum* SpdSyn.[193] A rapid screening technique to actually assay libraries of possible inhibitors has been described using monoclonal antibodies to MTA, the product of the amino-propyltransferase reactions.[194]

4.4.3 SpmSyn Inhibitors

Spermine is not required for growth of mammalian cells in culture, is not present in many micro-organisms[25,167] and has not been targeted for therapy. However, it should be noted that contrary to a number of reports in the literature, there are significant numbers of micro-organisms that do contain spermine, and these include bacteria in families that contain significant pathogens such as Clostridiales, Bacilliales and Actinobacteria.[25] The enzymes having SpmSyn activity not yet been characterized from such organisms, and the importance of spermine to their growth is unknown, but this might be a field worth exploring in a search for drug targets since, as described below, potent and specific inhibitors of SpmSyn are available, and these have little effect on mammalian growth.

The multisubstrate analog approach described above has also been applied to SpmSyn leading to the synthesis of *S*-adenosyl-1,12-diamino-3-thio-9-aza-dodecane (AdoDATAD, Figure 4.4) (IC$_{50}$ *ca.* 20 nM)[195] and (3-(*RS*)-(5′-deoxy-5′-carbaadenos-6′-yl)spermidine) (adenosylspermidine, Figure 4.4) (IC$_{50}$ *ca.* 14 nM).[196] AdoDATAD reduced spermine content, increased spermidine levels in cultured L1210 cells and transformed fibroblasts but had little effect on growth when its conversion to oxidative metabolites was prevented by aminoguanidine.[197]

Other structure/mechanism-based inhibitors of SpmSyn have been produced by adding hydrophobic groups to 1,3-diaminopropane.[179,180,198] These allow the inhibitors to bind strongly in the active site pocket.[168] The most active compounds were *N*-(3-aminopropyl)cyclohexylamine (APCHA, Figure 4.4) and *N*-butyl-1,3-diaminopropane (BDAP, Figure 4.4) with IC$_{50}$ values of 0.2–0.4 μM. These compounds are effective inhibitors of spermine synthesis in cultured cells leading to the same result as exposure to AdoDATAD of a decrease in spermine and an increase in spermidine. In most reported studies, no effect on growth was observed.[179,198,199] This is consistent with studies using embryonic fibroblasts from mice in which the gene encoding spermine synthase was inactivated leading to a total absence of spermine.[200,201] An exception was that treatment with APCHA had a significant antiproliferative effect in ZR-75-1 breast cancer cells.[202] It should be pointed out that the compensatory increase in spermidine may mask any effects, and this increase is quite limited in the breast-cancer cell line. Also, recently, APCHA has been used to show that spermine is able to contribute to microtubule assembly.[51]

In rats treated with N-(3-aminopropyl)cyclohexylamine, there was no effect on growth or physiology, but in many tissues, spermine was not fully depleted.[184] A cautionary note for the possible use of SpmSyn inhibitors is that the loss of SpmSyn activity that is seen in the rare inherited condition Snyder-Robinson syndrome causes a range of pathological effects including mental retardation, hypotonia and movement disorders as well as bone-related abnormalities.[203–205] Mice lacking SpmSyn activity have even more severe defects including a tendency to neurological defects, a propensity to sudden death, small size, sterility and deafness.[206,207]

4.5 Conclusions

There are now inhibitors for all steps of the polyamine biosynthetic pathway that have adequate specificity, potency and biological stability for studies of efficacy as drugs against a variety of conditions. Improvements may need to be made to provide approvable clinical drugs. Some of the known inhibitors have been so long in development that patent protections issues may also be a problem. Thus, second-generation compounds may have commercial advantages. Modifications may be possible to enhance species specificity for treatment of diseases caused by infectious agents. The rapidly increasing body of knowledge on the structures of the four key enzymes described in this review should aid in the design of such inhibitors.

It should also be pointed out that numerous species, including some micro-organisms causing human pathology, do not use ODC for putrescine production but instead rely on decarboxylation of arginine to form agmatine, which is converted to putrescine or to aminopropylagmatine and then to spermidine.[208–213] Arginine decarboxylase is readily inactivated by inhibitors designed in the same way as DFMO to act as enzyme-activated irreversible inhibitors.[214] Thus, compounds such as α-difluoromethylarginine were shown to block agmatine production, and it is surprising that there has been little interest in developing such compounds as therapeutics. Furthermore, many micro-organisms including *Vibrio cholerae* do not use AdoMetDC and aminopropyltransferases employing dcAdoMet for the synthesis of higher polyamines. These species use a condensation pathway in which aspartate β-semialdehyde forms the aminopropyl group producing carboxyspermidine or carboxynorspermidine, which is then decarboxylated.[212,215,216] The synthesis of *sym*-homospermidine, a polyamine present not in mammals but in many other species, also involves a condensation pathway.[217] These alternative pathways to essential metabolites offer the opportunity to generate potentially valuable species-specific inhibitors of polyamine production designed to inactivate their unique enzymes.

Modifications to provide better pharmacokinetic properties are potentially highly valuable improvements for the use of existing drugs both for parasitic diseases and for cancer treatment and chemoprevention. Some examples of such modifications include: the use of methyl ester of Δ-MFMO and related ornithine derivatives; the BOC-protected methyl ester derivatives of *N*-(5-phosphopyridoxyl)-ornithine; and the melamine derivative of 2-hydroxy-AOPA described above. These modifications may increase the difficulty of synthesis on a large scale, but this would be more than compensated for by the improved cellular uptake and retention leading to lower doses and more prolonged effects.

Another promising area for future development is the combination of polyamine synthesis inhibitors with other drugs. There are numerous examples, given above, in which this has significant potential for cancer chemotherapy and chemoprevention. This approach combined with targeting the inhibitors towards neoplasia in which activation of oncogenes known to stimulate polyamine synthesis is a causative feature may rejuvenate approaches that suffer from a lack of interest due to the current focus on highly targeted therapies for tumor therapy.

Combination therapy with other drugs against the polyamine pathway or other targets may also improve the chances that the inhibitors of polyamine production will have more widespread impact than the current use against *T. brucei gambiense*. The combination of DFMO with nifurtimox (a drug which increases oxidative stress) improves the efficacy and decreases the dose of DFMO needed for African sleeping sickness.[12,218] Nifurtimox has some activity against Chagas disease,[219] and its combination with the AdoMetDC inhibitors for treatment of *T. cruzi* would be of interest. Co-inhibition of ODC and AdoMetDC shows promise for treatment of *P. falciparum*, which contains a bifunctional enzyme that combines both activities in a single gene product.[60,220] The renewed interest of major drug companies in supporting studies against parasitic diseases present in economically disadvantaged countries exemplified by the development of Genz-644131[109] is an encouraging sign that such development of the polyamine biosynthetic inhibitors will continue.

References

1. A. S. Tyms, J. D. Williamson and C. J. Bacchi, *J. Antimicrobial Chemotherapy*, 1988, **22**, 403.
2. A. E. Pegg, P. P. McCann and A. Sjoerdsma, in *Enzymes as Targets for Drug Design*, ed. M. G. Palfreyman, P. P. McCann, W. Lovenberg, J. G. Temple and A. Sjoerdsma, Academic Press, Orlando, FL, 1990, p. 407.

3. P. P. McCann and A. E. Pegg, *Pharmac. Ther.*, 1992, **54**, 195.
4. A. E. Pegg and P. P. McCann, *Pharmac. Ther.*, 1992, **56**, 359.
5. L. J. Marton and A. E. Pegg, *Annu. Rev. Pharm.* 1995, **35**, 55.
6. O. Heby, S. C. Roberts and B. Ullman, *Biochem. Soc. Trans.*, 2003, **31**, 415.
7. E. W. Gerner and F. L. Meyskens, Jr, *Nat. Rev. Cancer*, 2004, **4**, 781.
8. U. K. Basuroy and E. W. Gerner, *J. Biochem. (Tokyo)*, 2006, **139**, 27.
9. R. A. Casero, Jr and L. J. Marton, *Nat. Rev. Drug Discov.*, 2007, **6**, 373.
10. O. Heby, L. Persson and M. Rentala, *Amino Acids*, 2007, **33**, 359.
11. H. M. Wallace, *Expert Opin. Pharmacother*, 2007, **8**, 2109.
12. C. J. Bacchi, *Interdiscip. Perspect. Infect. Dis.*, 2009, **2009**, 195040.
13. M. T. Hyvönen, T. A. Keinänen, M. Cerrada-Gimenez, R. Sinervirta, N. Grigorenko, A. R. Khomutov, J. Vepsäläinen, L. Alhonen and J. Jänne, *J. Biol. Chem.*, 2007, **282**, 34700.
14. G. Landau, Z. Bercovich, M. H. Park and C. Kahana, *J. Biol. Chem.*, 2010, **285**, 2474.
15. C. G. Nichols and A. N. Lopatin, *Annu. Rev. Physiol.*, 1998, **59**, 171.
16. K. Igarashi and K. Kashiwagi, *Int. J. Biochem. Cell Biol.*, 2010, **42**, 39.
17. A. E. Pegg, *IUBMB Life*, 2009, **61**, 880.
18. C. Bello-Fernandez, G. Packham and J. L. Cleveland, *Proc. Natl. Acad. Sci. U. S. A.*, 1993, **90**, 7804.
19. J. A. Nilsson, U. B. Keller, T. A. Baudino, C. Yang, S. Norton, J. A. Old, L. M. Nilsson, G. Neale, D. L. Kramer, C. W. Porter and J. L. Cleveland, *Cancer Cell*, 2005, **7**, 433.
20. A. E. Pegg, *J. Biol. Chem.*, 2006, **281**, 14529.
21. L. M. Shantz, *Biochem. J.*, 2004, **377**, 257.
22. N. S. Rial, J. Meyskens, F. L. and E. W. Gerner, *Essays Biochem.*, 2009, **46**, 111.
23. L. M. Shantz and V. A. Levin, *Amino Acids*, 2007, **33**, 213.
24. A. E. Pegg and P. P. McCann, *Am. J. Physiol.*, 1982, **243**, C212.
25. A. E. Pegg and A. J. Michael, *Cell. Mol. Life Sci.*, 2010, **67**, 113.
26. R. Poulin, L. Lu, B. Ackermann, P. Bey and A. E. Pegg, *J. Biol. Chem.*, 1992, **267**, 150.
27. A. L. Osterman, H. B. Brooks, L. Jackson, J. J. Abbott and M. A. Phillips, *Biochemistry*, 1999, **38**, 11814.
28. L. McConlogue and P. Coffino, *J. Biol. Chem.*, 1983, **258**, 12083.
29. T. Kameji, S. Hayashi, K. Hoshino, Y. Kakinuma and K. Igarashi, *Biochem. J.*, 1993, **289**, 581.
30. J. Jänne, L. Alhonen and P. Leinonen, *Ann. Med.*, 1991, **23**, 241.
31. S. Hanson, S. M. Beverley, W. Wagner and B. Ullman, *Mol. Cell. Biol.*, 1992, **12**, 5499.
32. M. A. Phillips, P. Coffino and C. C. Wang, *J. Biol. Chem.*, 1987, **262**, 8721.
33. A. H. Fairlamb, P. Blackburn, P. Ulrich, B. T. Chait and A. Cerami, *Science (Washington DC)*, 1985, **227**, 1485.

34. S. Wyllie, S. L. Oza, S. Patterson, D. Spinks, S. Thompson and A. H. Fairlamb, *Mol. Microbiol.*, 2009, **74**, 529.
35. C. J. Bacchi, *Parasitol. Today*, 1993, **9**, 190.
36. T. L. Byers, T. L. Bush, P. P. McCann and A. J. Bitonti, *Biochem. J.*, 1991, **274**, 527.
37. N. Yarlett, J. Garofalo, B. Goldberg, M. A. Ciminelli, V. Ruggiero, J. R. Sufrin and C. J. Bacchi, *Biochim. Biophys. Acta*, 1993, **1181**, 68.
38. C. J. Bacchi, J. Garofalo, M. Ciminelli, D. Rattendi, B. Goldberg, P. P. McCann and N. Yarlett, *Biochem. Pharmacol.*, 1993, **46**, 471.
39. R. J. Vaidya, R. M. Ray and L. R. Johnson, *Cell. Mol. Life Sci.*, 2006, **63**, 2871.
40. L. Xiao, J. N. Rao, T. Zou, L. Liu, B. S. Marasa, J. Chen, D. J. Turner, A. Passaniti and J. Y. Wang, *Biochem. J.*, 2007, **403**, 573.
41. R. M. Ray, S. Bhattacharya and L. R. Johnson, *Cell. Signal.*, 2007, **19**, 2519.
42. T. Zou, L. Liu, J. N. Rao, B. S. Marasa, J. Chen, L. Xiao, H. Zhou, M. Gorospe and J. Y. Wang, *Biochem. J.*, 2008, **409**, 389.
43. S. Bhattacharya, H. Guo, R. M. Ray and L. R. Johnson, *Biochem. J.*, 2007, **407**, 243.
44. D. L. Koomoa, L. P. Yco, T. Borsics, C. J. Wallick and A. S. Bachmann, *Cancer Res.*, 2008, **68**, 9825.
45. S. Bhattacharya, R. M. Ray and L. R. Johnson, *Cell Signal.*, 2009, **21**, 509.
46. R. J. Rounbehler, W. Li, M. A. Hall, C. Yang, M. Fallahi and J. L. Cleveland, *Cancer Res.*, 2009, **69**, 547.
47. P. Kucharzewska, J. E. Welch, K. J. Svensson and M. Belting, *Biochem. Biophys. Res. Commun.*, 2009, **380**, 413.
48. P. Y. Wang, J. N. Rao, T. Zou, L. Liu, L. Xiao, T. X. Yu, D. J. Turner, M. Gorospe and J. Y. Wang, *Biochem. J.*, 2010, **426**, 293.
49. C. X. Xu, Y. F. Yan, Y. P. Yang, B. Liu, J. X. Xin, S. M. Chen, W. Wang, C. Y. Jiang, Z. X. Lu and X. X. Liu, *Mol. Biol. Rep.*, 2011, **38**, 949.
50. A. Banan, S. A. McCormack and L. R. Johnson, *Am. J. Physiol.*, 1998, **274**, G879.
51. P. Savarin, A. Barbet, S. Delga, V. Joshi, L. Hamon, J. Lefevre, S. Nakib, J. P. De Bandt, C. Moinard, P. A. Curmi and D. Pastre, *Biochem. J.*, 2010, **430**, 151.
52. A. Sjoerdsma and P. J. Schechter, *Clin. Pharmacol. Ther.*, 1984, **35**, 287.
53. C. J. Bacchi and P. P. McCann, in *Inhibition of Polyamine Metabolism* ed. P. P. McCann, A. E. Pegg and A. Sjoerdsma, Academic Press, Orlando, FL, 1987, p. 317.
54. C. Burri and R. Brun, *Parasitol. Res.*, 2003, **90**(Supp 1), S49.
55. M. P. Barrett, D. W. Boykin, R. Brun and R. R. Tidwell, *Br. J. Pharmacol.*, 2007, **152**, 1155.
56. F. Milord, J. Pépin, L. Loko, L. Ethier and B. Mpia, *Lancet*, 1992, **340**, 652.
57. B. G. Kim, P. P. McCann and T. J. Byers, *J. Parasitol.*, 1987, **34**, 264.

58. R. Mukhopadhyay, P. Kapoor and R. Madhubala, *Pharmacol. Res.*, 1996, **34**, 43.
59. T. Coons, S. Hanson, A. J. Bitonti, P. P. McCann and B. Ullman, *Mol. Biochem. Parasitol.*, 1990, **39**, 77.
60. I. B. Muller, R. Das Gupta, K. Luersen, C. Wrenger and R. D. Walter, *Mol. Biochem. Parasitol.*, 2008, **160**, 1.
61. C. Maia, A. Lanfredi-Rangel, K. G. Santana-Anjos, M. F. Oliveira, W. De Souza and M. A. Vannier-Santos, *Parasitol. Res.*, 2008, **103**, 363.
62. J. M. Boitz, P. A. Yates, C. Kline, U. Gaur, M. E. Wilson, B. Ullman and S. C. Roberts, *Infect. Immun.*, 2009, **77**, 756.
63. S. Sarhan, B. Knödgen and N. Seiler, *Anticancer Res.*, 1989, **9**, 215.
64. V. Quemener, J. P. Moulinoux, R. Havouis and N. Seiler, *Anticancer Res.*, 1992, **12**, 1447.
65. A. Ask, L. Persson and O. Heby, *Cancer Lett.*, 1993, **66**, 29.
66. N. Seiler, S. Sarhan, C. Grauffel, R. Jones, Knödgen and J. P. Moulinoux, *Cancer Res.*, 1990, **50**, 5077.
67. V. Quemener, Y. Blancard, L. Chamaillard, R. Havouis, B. Cipolla and J. Moulinoux, *Anticancer Res.*, 1994, **14**, 443.
68. V. A. Levin, M. D. Prados, W. K. A. Yung, M. J. Gleason, S. Ictech and M. Malec, *J. Natl. Cancer Inst.*, 1992, **84**, 1432.
69. V. A. Levin, K. R. Hess, A. Choucair, P. J. Flynn, K. A. Jaeckle, A. P. Kyritsis, W. K. Yung, M. D. Prados, J. M. Bruner, S. Ictech, M. J. Gleason and H. W. Kim, *Clin. Cancer Res.*, 2003, **9**, 981.
70. V. A. Levin, J. L. Jochec, L. M. Shantz, P. E. Koch and A. E. Pegg, *J. Histochem. Cytochem.*, 2004, **52**, 1467.
71. V. A. Levin, J. L. Jochec, L. M. Shantz and K. D. Aldape, *Int. J. Cancer*, 2007, **121**, 2279.
72. C. J. Wallick, I. Gamper, M. Thorne, D. J. Feith, K. Y. Takasaki, S. M. Wilson, J. A. Seki, A. E. Pegg, C. V. Byus and A. S. Bachmann, *Oncogene*, 2005, **24**, 5606.
73. N. F. Evageliou and M. D. Hogarty, *Clin. Cancer Res.*, 2009, **15**, 5956.
74. M. D. Hogarty, M. D. Norris, K. Davis, X. Liu, N. F. Evageliou, C. S. Hayes, B. Pawel, R. Guo, H. Zhao, E. Sekyere, J. Keating, W. Thomas, N. C. Cheng, J. Murray, J. Smith, R. Sutton, N. Venn, W. B. London, A. Buxton, S. K. Gilmour, G. M. Marshall and M. Haber, *Cancer Res.*, 2008, **68**, 9735.
75. D. Geerts, J. Koster, D. Albert, D. L. Koomoa, D. J. Feith, A. E. Pegg, R. Volckmann, H. Caron, R. Versteeg and A. S. Bachmann, *Int. J. Cancer*, 2010, **126**, 2012.
76. G. J. Kelloff, C. W. Boone, J. A. Crowell, V. E. Steele, R. Lubet and C. C. Sigman, *Cancer Epidemiol. Biomark. Prev.*, 1994, **3**, 85.
77. D. J. Feith, L. M. Shantz, P. L. Shoop, K. A. Keefer, C. Prakashagowda and A. E. Pegg, *Mol. Carcinogenesis*, 2007, **46**, 453.
78. X. Tang, A. L. Kim, D. J. Feith, A. E. Pegg, J. Russo, H. Zhang, M. Aszterbaum, L. Kopelovich, E. H. Epstein Jr, D. R. Bickers and M. Athar, *J. Clin. Invest.*, 2004, **113**, 867.

79. D. J. Feith, S. Origanti, P. L. Shoop, S. Sass-Kuhn and L. M. Shantz, *Carcinogenesis*, 2006, **27**, 1090.
80. L. Y. Fong, D. J. Feith and A. E. Pegg, *Cancer Res.*, 2003, **63**, 3945.
81. Y. Guo, J. L. Cleveland and T. G. O'Brien, *Cancer Res.*, 2005, **65**, 1146.
82. F. L. Meyskens, Jr and E. W. Gerner, *Clin. Cancer Res.*, 1999, **5**, 945.
83. A. R. Simoneau, E. W. Gerner, R. Nagle, A. Ziogas, S. Fujikawa-Brooks, H. Yerushalmi, T. E. Ahlering, R. Lieberman, C. E. McLaren, H. Anton-Culver and F. L. Meyskens, Jr, *Cancer Epidemiol. Biomark. Prev.*, 2008, **17**, 292.
84. H. H. Bailey, K. Kim, A. K. Verma, K. Sielaff, P. O. Larson, S. Snow, T. Lenaghan, J. L. Viner, J. Douglas, N. E. Dreckschmidt, M. Hamielec, M. Pomplun, H. H. Sharata, D. Puchalsky, E. R. Berg, T. C. Havighurst and P. P. Carbone, *Cancer Prev. Res.* 2010, **3**, 35.
85. F. L. Meyskens, C. E. McLaren, D. Pelot, S. Fujikawa-Brooks, P. M. Carpenter, E. Hawk, G. Kelloff, M. J. Lawson, J. Kidao, J. McCracken, C. G. Albers, D. J. Ahnen, D. K. Turgeon, S. Goldschmid, P. Lance, C. H. Hagedorn, D. L. Gillen and E. W. Gerner, *Cancer Prev. Res.* 2008, **1**, 9.
86. E. W. Gerner and F. L. Meyskens, Jr, *Clin. Cancer Res.*, 2009, **15**, 758.
87. P. A. Thompson, B. C. Wertheim, J. A. Zell, W. Pin Chen, C. E. McLaren, B. J. Lafleur, F. L. Meyskens and E. W. Gerner, *Gastroenterology*, 2010, **139**, 797.
88. J. Jänne, L. Alhonen, M. Pietila, T. A. Keinanen, A. Uimari, M. T. Hyvonen, E. Pirinen and A. Jarvinen, *J. Biochem. (Tokyo)*, 2006, **139**, 155.
89. Y. Ramot, M. Pietila, G. Giuliani, F. Rinaldi, L. Alhonen and R. Paus, *Exp. Dermatol.*, 2010, **19**, 784.
90. S. R. Smith, D. J. Piacquadio, B. Beger and C. Littler, *Dermatol. Surg.*, 2006, **32**, 1237.
91. J. Jackson, J. J. Caro, G. Caro, F. Garfield, F. Huber, W. Zhou, C. S. Lin, D. Shander and K. Schrode, *Int. J. Dermatol.*, 2007, **46**, 976.
92. M. Lapidoth, C. Dierickx, S. Lanigan, U. Paasch, A. Campo-Voegeli, S. Dahan, L. Marini and M. Adatto, *Dermatology*, 2010, **221**, 34.
93. P. Bey, C. Danzin and M. Jung, in *Inhibition of Polyamine Metabolism. Biological Significance and Basis for New Therapies*, ed. P. P. McCann, A. E. Pegg and A. Sjoerdsma, Academic Press, Orlando, FL, 1987, p. 1.
94. N. Claverie and P. S. Mamont, *Cancer Res.*, 1989, **49**, 4466.
95. T. L. Bowlin, B. J. Hoeper, A. L. Rosenberger, G. F. Davis and P. S. Sunkara, *Cancer Res.*, 1990, **50**, 4510.
96. T. Hyvönen, L. Alakuijala, L. Andersson, A. R. Khomutov, R. M. Khomutov and T. O. Eloranta, *J. Biol. Chem.*, 1988, **263**, 11138.
97. R. Poulin, J. A. Secrist III and A. E. Pegg, *Biochem. J.*, 1989, **263**, 215.
98. J. Stanek, J. Frei, H. Mett, P. Schnieder and U. Regenass, *J. Med. Chem.*, 1992, **35**, 1339.
99. H. Mett, J. Stanek, J. A. Lopez-Ballester, J. Jänne, L. Alhonen, R. Sinervirta, J. Frei and U. Regenass, *Cancer Chemother. Pharmacol.*, 1993, **32**, 39.

100. R. Das Gupta, T. Krause-Ihle, B. Bergmann, I. B. Muller, A. R. Khomutov, S. Muller, R. D. Walter and K. Luersen, *Antimicrob. Agents Chemother.*, 2005, **49**, 2857.
101. V. T. Dufe, D. Ingner, O. Heby, A. R. Khomutov, L. Persson and S. Al-Karadaghi, *Biochem. J.*, 2007, **405**, 261.
102. N. Klee, P. E. Wong, B. Baragana, F. E. Mazouni, M. A. Phillips, M. P. Barrett and I. H. Gilbert, *Bioorg. Med. Chem. Lett.*, 2010, **20**, 4364.
103. J. S. Heller, E. S. Canellakis, D. L. Bussolotti and J. K. Coward, *Biochem. Biophys. Acta*, 1975, **403**, 197.
104. F. Wu, D. Grossenbacher and H. Gehring, *Mol. Cancer Ther.*, 2007, **6**, 1831.
105. F. Wu and H. Gehring, *FASEB J.*, 2009, **23**, 565.
106. I. B. Muller, F. Wu, B. Bergmann, J. Knockel, R. D. Walter, H. Gehring and C. Wrenger, *PLoS ONE*, 2009, **4**, e4406.
107. D. C. Smithson, J. Lee, A. A. Shelat, M. A. Phillips and R. K. Guy, *J. Biol. Chem.*, 2010, **285**, 16771.
108. D. E. McCloskey, S. Bale, J. A. Secrist, A. Tiwari, T. H. Moss, J. Valiyaveettil, W. H. Brooks, W. C. Guida, A. E. Pegg and S. E. Ealick, *J. Med. Chem.*, 2009, **52**, 1388.
109. R. H. Barker Jr, H. Liu, B. Hirth, C. A. Celatka, R. Fitzpatrick, Y. Xiang, E. K. Willert, M. A. Phillips, M. Kaiser, C. J. Bacchi, A. Rodriguez, N. Yarlett, J. D. Klinger and E. Sybertz, *Antimicrob. Agents Chemother.*, 2009, **53**, 2052.
110. F. Svensson, L. Persson and H. Mett, *Biochem. J.*, 1997, **322**, 297.
111. F. Svensson, I. Kockum and L. Persson, *Mol. Cell. Biochem.*, 1993, **124**, 141.
112. A. E. Pegg, D. B. Jones and J. A. Secrist III, *Biochemistry*, 1988, **27**, 1408.
113. A. E. Pegg, J. A. Secrist III and R. Madhubala, *Cancer Res.*, 1988, **48**, 2678.
114. U. Regenass, H. Mett, J. Stanek, M. Mueller, D. Kramer and C. W. Porter, *Cancer Res.*, 1994, **54**, 3210.
115. M. H. Park, *J. Biochem. (Tokyo)*, 2006, **139**, 161.
116. H. G. Williams-Ashman and A. Schenone, *Biochem. Biophys. Res. Commun.*, 1972, **46**, 288.
117. H. G. Williams-Ashman and J. Seidenfeld, *Biochem. Pharmacol.*, 1986, **35**, 1217.
118. H. Elo, I. Mutikainen, L. Alhonen-Hongisto, R. Laine, J. Jänne and P. Lumme, *Z. Naturforsch.*, 1986, **41c**, 851.
119. J. Jänne, L. Alhonen-Hongisto, P. Nikula and H. Elo, *Adv. Enzyme Regul.*, 1985, **24**, 125.
120. U. Regenass, G. Caravatti, H. Mett, J. Stanek, P. Schneider, M. Müller, A. Matter, P. Vertino and C. W. Porter, *Cancer Res.*, 1992, **52**, 4712.
121. J. Stanek, G. Caravatti, H.-G. Caprano, P. Furet, H. Mett, P. Schneider and U. Regenass, *J. Med. Chem.*, 1993, **36**, 46.
122. J. Stanek, G. Caravatti, J. Frei, P. Furet, H. Mett, P. Schneider and U. Regenass, *J. Med. Chem.*, 1993, **36**, 2168.

123. T. C. Beswick, E. K. Willert and M. A. Phillips, *Biochemistry*, 2006, **45**, 7797.
124. C. J. Bacchi, R. Brun, S. L. Croft, K. Alicea and Y. Buhler, *Antimicrobial Agents & Chemotherapy*, 1996, **40**, 1448.
125. R. Brun, Y. Bühler, U. Sandmeier, R. Kaminsy, C. J. Bacchi, D. Rattendi, S. Lane, S. L. Croft, D. Snowdon, V. Yardley, G. Caravatti, J. Frei, J. Stanek and H. Mett, *Antimicrobial Agents & Chemotherapy*, 1996, **40**, 1442.
126. D. W. Tolbert, J. L. Ekstrom, I. I. Mathews, J. A. I. Secrist, P. Kapoor, A. E. Pegg and S. E. Ealick, *Biochemistry*, 2001, **40**, 9484.
127. A. E. Pegg, A. Corti and H. G. Williams-Ashman, *Biochem. Biophys. Res. Commun.*, 1973, **52**, 696.
128. A. Yerlikaya and B. A. Stanley, *J. Biol. Chem.*, 2004, **279**, 12469.
129. E. Y. Artamonova, L. L. Zavalova, R. M. Khomutov and A. R. Khomutov, *Biorg. Khim.*, 1986, **12**, 206.
130. J. A. Secrist III, *Nucleosides & Nucleotides*, 1987, **6**, 73.
131. B. L. Tekwani, C. J. Bacchi, J. A. Secrist and A. E. Pegg, *Biochem. Pharmacol.*, 1992, **44**, 905.
132. P. Casara, P. Marchal, J. Wagner and C. Danzin, *J. Am. Chem. Soc.*, 1989, **111**, 9111.
133. C. Danzin, P. Marchal and P. Casara, *Biochem. Pharmacol.*, 1990, **40**, 1499.
134. C. Danzin, P. Marchal and P. Casara, in *Enzymes Dependent on Pyridoxal Phosphate and Other Carbonyl Compounds as Cofactors*, ed. T. Fukui, H. Kagamiyama, K. Soda and H. Wada, Pergamon Press, Oxford, 1991, p. 445.
135. A. J. Bitonti, T. L. Byers, T. L. Bush, P. J. Casara, C. J. Bacchi, A. B. Clarkson, P. P. McCann and A. Sjoerdsma, *Antimicrobial Agents Chemother.*, 1990, **34**, 1485.
136. L. M. Shantz, B. A. Stanley, J. A. Secrist and A. E. Pegg, *Biochemistry*, 1992, **31**, 6848.
137. C. J. Marasco Jr, D. L. Kramer, J. Miller, C. W. Porter, C. J. Bacchi, D. Rattendi, L. Kucera, N. Iyer, R. Bernacki, P. Pera and J. R. Sufrin, *J. Med. Chem.*, 2002, **45**, 5112.
138. C. J. Bacchi, R. H. Barker Jr, A. Rodriguez, B. Hirth, D. Rattendi, N. Yarlett, C. L. Hendrick and E. Sybertz, *Antimicrob. Agents Chemother.*, 2009, **53**, 3269.
139. B. Hirth, R. H. Barker Jr, C. A. Celatka, J. D. Klinger, H. Liu, B. Nare, A. Nijjar, M. A. Phillips, E. Sybertz, E. K. Willert and Y. Xiang, *Bioorg. Med. Chem. Lett.*, 2009, **19**, 2916.
140. Y. Wu and P. M. Woster, *J. Med. Chem.*, 1992, **35**, 3196.
141. Y. Q. Wu and P. M. Woster, *Biochem. Pharmacol.*, 1995, **49**, 1125.
142. W. H. Brooks, D. E. McCloskey, K. G. Daniel, S. E. Ealick, J. A. Secrist, 3rd, W. R. Waud, A. E. Pegg and W. C. Guida, *J. Chem. Inf. Model.*, 2007, **47**, 1897.
143. A. E. Pegg, *Essays Biochem.*, 2009, **46**, 25.
144. S. Bale and S. E. Ealick, *Amino Acids*, 2010, **38**, 451.

145. G. A. Wells, L. M. Birkholtz, F. Joubert, R. D. Walter and A. I. Louw, *J. Mol. Graph. Model.*, 2005, **24**, 307.

146. D. Kramer, H. Mett, A. Evans, U. Regenass, P. Diegelman and C. W. Porter, *J. Biol. Chem.*, 1995, **270**, 2124.

147. X. Hu, S. Washington, M. F. Verderame, L. M. Demers, D. Mauger and A. Manni, *Int. J. Oncol.*, 2004, **25**, 1831.

148. R. Paridaens, D. R. A. Uges, N. Barbet, L. Choi, M. Seeghers, W. T. A. van der Graaf and J. J. M. Groen, *Br. J. Cancer*, 2000, **83**, 594.

149. F. A. Eskens, G. A. Greim, C. van Zuylen, I. Wolff, L. J. Denis, A. S. Planting, F. A. Muskiet, J. Wanders, N. C. Barbet, L. Choi, R. Capdeville, J. Verweij, A. R. Hanauske and U. Bruntsch, *Clin. Cancer Res.*, 2000, **6**, 1736.

150. M. Pless, K. Belhadj, W. Kern, C. Dumontet, J. Chemnitz, H. D. Menssen, R. Herrmann, N. C. Barbet and R. Capdeville, *Proc. Am. Soc. Clin. Oncol.*, 2000, **36**, 62.

151. L. van Zuylen, F. Eskens, J. Bridgewater, A. Sparreboom, I. Sklenar, A. Planting, L. Choi, C. Mueller, R. Capdeville, J. Ledermann and J. Verweij, *Proc. Am. Soc. Clin. Oncol.*, 2000, **36**, 751.

152. L. L. Siu, E. K. Rowinsky, L. A. Hammond, G. R. Weiss, M. Hidalgo, G. M. Clark, J. Moczygemba, L. Choi, R. Linnartz, N. Barbet, I. T. Sklenar, R. Capdeville, G. Gan, C. W. Porter, D. D. Von Hoff and S. G. Eckhardt, *Clin. Cancer Res.*, 2002, **8**, 2157.

153. M. Pless, K. Belhadj, H. D. Menssen, W. Kern, B. Coiffier, J. Wolf, R. Herrmann, E. Thiel, D. Bootle, I. Sklenar, C. Muller, L. Choi, C. Porter and R. Capdeville, *Clin. Cancer Res.*, 2004, **10**, 1299.

154. M. J. Millward, A. Joshua, R. Kefford, S. Aamdal, D. Thomson, P. Hersey, G. Toner and K. Lynch, *Invest. New Drugs*, 2005, **23**, 253.

155. D. L. Koomoa, T. Borsics, D. J. Feith, C. C. Coleman, C. J. Wallick, I. Gamper, A. E. Pegg and A. S. Bachmann, *Mol. Cancer Ther.*, 2009, **8**, 2067.

156. T. L. Byers, B. Ganem and A. E. Pegg, *Biochem. J.*, 1992, **287**, 717.

157. T. L. Byers, R. Wechter, R. Hu and A. E. Pegg, *Biochem. J.*, 1994, **303**, 89.

158. N. Seiler, S. Sarhan, P. Mamont, P. Casara and C. Danzin, *Life Chem. Reports*, 1991, **9**, 151.

159. R. Blavid, P. Kusch, J. Hauber, U. Eschweiler, S. R. Sarite, S. Specht, S. Deininger, A. Hoerauf and A. Kaiser, *Amino Acids*, 2010, **38**, 461.

160. J. Guo, Y. Q. Wu, D. Rattendi, C. J. Bacchi and P. M. Woster, *J. Med. Chem.*, 1995, **38**, 1770.

161. C. J. Bacchi, H. C. Nathan, N. Yarlett, B. Goldberg, P. P. McCann, A. J. Bitonti and A. Sjoerdsma, *Antimicrob. Agents Chemother.*, 1992, **36**, 2736.

162. M. A. Yakubu, S. Majumder and F. Kierszenbaum, *J. Parasitol.*, 1993, **79**, 525.

163. L. Persson, *Biochem. Soc. Trans.*, 2007, **35**, 314.

164. J. A. Urbina, *Acta Trop.*, 2010, **115**, 55.

165. B. Schafer, I. Hauber, A. Bunk, J. Heukeshoven, A. Dusedau, D. Bevec and J. Hauber, *J. Infect. Dis.*, 2006, **194**, 740.

166. D. Bevec and J. Hauber, *Biological Signals*, 1997, **6**, 124.

167. Y. Ikeguchi, M. Bewley and A. E. Pegg, *J. Biochem.*, 2006, **139**, 1.
168. H. Wu, J. Min, H. Zeng, D. E. McCloskey, Y. Ikeguchi, P. Loppnau, A. J. Michael, A. E. Pegg and A. N. Plotnikov, *J. Biol. Chem.*, 2008, **283**, 16135.
169. A. E. Pegg, *Biochem. J.*, 1986, **234**, 249.
170. M. A. Avila, E. R. Garcia-Trevijano, S. C. Lu, F. J. Corrales and J. M. Mato, *Int. J. Biochem. Cell Biol.*, 2004, **36**, 2125.
171. E. Albers, *IUBMB Life*, 2009, **61**, 1132.
172. I. Basu, G. Cordovano, I. Das, T. J. Belbin, C. Guha and V. L. Schramm, *J. Biol. Chem.*, 2007, **282**, 21477.
173. G. B. Evans, R. H. Furneaux, D. H. Lenz, G. F. Painter, V. L. Schramm, V. Singh and P. C. Tyler, *J. Med. Chem.*, 2005, **48**, 4679.
174. A. I. Longshaw, F. Adanitsch, J. A. Gutierrez, G. B. Evans, P. C. Tyler and V. L. Schramm, *J. Med. Chem.*, 2010, **53**, 6730.
175. G. Colotti and A. Ilari, *Amino Acids,* 2011, **40**, 269.
176. J. V. Becker, L. Mtwisha, B. G. Crampton, S. Stoychev, A. C. van Brummelen, S. Reeksting, A. I. Louw, L. M. Birkholtz and D. T. Mancama, *BMC Genomics*, 2010, **11**, 235.
177. M. C. Taylor, H. Kaur, B. Blessington, J. M. Kelly and S. R. Wilkinson, *Biochem. J.*, 2008, **409**, 563.
178. Y. Xiao, D. E. McCloskey and M. A. Phillips, *Eukaryot Cell*, 2009, **8**, 747.
179. A. E. Pegg, R. Poulin and J. K. Coward, *Int. J. Biochem.*, 1995, **27**, 425.
180. A. Shirahata, T. Morohohi, M. Fukai, F. Akatsu and K. Samejima, *Biochem. Pharmacol.*, 1991, **41**, 205.
181. H. Goda, T. Watanabe, N. Takeda, M. Kobayashi, M. Wada, H. Hosoda, A. Shirahata and K. Samejima, *Biol. Pharm. Bull.*, 2004, **27**, 1327.
182. H. Wu, J. Min, Y. Ikeguchi, H. Zeng, A. Dong, P. Loppnau, A. E. Pegg and A. N. Plotnikov, *Biochemistry*, 2007, **46**, 8331.
183. V. T. Dufe, W. Qiu, I. B. Muller, R. Hui, R. D. Walter and S. Al-Karadaghi, *J. Mol. Biol.*, 2007, **373**, 167.
184. A. Shirahata, N. Takahashi, T. Beppu, H. Hosoda and K. Samejima, *Biochem. Pharmacol.*, 1993, **45**, 1897.
185. M. Kobayashi, T. Watanabe, Y. J. Xu, M. Tatemori, H. Goda, M. Niitsu, A. Shirahata and K. Samejima, *Biol. Pharm. Bull.*, 2005, **28**, 569.
186. K. C. Tang, A. E. Pegg and J. K. Coward, *Biochem. Biophys. Res. Commun.*, 1980, **96**, 1371.
187. K. C. Tang, R. Mariuzza and J. K. Coward, *J. Med. Chem.*, 1981, **24**, 1277.
188. J. K. Coward and A. E. Pegg, *Adv. Enzyme Regul.*, 1987, **26**, 107.
189. S. Korolev, Y. Ikeguchi, T. Skarina, S. Beasley, C. Arrowsmith, A. Edwards, A. Joachimiak, A. E. Pegg and A. Savchenko, *Nature Struct. Biol.*, 2002, **9**, 27.
190. A. E. Pegg, K. C. Tang and J. K. Coward, *Biochemistry*, 1982, **21**, 5082.
191. M. L. Sherman, T. D. Shafman, J. K. Coward and D. W. Kufe, *Biochem. Pharmacol.*, 1986, **35**, 2633.
192. T. P. Forshell, S. Rimpi and J. A. Nilsson, *Cancer Prev. Res.*, 2010, **3**, 140.

193. M. Jacobsson, M. Garedal, J. Schultz and A. Karlen, *J. Med. Chem.*, 2008, **51**, 2777.
194. K. Enomoto, T. Nagasaki, A. Yamauchi, J. Onoda, K. Sakai, T. Yoshida, K. Maekawa, Y. Kinoshita, I. Nishino, S. Kikuoka, T. Fukunaga, K. Kawamoto, Y. Numata, H. Takemoto and K. Nagata, *Anal. Biochem.*, 2006, **351**, 229.
195. P. M. Woster, A. Y. Black, K. J. Duff, J. R. Coward and A. E. Pegg, *J. Med. Chem.*, 1989, **32**, 1300.
196. J. R. Lakanen, A. E. Pegg and J. K. Coward, *J. Med. Chem.*, 1995, **38**, 2714.
197. A. E. Pegg, R. Wechter, R. Poulin, P. M. Woster and J. K. Coward, *Biochemistry*, 1989, **28**, 8446.
198. J. G. Baillon, M. Kolb and P. S. Mamont, *Eur. J. Biochem.*, 1989, **179**, 17.
199. A. E. Pegg and J. K. Coward, *Biochem. Biophys. Res. Commun.*, 1985, **133**, 82.
200. C. A. Mackintosh and A. E. Pegg, *Biochem. J.*, 2000, **351**, 439.
201. J. Nilsson, A. Gritli-Linde and O. Heby, *Biochem. J.*, 2000, **352**, 381.
202. M. Huber and R. Poulin, *Cancer Res.*, 1995, **55**, 934.
203. A. L. Cason, Y. Ikeguchi, C. Skinner, T. C. Wood, H. A. Lubs, F. Martinez, R. J. Simensen, R. E. Stevenson, A. E. Pegg and C. E. Schwartz, *Eur. J. Human Genet.*, 2003, **11**, 937.
204. G. de Alencastro, D. E. McCloskey, S. E. Kliemann, C. M. Maranduba, A. E. Pegg, X. Wang, D. R. Bertola, C. E. Schwartz, M. R. Passos-Bueno and A. L. Sertie, *J. Med. Genet.*, 2008, **45**, 539.
205. L. E. Becerra-Solano, J. Butler, G. Castañeda-Cisneros, D. E. McCloskey, X. Wang, A. E. Pegg, C. E. Schwartz, J. Sánchez-Corona and J. E. Garcia-Ortiz, *Am. J. Med. Genet. A*, 2009, **149A**, 328.
206. A. E. Pegg and X. Wang, *Commun. Integr. Biol.*, 2009, **2**, 271.
207. X. Wang, S. Levic, M. A. Gratton, K. J. Doyle, E. N. Yamoah and A. E. Pegg, *J. Biol. Chem.*, 2009, **284**, 930.
208. A. J. Kahn and S. C. Minocha, *Life Sci.*, 1989, **44**, 1215.
209. W. D. Tolbert, D. E. Graham, R. H. White and S. E. Ealick, *Structure*, 2003, **11**, 285.
210. H. Kidron, S. Repo, M. S. Johnson and T. A. Salminen, *Mol. Biol. Evol.*, 2007, **24**, 79.
211. T. Oshima, *Amino Acids*, 2007, **33**, 367.
212. X. Deng, J. Lee, A. J. Michael, D. R. Tomchick, E. J. Goldsmith and M. A. Phillips, *J. Biol. Chem.*, 2010, **285**, 25708.
213. N. Morimoto, W. Fukuda, N. Nakajima, T. Masuda, Y. Terui, T. Kanai, T. Oshima, T. Imanaka and S. Fujiwara, *J. Bacteriol.*, 2010, **192**, 4991.
214. A. J. Bitonti, P. J. Casara, P. P. McCann and P. Bey, *Biochem. J.*, 1987, **242**, 69.
215. G. H. Tait, *Biochem. Soc. Trans.*, 1976, **4**, 610.
216. J. Lee, V. Sperandio, D. E. Frantz, J. Longgood, A. Camilli, M. A. Phillips and A. J. Michael, *J. Biol. Chem.*, 2009, **284**, 9899.

217. F. L. Shaw, K. A. Elliott, L. N. Kinch, C. Fuell, M. A. Phillips and A. J. Michael, *J. Biol. Chem.*, 2011, **285**, 14711.
218. G. Priotto, S. Kasparian, D. Ngouama, S. Ghorashian, U. Arnold, S. Ghabri and U. Karunakara, *Clin. Infect. Dis.*, 2007, **45**, 1435.
219. A. M. Canavaci, J. M. Bustamante, A. M. Padilla, C. M. Perez Brandan, L. J. Simpson, D. Xu, C. L. Boehlke and R. L. Tarleton, *PLoS Negl Trop Dis*, 2010, **4**, e740.
220. A. C. van Brummelen, K. L. Olszewski, D. Wilinski, M. Llinas, A. I. Louw and L. M. Birkholtz, *J. Biol. Chem.*, 2008, **7**, 4635.

CHAPTER 5

Symmetrical- and Unsymmetrical Terminally Alkylated Polyamines

PATRICK M. WOSTER*[a] AND ROBERT A. CASERO, JR[b]

[a] Department of Pharmaceutical and Biomedical Sciences, Medical University of South Carolina, 70 President St., Charleston, SC 29425, USA; [b] The Sidney Kimmel Comprehensive Cancer Center, Bunting-Blaustein Cancer Research Building, Johns Hopkins University, 1650 Orleans St., Baltimore, MD 21231, USA

5.1 Introduction

The polyamines putrescine (1,4-diaminobutane), spermidine (1,8-diamino-4-azaoctane, **2**) and spermine (1,12-diamino-4,9-diazadodecane, **3**) (Figure 5.1) are ubiquitous, naturally occurring polycationic compounds that are found in significant amounts in nearly every prokaryotic and eukaryotic cell type. Under physiological conditions, spermidine and spermine primarily exist as fully protonated polycations with the pK_a values indicated in Figure 5.1.[1] This high degree of positive charge is an important factor in the biological functions of these molecules, and as will be discussed below, alterations in the pK_a of polyamine nitrogens can significantly disrupt their cellular function. Polyamines are widely distributed in nature, and are known to be required in micromolar to millimolar concentrations to support a wide variety of cellular functions. However, data establishing the precise role of the polyamines and their analogs in cellular processes are incomplete. The polyamine pathway

RSC Drug Discovery Series No. 17
Polyamine Drug Discovery
Edited by Patrick M. Woster and Robert A. Casero, Jr.
© Royal Society of Chemistry 2012
Published by the Royal Society of Chemistry, www.rsc.org

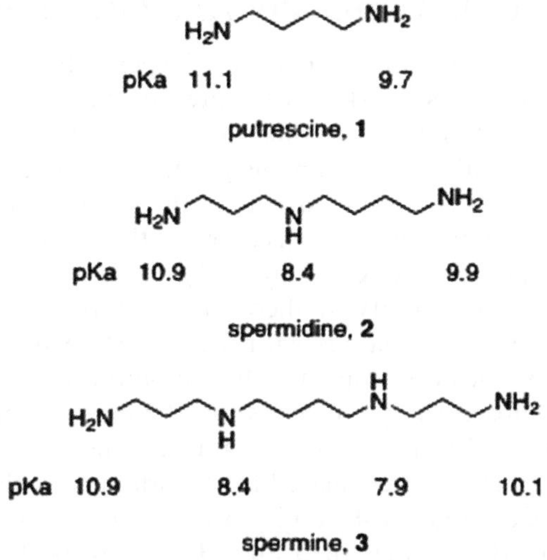

Figure 5.1 Structures of putrescine (**1**), spermidine (**2**) and spermine (**3**), and the pKa values for each of their nitrogens.

represents an important target for chemotherapeutic intervention, since depletion of polyamines results in the disruption of a variety of cellular functions and may in specific cases result in cytotoxicity.[2,3] The human and mammalian pathways for polyamine metabolism have been studied extensively, and analogous pathways have now been elucidated for a relatively large number of organisms. There are important interspecies differences in polyamine metabolism, especially among eukaryotic cells, plants, and some bacteria and protozoa. Some pathogenic organisms possess enzymes involved in polyamine metabolism that are not present in the host, and thus provide potential targets for the design of specific therapeutic agents. The enzymes involved in human and mammalian polyamine metabolism are reasonably similar, and inhibitors targeted to these enzymes rely on the observation that polyamine metabolism is accelerated, and polyamines are required in higher quantities, in target cell types. The rational design of polyamine antimetabolites such as those discussed below had its origins in the theory that the terminal nitrogens of the linear polyamines are most important for conserving the biological function of the natural polyamines.[1] It was proposed that carefully designed polyamine analogs with altered pK_a values at these terminal nitrogens would have the potential to disrupt polyamine metabolism, and thus such agents have been investigated as potential therapeutic agents *in vitro* and *in vivo*. This chapter will summarize the development of synthetic derivatives of the polyamines, and describe their use as potential chemotherapeutic agents. Comprehensive reviews of polyamine biochemistry, polyamine biosynthesis inhibitors and the role of polyamines in normal and tumor cell metabolism have recently been published.[2,4–10]

5.2   Symmetrical, Terminally Alkylated Polyamines

The development of unnatural analogs of spermidine and spermine as potential antitumor agents was first attempted in the late 1980s. The first of these analogs featured simple modifications of the natural polyamines, i.e they possessed terminal primary amine groups or secondary amines with small alkyl substituents, and differed only in the length of the central carbon chain.[11] Compounds with eight carbons between the central nitrogens, such as tetraamines A–C (**4a**, **b** and **c**, Figure 5.2), generally showed significant antitumor activity. The activity of these first unnatural polyamine analogs supported the theory that analogs with altered pK_a values at the terminal nitrogens could disrupt polyamine metabolism and potentially provide therapeutic benefit. Subsequently, a series of (bis)benzyl analogs, the most effective of which was MDL 27695 (**5**, Figure 2), and an additional series of substituted tetraamines were evaluated for the ability to inhibit proliferation of HeLa cells.[12] Compounds **4a** and **5** were active antiproliferative compounds, exhibiting IC_{50} values of 5 and 50 μM, respectively. Because the natural polyamines are known to bind to DNA, the question was raised as to whether their activity was due to DNA binding; however, no correlation between the DNA binding of these analogs and antitumor activity was observed. (Bis)benzyl polyamines, and in particular MDL 27695, were subsequently found to possess excellent antimalarial activity *in vitro* and *in vivo*,[13,14] and were effective against *Leishmania donovanii*,[15] but these promising lead compounds were not developed further. Interestingly, the *in vivo* antimalarial activity of MDL 27695 was later attributed to the fact that it served as a substrate for the plasmodial form of polyamine oxidase.[16] More will be said about the antiparasitic properties of analogs of **5** later in this chapter.

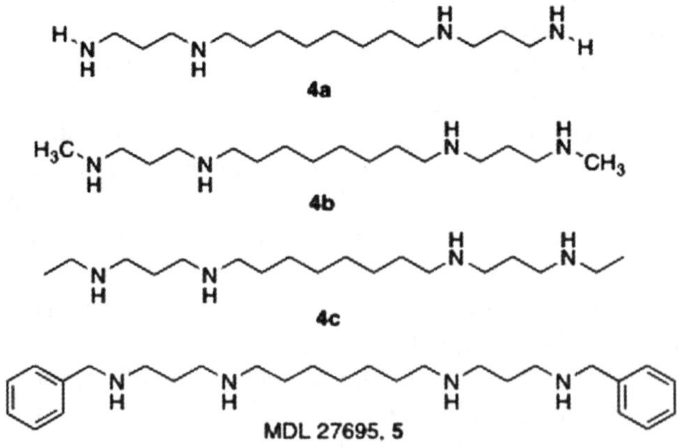

Figure 5.2   Early examples of polyamine analogs **4a–c** and **5** that exhibit biological activity *in vitro*.

Subsequent early attempts to develop analogs of the natural polyamines that could modulate their functions focused on symmetrical, terminally substituted bis(alkyl)polyamines. Their design was based on the finding that natural polyamines closely regulate their cellular levels using specific feedback mechanisms that regulate their synthesis,[17–19] and that they can be taken into cells by an energy-dependent polyamine transport system. These analogs also featured terminal secondary amines to explore the relationship between altered pK_a and cellular function. A variety of symmetrically substituted polyamine analogs have been synthesized which enter the cell using the polyamine transport system, the most prominent of which include compounds **6–11** (Figure 5.3). These analogs slow the synthesis of polyamines through downregulation of the biosynthetic enzymes ornithine decarboxylase (ODC) and *S*-adenosylmethionine decarboxylase (AdoMet-DC), but cannot support the cell growth and survival functions that polyamines mediate.[2,20] Depending on the structure of the terminal alkyl substituents, the cellular effects of these compounds can be variable. For example, some alkylpolyamine analogs are potent inducers of the polyamine catabolic

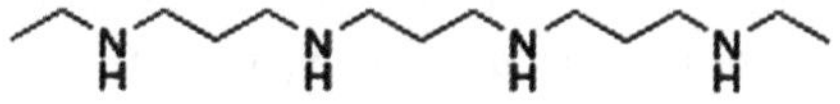

BENSpm, 6

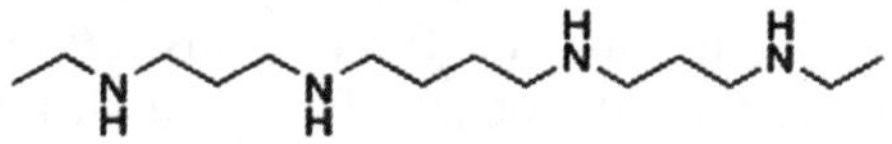

BESpm, 7

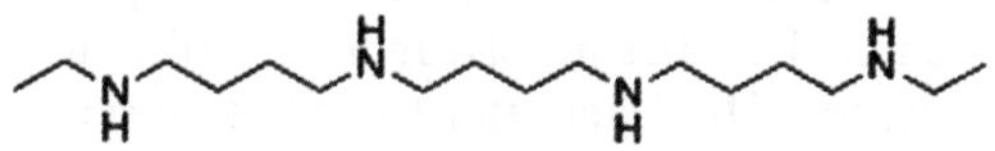

BEHSpm, 8

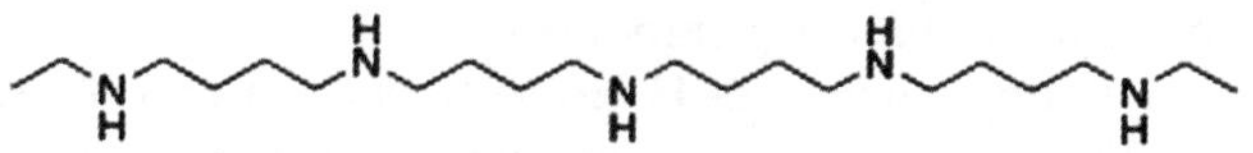

BE-4444, 9

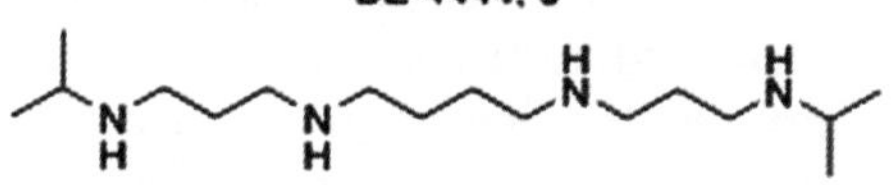

BIPSpm, 10

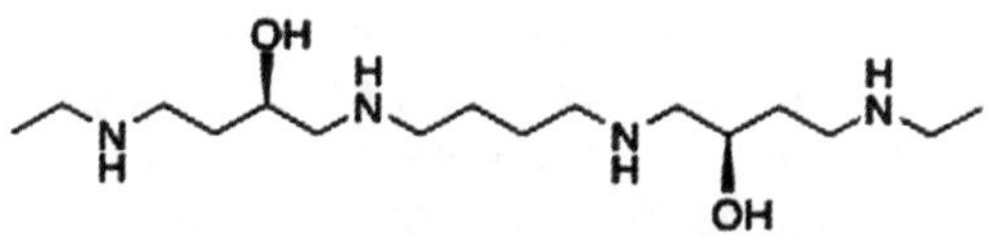

3,12-dihydroxy-BEHSpm, 11

Figure 5.3 Structures of bis(ethyl)polyamines **6–11**.

enzymes spermidine/spermine-N^1-acetyltransferase (SSAT) and/or spermine oxidase (SMO) in cultured tumor cells, while others are not.[2,21] However, following treatment of tumor cells with μM concentrations of these analogs, active compounds (whether the induce polyamine catabolic enzymes or not) initiate apoptosis as a common mechanism of cell death. The most successful of the symmetrically substituted polyamines analogs to date are the *N,N'*-bis(ethyl)polyamines shown in Figure 5.3: bis(ethyl)norspermine (BENSpm, **6**), bis(ethyl)spermine (BESpm, **7**),[20,22,23] bis(ethyl)homospermine (BEHSpm, **8**)[20,22,23] and 1,20-(ethylamino)-5,10,15-triazanonadecane (BE-4444, **9**).[24,25] These compounds have been shown to possess a wide variety of therapeutic effects[2,3,6] and illustrate the fact that small structural changes in alkylpolyamine analogs can result in surprisingly significant changes in biological activity.

A major advantage of the bis(ethyl)polyamines lies in the fact that their synthesis is extremely straightforward and depends only on the availability of the appropriate parent polyamine backbone.[20,26,27] These syntheses can also be readily scaled up to pilot-plant quantities. For this reason, biological data were collected using only symmetrically substituted (bis)alkylpolyamines, and as a result, only bis(ethyl)-substituted analogs were advanced to clinical trials. Among these analogs, the most promising were **6**, which has a 3-3-3 backbone, **8**, which has a 4-4-4 carbon skeleton, and **9**, which has a 4-4-4-4 architecture. Interestingly, BEHSpm proved to be useful as an antidiarrheal agent[28–30] and was advanced to clinical trials for this indication. Phase I and II clinical studies involving **6** revealed that it was safe for administration, and had minimal toxicity, but the compound did not show any significant clinical effects in patients with breast or lung malignancies.[31–33] However, compelling pre-clinical data suggested that the study of additional polyamine analogs was warranted. Phase I clinical studies involving **8** were discontinued due to neuro- and hepatotoxicity issues. However, the toxicities observed with **8** could be reduced by the introduction of hydroxyl groups into the intermediate chain, such as in **11**, probably due to the fact that these hydroxylated analogs are more rapidly cleared through phase 2 metabolism.[28,34]

The success of analogs such as **6–11** led to the design of a series of 'second-generation' bis(ethyl)polyamines with enhanced antitumor and antiparasitic activity. In an attempt to confer a degree of structural rigidity, analogs of **8** were synthesized that featured restricted rotation in the central polyamine chain. This was accomplished by incorporating a *cis*- and *trans*-cyclopropyl (**12**) or -cyclobutyl (**13**) ring, a *cis*- and *trans* double bond (**14**), a triple bond (**15**), or a 1,2-disubstituted aromatic ring (**16**) into the polyamine skeleton (Figure 5.4).[35–37] The resulting analogs exhibited varying antitumor activity against a panel of human tumor cell lines (A549, HT-29, U251MG, DU145, PC-3 and MCF7).[38] It is of interest to note that a subset of terminally alkylated polyamine analogs (such as MDL 27695, **5**) exhibit potent antiparasitic activity *in vitro* and *in vivo*. Such activity is most often observed in analogs with central chains that are seven or eight carbons in length, but antiparasitic properties have also been observed in 3-3-3 or 3-4-3 compounds.[2] Along these lines, *cis*-**14**

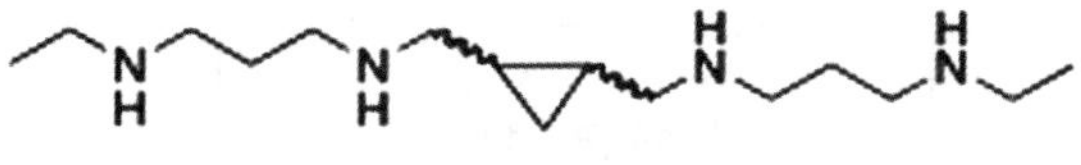

Figure 5.4 Conformationally restricted analogs of BESpm **12–16**.

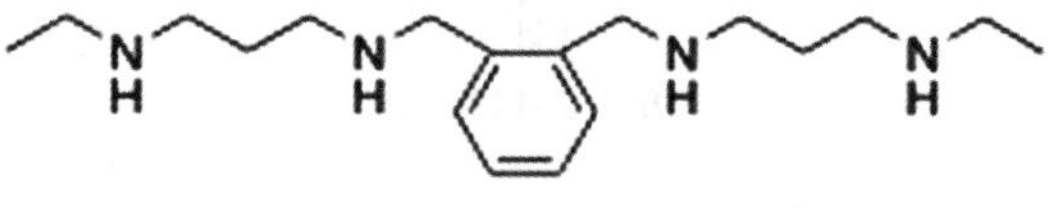

(PG-11047, Figure 5.5) was found to be a curative treatment for *Cryptosporidum parvum* infection in a mouse model.[39] By contrast, the 4-4-4 (homospermine) analog PG-11093 (**17**, Figure 5), which contains a *trans*-cyclopropyl moiety in the central region, was an effective antitumor agent *in vitro*, and *in vivo* against DU-145 nude mouse xenografts,[36] but was far less effective against *Cryptosporidia*. These data underscore the observation that small changes in chemical structure within this class of compounds can result in significant changes in biological activity. In general, structural modifications to homospermine-like (4-4-4) backbones such as the addition of *cis*- and *trans*-cyclopropyl, -cyclobutyl and *cis*-olefin moieties afforded analogs with enhanced antitumor activity and reduced systemic toxicity. Compound **17** (PG-11093) has recently been found to be effective in combination with the proteosome inhibitor bortezomib for the treatment of myeloma.[40] Compounds with a 4-4-4 or 4-4-4-4 backbone that featured *trans*-cyclopropyl, *trans*-cyclobutyl or *cis* olefin moieties in non-central regions of the chain were more active in prostate-tumor cell lines, an effect attributed to enhanced binding of the analogs to DNA.[37,40] The *trans*-bis(cyclopropyl) analogs **18** and **19** (Figure 5.5) were effective antitumor agents against prostate-tumor lines *in vitro*.[35] Insertion of a *cis* double bond into the terminal aminobutyl moieties of BE-4-4-4-4 **9** (*i.e.* as in PG-11121, **20** and PG-11128, **21**, Figure 5.5) produced analogs that are equipotent to **9** with respect to ID_{50} values but that are an order of magnitude more

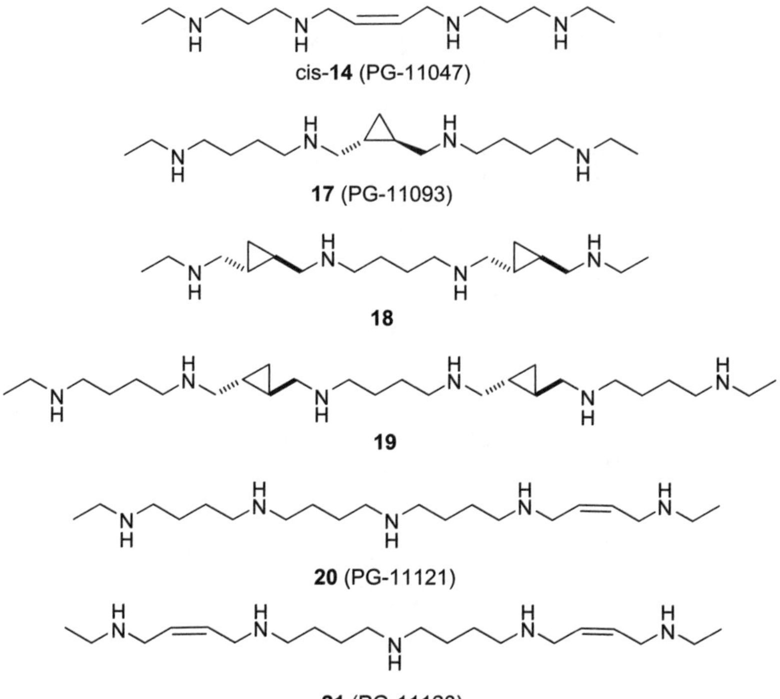

Figure 5.5 Conformationally restricted bis(ethyl) polyamine analogs **14** and **17–21** that exhibit enhanced antitumor activity against prostate tumor cells in culture.

cytotoxic in a dose–response study.[41] More recently, *cis*-**14** was shown to inhibit the growth of both small-cell and non-small-cell lung cancer cells effectively *in vitro*,[42] and significantly delayed the progression of established tumors in an A549 xenograft model. Finally, **14** and **17** were found to cause significant suppression and regression of laser-induced choroidal neovascularization following periocular injection, most likely by initiating apoptosis in proliferating vascular cells.[43] These data suggest a potential role for these analogs in the treatment of macular degeneration. Thus, *cis*-**14** represents a promising new polyamine analog that warrants further clinical evaluation, and the data suggest that polyamine analogs like *cis*-**14** could also be useful in diseases other than cancer.

In 2003, Valasinas *et al.* described a series of bis(ethyl)oligoamine analogs that show promise as potential chemotherapeutic agents.[44] The decamine PG-11144 (**22**) and the octamine PG-11158 (**23**) (Figure 5.6) proved to be most growth-inhibitory against a panel of prostate tumor cells *in vitro*. These analogs displayed IC$_{50}$ values an order of magnitude lower than the shorter analogs described above, with activity that correlated with their ability to aggregate DNA.[44] Compound **22** had good activity against human breast cancer cells *in vitro* and *in vivo*, most likely exerting these effects through multiple apoptotic mechanisms.[45] In addition, oligoamine-induced cell death appears to be mediated through the c-Jun and c-Fos pathway in MDA-MB-435cells.[46] Also, **22** and two related oligoamines specifically suppressed the mRNA and protein expression of estrogen-receptor α and estrogen-receptor target genes in the estrogen-receptor-positive human breast-cancer cell lines, whereas neither estrogen receptor β nor other steroid hormonal receptors were affected. These results suggest a novel antiestrogenic mechanism for specific polyamine analogs in human breast-cancer cells.[47] As mentioned above, some polyamine analogs exhibit potent antiparasitic activity, and compound **23** and related analogs have been shown to have significant inhibitory activity against the microsporidia *Enterocytozoon bieneusi*, the most common cause of chronic diarrhea in HIV/AIDS patients.[48] As will be described below, some of the antitumor activity observed following treatment with **22** and its analogs can be attributed to an epigenetic effect that will be described later in this chapter.

Although, strictly speaking, they are not members of this structural class, macrocyclic polyamines[49] act as potent antitumor agents by virtue of their ability to selectively deplete ATP. Based on this observation, a series of five macrocyclic polyamines typified by **24** (Figure 5.6) were synthesized and evaluated for antitumor activity in multiple human prostate cell lines.[50] All five analogs were readily imported by cells and caused a dramatic depletion of cellular polyamines accompanied by cytotoxicity that roughly correlated to their ability to deplete ATP.

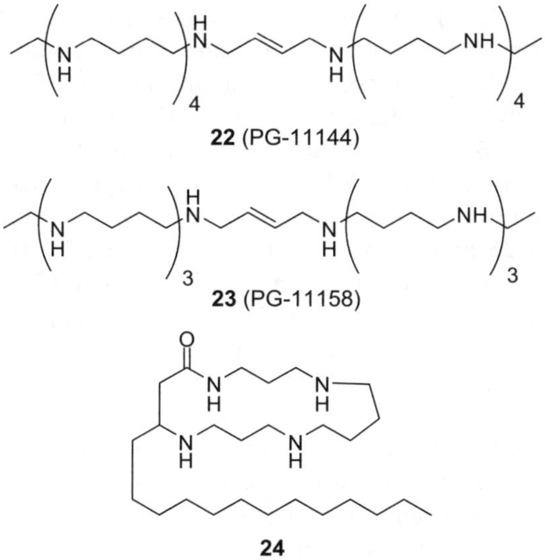

22 (PG-11144)

23 (PG-11158)

24

Figure 5.6   Structures of oligamines **22** and **23** and macrocyclic polyamines **24** that possess antitumor activity.

5.3 Unsymmetrical, Terminally Alkylated Polyamines

Studies involving the bis(alkyl)polyamines were instrumental in beginning to understand the influence of charge, length and flexibility of the polyamine backbone structure on biological activity. However, prior to 1993 the most promising polyamine analogs were symmetrically substituted with ethyl groups at the terminal nitrogens, leading to the hypothesis that substituents of a greater size than ethyl would reduce activity dramatically. These studies also suggested that compounds with a 3-3-3 carbon skeleton were more effective than the corresponding 3-4-3 or 3-7-3 analogs and that spermine-like compounds are more effective that spermidine-like analogs. In order to test this hypothesis, unsymmetrically substituted alkylpolyamines had to be synthesized to determine the optimal substituent pattern for the terminal nitrogens and to explore the chemical space surrounding the terminal alkyl groups. Despite the success of the bis(ethyl)polyamines **6–10** and their homologs as antitumor agents, no efficient syntheses leading to unsymmetrically substituted polyamines were described prior to 1993. The synthesis of these analogs is less straightforward and requires selective protection and deprotection of the internal and external nitrogens coupled with assembly of compounds in aminopropyl- or aminobutyl subunits. Saab *et al.*[51] were the first to describe a synthesis for unsymmetrically substituted alkylpolyamines, and an almost identical route was published independently by another group in 1994.[23] From this literature, a large number of synthetic techniques emerged that can be used to produce selectively functionalized mono- and bis-alkyl substituted alkylpolyamines.[11,12,20,23,34,51–57]

The first examples of unsymmetrically substituted alkylpolyamines, N^1-propargyl-N^{11}-ethylnorspermine (PENSpm, **25**) and N^1-cyclopropylmethyl-N^{11}-ethylnorspermine (CPENSpm, **26**), were described in 1993 (Figure 5.7).[51] Both **25** and **26** were more active than BESpm **7** and produced cytotoxicity in H157 non-small cell lung carcinoma cells in culture with IC_{50} values of 0.7 and 0.4 µM, respectively. It is interesting to note that treatment with bis(ethyl)-polyamines such as **6** and **7** produced a phenotype-dependent induction of SSAT in the H157 non-small cell line, resulting in cytotoxicity, but not in the H82 small-cell lung-tumor line, where cytostasis was observed.[58] As was the case with **7**, both **25** and **26** produced a significant induction of SSAT and retained the cell-type-specific cytotoxicity typical of **6** and **7**, wherein the H157 cell line was rapidly killed, while the H82 small-cell lung-carcinoma cell line responded in a cytostatic manner. In each case, analog cytotoxicity was directly correlated to the level of induction of SSAT, and this increase in SSAT activity was accompanied by a cell-specific increase in steady-state SSAT mRNA in H157 cells. These data suggest a similar mechanism of induction of SSAT for **6**, **7**, **25** and **26**, and in theory, the resulting cellular increase in hydrogen peroxide (and hence reactive oxygen species) was responsible for the induction of apoptosis. This contention is supported by the fact that apoptosis following treatment with **26** can be ablated by the addition of catalase or the polyamine oxidase inhibitor MDL-72527.[59] However, more recent data suggest that the

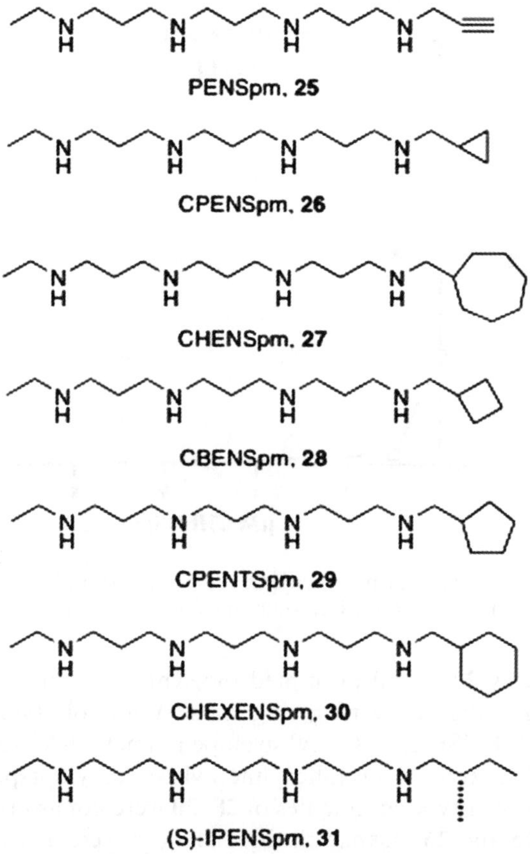

PENSpm, **25**

CPENSpm, **26**

CHENSpm, **27**

CBENSpm, **28**

CPENTSpm, **29**

CHEXENSpm, **30**

(S)-IPENSpm, **31**

Figure 5.7 Unsymmetrically substituted alkylpolyamines PENSpm **25**, CPENSpm **26**, CBENSpm **27**, CPENTSpm **28**, CHEXENSpm **29**, CHENSpm **30** and (S)-IPENSpm **31**.

source of reactive oxygen species is actually from spermine oxidase and not the SSAT/acetylpolyamine oxidase (APAO) pathway (see also Chapter 6 of this volume).[60-62] Data gathered from the unsymmetrically substituted alkylpolyamines **25** and **26** supported the hypothesis that there may be a functional relationship between cytotoxicity and SSAT induction in the H157 and H82 lung-tumor cell lines. A third compound in this series, N^1-cycloheptylmethyl-N^{11}-ethylnorspermine (CHENSpm, **27**, Figure 5.7), was subsequently synthesized and evaluated, and produced a potent cytotoxic effect ($IC_{50} = 0.5 \ \mu M$) in the H157 line, as seen with **7**, **25** and **26**, while producing a less pronounced cell-type specificity, as shown in Figure 5.8 (H82 $IC_{50} = 1.5 \ \mu M$). Surprisingly, treatment with 10 μM **22** had almost no effect on the levels of putrescine, spermidine and spermine, and caused only a 15-fold induction of SSAT activity. These data suggested that the cytotoxic effects produced by **6**, **7**, **25** and **26** in H157 cells could be mediated by different cellular mechanisms than the

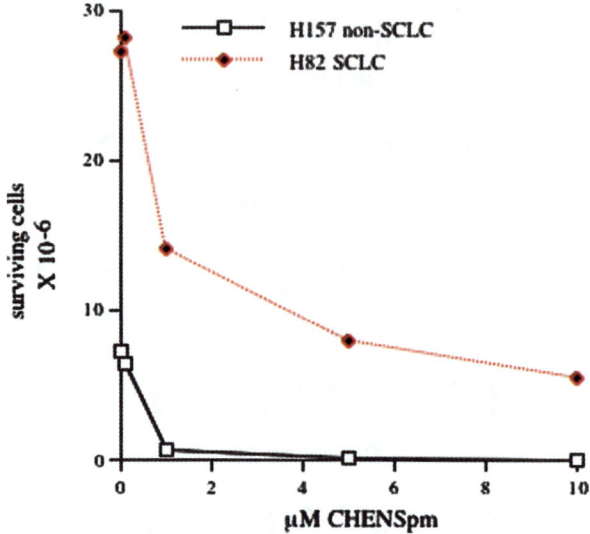

Figure 5.8 Effect of 96 h treatment with CHENSpm (**27**) on H157 non-small cell and
H82 small-cell lung-carcinoma lines *in vitro*.

effects produced by **27**,[63] and prompted the synthesis and evaluation of compounds containing the intervening ring sizes: N^1-cyclobutylmethyl-N^{11}-ethyl-
norspermine (CBENSpm, **28**), N^1-cyclopentylmethyl-N^{11}-ethylnorspermine
(CPENTSpm, **29**) and N^1-cyclohexylmethyl-N^{11}-ethylnorspermine (CHEX-
ENSpm, **30**). The cell-type specificities of **28–30** were not as pronounced as that
seen with **6**, **7**, **25** and **26**, in that all three analogs were generally cytotoxic in
both cell lines. However, there was no correlation between the induction of
SSAT and the IC_{50} value, as shown in Figure 5.9. In the cycloalkyl series, the
induction of SSAT decreased dramatically as a function of ring size, while the
IC_{50} values were remarkably constant (0.4–0.7 μM). These data clearly support
the contention that there are at least two mechanisms by which unsymmetrically substituted alkylpolyamines produce cytotoxicity in H157 non-SCLC
cells. In addition to production of apoptosis by at least two mechanisms,[63]
unsymmetrically substituted alkylpolyamines can have dramatically different
effects on the cell cycle.[64] Symmetrically substituted alkylpolyamines such as
6–11, and unsymmetrical alkylpolyamines with small substituents such as **25**
and **26**, either produce no effect on cell cycle or induce a G_1 cell-cycle arrest,
while analogs with one large substituent (*e.g.* **27** and **31**) produce a G_2/M cell-
cycle arrest.[64] In conjunction with a G_2/M cell-cycle block, the unsymmetrically
substituted polyamine analog (S)-N^1-(3-methyl)butyl-N^{11}-ethylnorspermine
(IPENSpm) alters tubulin polymerization, thus providing a lead compound
that may have a similar spectrum of activity to more difficult to synthesize
compounds typified by paclitaxel. Finally, it is now known that polyamine
analogs such as **26** have the ability to induce spermine oxidase,[62] a flavin-
dependent amine oxidase that oxidizes spermine directly to spermidine.[60]

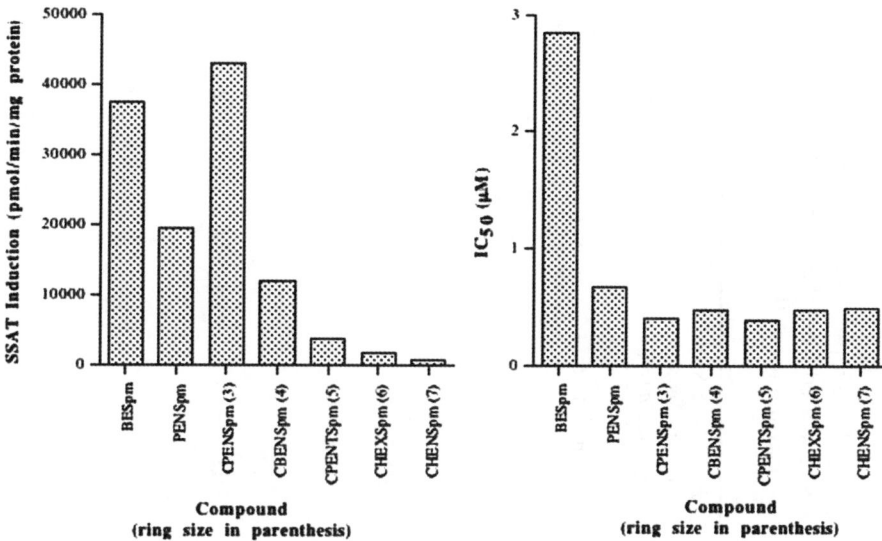

Figure 5.9 SSAT induction and IC$_{50}$ data for selected alkylpolyamine analogs.

To date, more than 120 unsymmetrical alkylpolyamines have been synthesized and evaluated as antitumor agents. These analogs, as well as the symmetrically substituted alkylpolyamines described above, have been of great value for developing an SAR model for alkylpolyamine analogs with antitumor activity. The library of unsymmetrically substituted alkylpolyamine analogs was initially evaluated using a high-throughput MTT-based cell-viability determination in NCI H157 lung-tumor cells. To increase chemical diversity, individual analogs in the library included entries with varying polyamine backbone structure (3-4, 3-3-3, 3-4-4 and 3-7-3), and both symmetrically and unsymmetrically substituted analogs with a variety of terminal alkyl substituents. As previously described,[51,56,64–69] these alkylpolyamine analogs have a variety of cellular effects, including cytotoxicity, induction of apoptosis, disruption of tubulin polymerization, induction of spermidine/spermine-N^1-acetyltransferase (SSAT) and spermine oxidase (SMO), and inhibition of the parasitic enzyme trypanothione reductase. A total of 27 alkylpolyamine analogs (**32–58**) that exhibited IC$_{50}$ values of 5.0 μM or less against the NCI H157 non-small cell lung tumor line *in vitro* were selected for further study, as shown in Table 5.1. Like the analogs described above, the compounds in Table 5.1 were taken up into tumor cells by the polyamine transporter. However, the analogs produced widely varying effects on SSAT and on cellular polyamines. In some cases (*e.g.* **36**, **43–46** and **51**), the compounds retained the cell-type specificity observed with **6** and **26** (*i.e.* they were significantly more cytotoxic against the H157 line than against the H82 line), while in other cases they did not. Analogs of spermidine exemplified by compounds **32–35** (3-4 architecture, Table 5.1) are capable of producing potent antitumor effects, although in this series symmetrically substituted compounds with terminal aralkyl substituents

Table 5.1 Structures, physical data and IC_{50} values from NCI H157 cell line high-throughput screen of alkylpolyamine analogues. IC_{50} values were determined from 96 hour dose–response curves generated from the MTS high-throughput screen. Each data point on the dose–response curve was the average of two determinations that differed by <5% in all cases. ND: value was not determined.

No.	Structure and Sequence Number	Effect on polyamine content in H157 cells	H157 24 h SSAT Activity (pmol/min/mg protein)	H82 IC_{50} (μM)	H157 IC_{50} (μM)
3-4 carbon skeleton					
32	42-TDW-35C	25% reduction in spermidine	No induction	ND	4.0
33	46-TDW-22	no effect	ND	ND	4.2
34	46-TDW-23C	no effect	ND	7.6	4.6

	Structure	Effect			
35	46-TDW-39	no effect	ND	ND	4.2
3-3-3 carbon skeleton					
36	39-TDW-12C (PG-11400)	2 fold increase in spermine	2533.81	27.2	1.8
37	39-TDW-3	no effect	No induction	3.9	4.2
38	39-TDW-47C	20% increase in spermidine	No induction	1.9	4.1
39	39-TDW-43C	No effect	10.11	ND	4.2
40	40-TDW-31C	95% reduction in spermidine, 60% reduction in spermine	2384.73	4.2	0.6
41	40-TDW-30	90% reduction in spermidine, 50% reduction in spermine	705.70	4.5	0.8

Table 5.1 (Continued)

No.	Structure and Sequence Number	Effect on polyamine content in H157 cells	H157 24 h SSAT Activity (pmol/min/ mg protein)	H82 IC$_{50}$ (μM)	H157 IC$_{50}$ (μM)
42	 42-TDW-20C	10 fold increase in putrescine	13.61	ND	4.1
43	 46-TDW-34C (PG-11402)	ND	10435.87	>100	2.2
44	 49-TDW-17C	ND	ND	18.2	0.6
45	 49-TDW-15	ND	ND	26.4	1.4
46	 49-TDW-29C	ND	ND	19.9	0.3
47	 ZQW-35	ND	ND	ND	2.8
48	 ZQW-35-8C	ND	ND	ND	0.7
49	 44-DHEJ-8C	ND	22.32	ND	4.0

#	Compound	Structure				
50	51-DHEJ-45		ND	ND	ND	3.6
51	DG-52-27C (PG-11401)		75% reduction in spermine, 95–99% reduction of spermidine and putrescine	697-fold induction	86.6	1.3
52	DG-52-28		95–99% reduction of spermine, spermidine and putrescine	1907-fold induction	ND	1.4
3-4-3 and other carbon skeletons						
53	Bis-CH-3-7-3		ND	ND	ND	2.9
54	42-TDW-38		20% reduction in spermidine	18.42	ND	3.8
55	42-TDW-45C		33.9% increase in spermine, 44.5% reduction of spermidine, 70% reduction of putrescine	no induction	ND	1.7
56	46-TDW-9		ND	ND	ND	4.1
57	46-TDW-35C		92% increase in putrescine, 50% reduction of spermidine, 50% reduction in spermine	216-fold induction	4.7	3.1
58	YZZ33049c		ND	ND	ND	2.2

were much more potent against the H157 non-small-cell lung-tumor line than symmetrically substituted alkylpolyamines or unsymmetrically substituted alkyl- or aralkylpolyamines.

The most effective antitumor alkylpolyamines were found among the large number of norspermine (3-3-3) analogs related to the parent compound **6**. The library contains numerous analogs of **6** that incorporate chemical diversity into the terminal alkyl substituent as a means to elucidate the structure–activity relationships for alkylpolamine analogs with antitumor activity. As shown in Table 5.1, 17 derivatives based on the norspermine (3-3-3) backbone (compounds **36–52**) with IC_{50} values of 5 µM or less have been identified. It is apparent from the data in Table 5.1 that slight changes in chemical structure often effect large changes in biological response in tumor cells *in vitro*. It is noteworthy that the antitumor alkylpolyamine analogs shown in Table 5.1 feature functional groups that have not previously been reported in analogs of this type. Active analogs were identified possessing aralkyl and alkyl substituents on the terminal nitrogens, both symmetrical and unsymmetrical, that in some cases contain sulfur, oxygen or fluorine atoms. Although somewhat less potent overall, analogs with 3-4-3, 3-7-3 and 3-4-4 carbon backbones (*e.g.* **53–58**, Table 5.1) also possess significant antitumor activity.

Based on a number of factors that are outlined in Table 5.2, compounds **36** (PG-11400), **51** (PG11401) and **43** (PG-11402) were selected for scale-up and more extensive pre-clinical trials *in vivo*. All three analogs were taken up into cells using the polyamine transporter system, and each could be detected at levels higher than any of the natural polyamines by HPLC.[70] Despite having similar structures, **36**, **51** and **43** show different dose–response profiles against four lung-tumor cell lines (H157, H82, H69 and A549) over a range of doses between 0.1 and 100 µM. All three analogs retained the cell-type selectivity seen with **6**, exhibiting IC_{50} values in the H82 cell line that were 12 to 50 times higher than in the H157 cell line. All three analogs were potent antitumor compounds in the H69 small-cell lung-carcinoma line, and showed good activity against the A549 line of human alveolar basal epithelial cells *in vitro*. Importantly, **36** does not produce significant SSAT induction in any of these lung-tumor cell lines, while **51** and **43** produce SSAT induction in the H157 line of the same magnitude as **14**. With regard to the inducible oxidase SMO, significant induction was seen only in the A549 cell line, initiated by **36** and **43** but not **51**. The *in vivo* implications of the variation in these cellular effects between analogs and among different cell lines remain to be determined. Compounds **36**, **51** and **43** have recently been evaluated in an A549 human lung-tumor xenograft model in athymic nu/nu mice. Mice were treated with either 75, 100 or 150 mg kg^{-1} of each agent twice a week over a period of 70 days. Compounds **36** and **43** proved to be most effective at inhibiting tumor growth, although **36** produced toxicity at the 100 mg kg^{-1} dose. In all three cases, the mean body weights of the animals were comparable to those of control animals. The toxicity observed during treatment with **36** was not observed during treatment with **43**, which was effective in limiting tumor growth by 65% at both the 75 and 150 mg kg^{-1} doses. The reduced *in vivo* toxicity observed during treatment with **43** may be

Table 5.2 Summary of the effects of compounds 36, 51 and 43 against four tumor cell lines, SSAT and SMO *in vitro*.

Cmpd	Structure	IC_{50} H157 (μM)	IC_{50} H82 (μM)	IC_{50} H69 (μM)	IC_{50} A549 (μM)	SSAT Induction (fold)	SMO Induction (fold)
36	PG-11400	1.8	27.2	0.7	10.6	57 (H157) 61 (A549) 15 (H69) 3 (H82)	1.45 (H157) 10.0 (A549) 0 (H69) 0 (H82)
51	PG-11401	1.3	86.6	0.7	2.2	452 (H157) 116 (A549) 22 (H69) 1.3 (H82)	0 (H157) 1.1 (A549) 0 (H69) 0 (H82)
43	PG-11402	2.2	>100	0.8	6.3	556 (H157) 131 (A549) 13 (H69) 4.3 (H82)	1.9 (H157) 16.5 (A549) 0 (H69) 0 (H82)

due to the presence of a hydroxyl moiety in one terminal alkyl substituent, consistent with what was observed for compound **11**.[34] Additional *in vivo* experiments are being conducted using compounds **36**, **51** and **43**, and additional analogs in this series are being synthesized and used to determine the structural requirements for binding at the various alkylpolyamine effector sites.

5.4   Polyamine Isosteres and Epigenetic Activity

Chromatin architecture is a key determinant in the regulation of gene expression, and this architecture is strongly influenced by post-translational modifications of histones.[71,72] Histone protein tails contain lysine residues that interact with the negative charges on the DNA backbone. These lysine-containing tails, consisting of up to 40 amino acid residues, protrude through the DNA strand, and act as a site for post-translational modification of chromatin, allowing alteration of the higher-order nucleosome structure.[73] Multiple post-translational modifications of histones can mediate epigenetic remodeling of chromatin, with acetylation being the best characterized process.[74] Normal mammalian cells exhibit an exquisite level of control of chromatin architecture by maintaining a balance between histone acetyltransferase (HAT) and histone deacetylase (HDAC) activity.[75] In 2004, it was discovered that lysine methylation, once thought to be a stable modification, can be reversed by an oxidative demethylase known as lysine-specific demethylase 1 (LSD1).[76] Methylation/demethylation at specific lysine residues is now known to be a dynamic process regulated by the addition of methyl groups by histone methyltransferases and removal of methyl groups from mono- and dimethyllysines by LSD1 and lysine-specific demethylase 2 (LSD2)[77] and from mono-, di, and trimethyllysines by specific Jumonji C (JmjC) domain-containing demethylases.[76,78–80] Additional demethylases in the JmjC demethylase class are continuing to be identified.[81,82] Thus, histone lysine methylation/demethylation is an important general mechanism for transcriptional control. To date, 17 lysine residues and seven arginine residues on histone proteins have been shown to undergo methylation,[83] and lysine methylation on histones can signal transcriptional activation or repression, depending on the specific lysine residue involved.[84–86]

A key positive chromatin mark found associated with promoters of active genes is histone 3 dimethyllysine 4 (H3K4).[87,88] LSD1, also known as BHC110 and KDM1,[76,89] catalyzes the oxidative demethylation of histone 3 methyllysine 4 (H3K4me1) and histone 3 dimethyllysine 4 (H3K4me2), and is associated with transcriptional repression (Figure 5.10).[76] H3K4me2 is a transcription-activating chromatin mark at gene promoters, and demethylation of this mark by LSD1 may prevent expression of tumor-suppressor genes important in human cancer.[90] Thus, LSD1 is emerging as an important new target for the development of specific inhibitors as a new class of antitumor drugs.[91]

The C-terminal domain of LSD1 shares significant sequence homology with the polyamine oxidases APAO and SMO.[76,92] Multiple groups have identified polyamine analogs that act as inhibitors of these two polyamine oxidases. It has

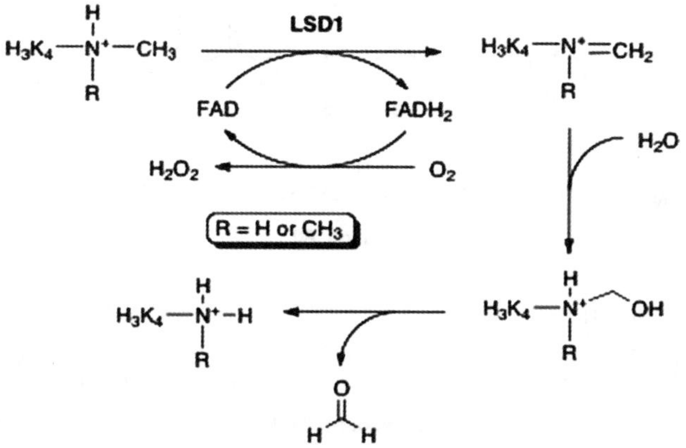

Figure 5.10 Catalytic mechanism for histone 3 mono- and dimethyl lysine 4 demethylation by LSD1.

recently been demonstrated that diamines related to BPOD[93] (**59**, Figure 5.11) act as inhibitors of plant amine oxidases, and the polyaminoguanidine guazatine (**60**, Figure 5.11) is a non-competitive inhibitor of maize polyamine oxidase.[94] Because of the sequence homology between LSD1 and the polyamine oxidases APAO and SMO, it is reasonable to conclude that potent and selective inhibitors for the homologous flavin-dependent amine oxidase LSD1 can also be designed and synthesized that contain substituted terminal amino or guanidino groups.

In 2006, a novel series of polyamino(bis)guanidines and polyaminobiguanides[65] were reported that act as potent antitrypanosomal agents *in vitro*, with 48 h IC_{50} values against *Trypanosoma brucei* as low as 90 nM. Based on the observation that polyaminoguanidines such as guazatine are potent inhibitors of amine oxidases, these compounds were evaluated for the ability to inhibit LSD1. A total of nine of the 13 compounds tested reduced LSD1 activity by greater than 50% at a concentration of 1 μM. Among the most effective compounds were the substituted (bis)biguanides 2a (**61**) and 2d (**62**) (Figure 5.11), and compound **62** was selected for further study. Like guazatine, **62** exhibited noncompetitive inhibition kinetics at concentrations of up to 2.5 μM, suggesting that it may or may not compete with H3K4me2 at the LSD1 active site. The effects on global H3K4 methylation were examined after exposure of HCT116 human colon carcinoma cells to increasing concentrations of **62** for 48 h. This exposure produced significant increases in both H3K4me1 and H3K4me2, while not affecting the global H3K9me2 levels, since it is a substrate for the JmjC domain demethylases.[90] By contrast, homologous compounds 1d and 2b (not shown), which are poor inhibitors of purified LSD1, were significantly less effective than **62** in increasing global levels of H3K4me2 in treated HCT116 cells. Levels of H3K4me3, which is not a substrate for LSD1,[79] were not affected by **62**. These data strongly suggest that the observed

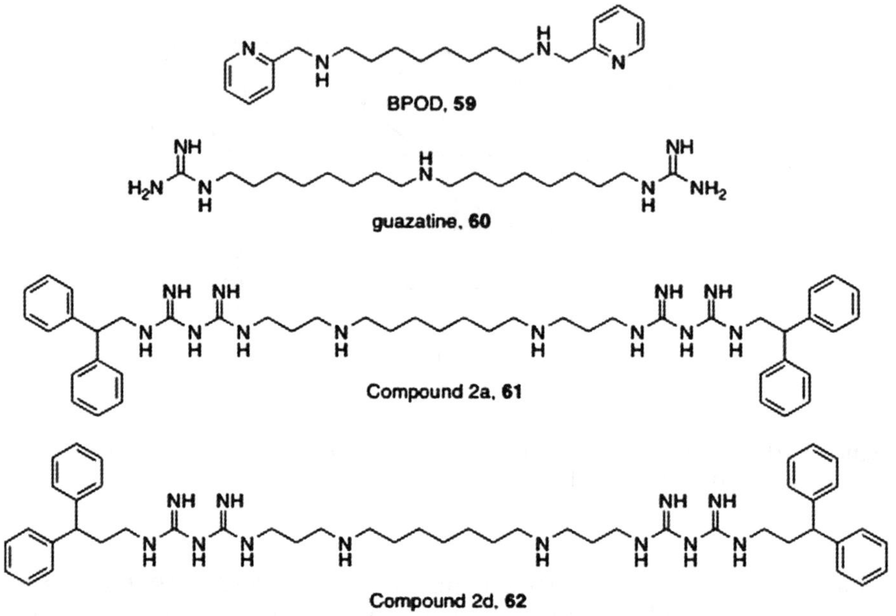

Figure 5.11 Structures for polyamine oxidase inhibitors BPOD (**59**), guazatine (**60**), compound 2a (**61**) and compound 2d (**62**).

increases in H3K4me1 and H3K4me2 produced by **62** are a direct result of LSD1 inhibition.

In cancer cells, H3K4me2 is depleted in the promoters of several aberrantly silenced DNA-hypermethylated genes that are important in tumorigenesis,[95] most notably four members of the secreted frizzle-related protein family (*SFRP1, SFRP2, SFRP4* and *SFRP5*)[96] and two *GATA* family transcription factors (*GATA4* and *GATA5*).[97] The *SFRP1, SFRP4, SFRP5* and *GATA5* factors were re-expressed following 48 h treatment with 1.0 µM **62**.[90] In HCT116 human colon-tumor cells, **62** exhibited a higher potency than other homologs such as **61,** inducing higher re-expression levels at 1.0 µM. Similar effects of **62** were observed in RKO colon cancer cells with respect to re-expression of *SFRP4* and *SFRP5*, and global H3K4me2 levels.[90] Treatment with 1 µM **62** resulted in ∼20–35% of the gene re-expression produced by the DNA methyltransferase inhibitor 5-aza-2′-deoxycytidine (DAC),[96] whereas no measurable re-expression was seen after treatment with the HDAC inhibitor trichostatin A. These results demonstrate that **62**, although not as potent as DAC, is effective at re-expressing epigenetically silenced genes. Importantly, compounds **61, 62** and their homologs are the first examples of synthetic LSD1 inhibitors that produce epigenetic changes leading to re-expression of tumor-suppressor genes. As in the case of HDAC inhibitors, these inhibitor-induced cellular effects have obvious therapeutic potential. *In vivo* studies were initiated to determine the maximum tolerated dose (MTD) for **62** in nu/nu nude mice,

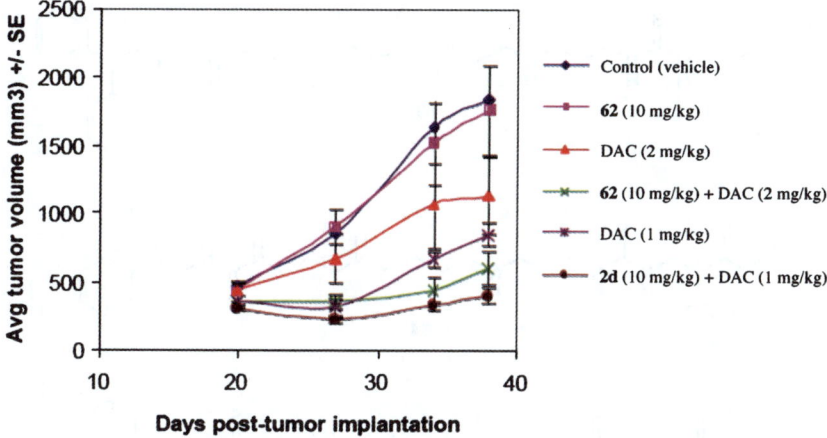

Figure 5.12 Effects of 5-azacytidine and **2d** alone or in combination in an HCT116 human colon tumor xenograft in athymic nu/nu Fox Chase mice.

and a single dose MTD was estimated to be between 10 and 25 mg kg^{-1}. *In vivo* antitumor activity was then determined in an HCT116 human colon tumor cells xenograft study, wherein mice were treated with **62**, DAC or both drugs in combination.[98] As shown in Figure 5.12, **62** alone at 10 mg kg^{-1} was not effective at limiting tumor growth over a 38-day period, and DAC (2 mg kg^{-1}) produced a 50% reduction in tumor growth. However, in combination, **62** (10 mg kg^{-1}) plus DAC (1 mg kg^{-1} or 2 mg kg^{-1}) produced an additive effect and limited tumor growth almost completely for 38 days with no concomitant loss of body weight. An even more dramatic effect was seen when the oligoamine PG-11144 (**22**) was administered in combination with DAC, indicating that the use of LSD1-inhibiting polyamine analogs in combination with DNA methyltransferase inhibitors represents a highly promising and novel approach for epigenetic therapy of cancer.

Because of the promising cellular and *in vivo* effects of **62**, the synthesis and evaluation of additional analogs were proposed. To access a library of more diverse analogs, previously published syntheses[65] were adapted and used to produce a series of (bis)alkylureas or (bis)alkylthioureas that are isosteric to **62**, and these analogs were evaluated for the ability to inhibit LSD1 and induce increases in global H3K4me2 *in vitro*. As previously observed, compound **2d** at 10 μM reduced LSD1 activity by 82.9%. Among the 30 urea and thiourea isosteres examined, six compounds were essentially inactive (*i.e.* produced <20% inhibition), while 11 analogs reduced LSD1 activity by 50% or greater at 10 μM concentration. The most effective LSD1 inhibitors proved to be thioureas **63–65** and the substituted carbamoylurea **66** (Figure 5.13) Compounds **63–65** produced levels of LSD1 inhibition that were comparable to that of **62** (Figure 5.14(A)), and thus were chosen for additional studies. These studies were conducted in the Calu-6 human anaplastic non-small-cell

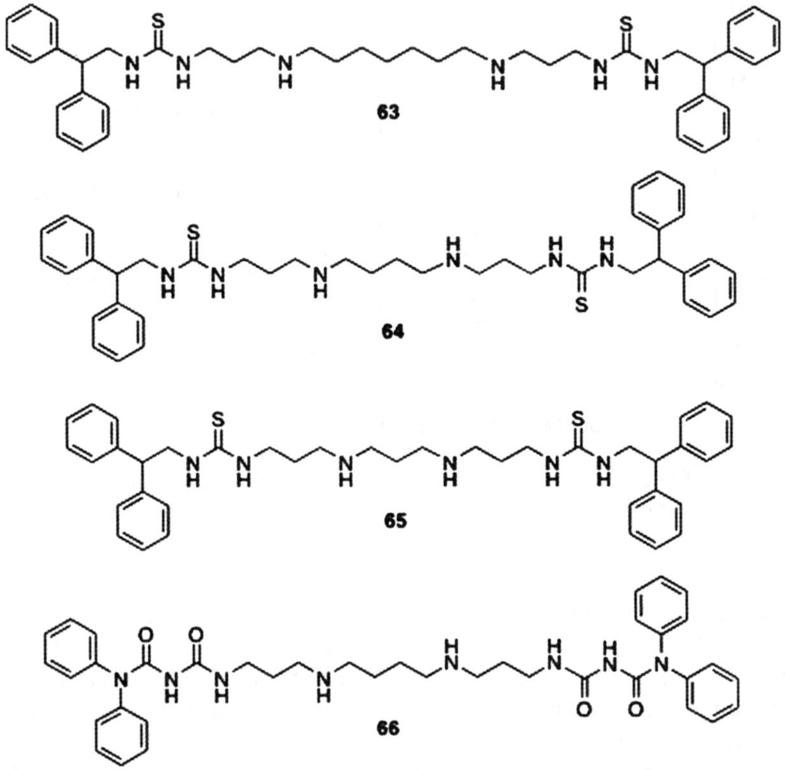

Figure 5.13 Structures of thiourea-based polyamine analogs **63-65** and (bis)carba-
moylurea **66** that act as potent inhibitors of LSD1.

lung-carcinoma line, which was chosen because it has a highly reproducible
response to epigenetic modulation, and because it is known that various tumor
suppressor genes are silenced in this line. In order for synthetic analogs to be
effective at the cellular level, any observed decreases in cellular LSD1 activity
should be accompanied by an increase in global H3K4me1 and H3K4me2
content. At 24 h, analogs **63** and **65** produced significant increases in H3K4me2
(Figure 5.14(B)), while analog **64** induced a significant increase in H3K4me1
but decreased the relative amount of H3K4me2. A similar pattern was observed
at 48 h. Compound **63** produced the most dramatic increases in H3K4me1 and
H3K4me2 at both 24 and 48 h. The reduction in H3K4me2 and corresponding
increase in H3K4me1 by **64** at both 24 and 48 h cannot be readily explained and
is the subject of continuing investigation. However, this anomolous finding
seems to correlate with the observed cytotoxicity of **64** (see below). These data
strongly suggest that intracellular inhibition of LSD1 by **63–65** leads to significant increases in methylation at the H3K4 chromatin mark. It is noteworthy
that in HCT116 human colon-tumor cells, compounds **63–65** all produced at
least a twofold increase in global H3K4me2 (data not shown), with the most
effective analog being compound **63** (17.4-fold increase).

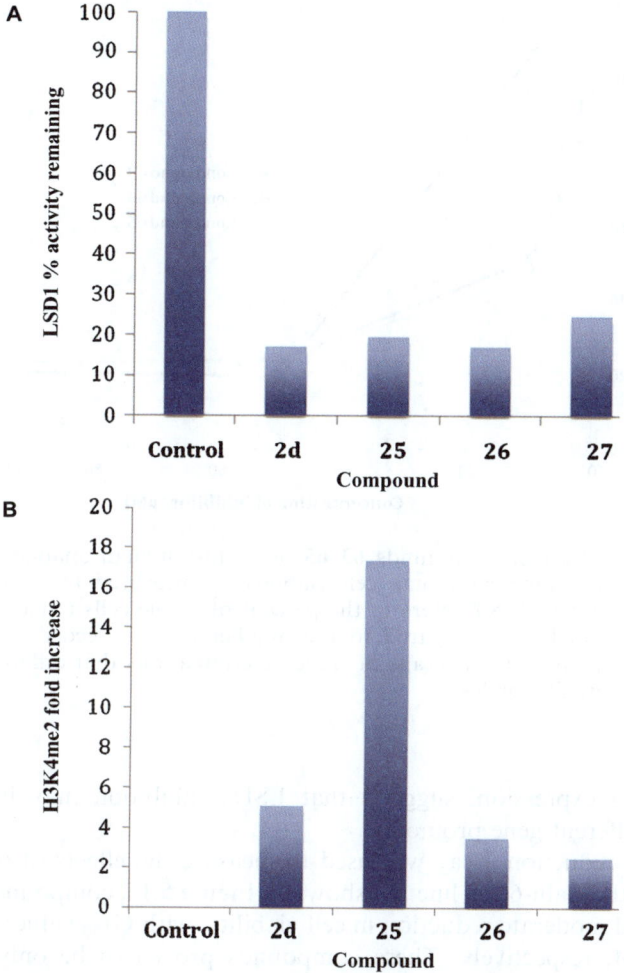

Figure 5.14   Effect of thiourea-based polyamine analogs **63–65** on *in vitro* LSD1 activity (A) and global levels of H3K4me2 (B).

The ability of compounds **63–65** to induce the re-expression of aberrantly silenced tumor suppressor genes *in vitro* was also measured in the Calu-6 human lung-carcinoma cell line. The tumor-suppressor genes SFRP2 and GATA4 were chosen because they are underexpressed in human lung cancer and play a role in tumorigenesis when silenced, making them well-documented LSD1 targets. All three compounds produced increases in SFRP2 expression that appeared to be dose-dependent for **63** and **65**. Compound **65** produced the largest increase in SFRP2 expression at 10 µM (4.8-fold increase). Compounds **63–65** did not produce any significant increases in GATA4 levels at either 5 or 10 µM. The disparity in the ability of **63–65** to induce SFRP2 expression, but

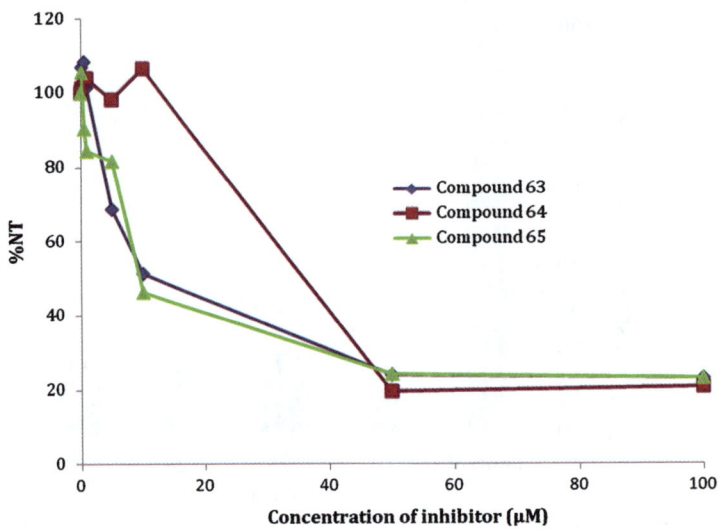

Figure 5.15 Effect of compounds **63–65** on Calu-6 human anaplastic non-small-cell lung-carcinoma cell viability as measured by a standard MTS assay. %NT refers to the percent of viable cells remaining at time T (96 h) as compared to the number of cells seeded, N_0. Each data point is the average of three determinations that differed in all cases by 5% or less.

not GATA4 expression, suggests that LSD1 inhibition may have variable effects at different gene promoters.

An MTS reduction assay was used to measure the effects of **63–65** on cell viability in the Calu-6 cell line. As shown in Figure 5.15, compounds **63**, **64** and **65** produced moderate reduction in cell viability, with GI_{50} values of 10.3, 38.3 and 9.4 μM, respectively. These compounds proved to be only moderately cytotoxic in the Calu-6 non-small cell lung carcinoma line *in vitro*. Compounds **63** and **65** produced the most prominent reduction in cell viability, exhibiting GI_{50} values of 10.3 and 9.4 μM, respectively, and these values are comparable to the GI_{50} value for other epigenetic modulators, including the parent compound **62**. In addition, these GI_{50} values are in the range of the histone deacetylase (HDAC) inhibitor MS-275, as measured in three colon tumor cell lines.[99] Compound **64** was significantly less cytotoxic, exhibiting a GI_{50} value of 38.3 μM. Our data suggest that decreases in H3K4me2 at 24 and 48 h and/or minimal effects on the re-expression of SFRP2 and GATA4 by **64** could account for this reduced cytotoxicity. It is important to note that epigenetic modulators such as those mentioned above are generally used in combination with traditional cytotoxic agents, and serve to restore the ability of transformed cells to undergo apoptosis.[100] As such, cytotoxicity is less of an issue, as long as the compound produces epigenetic effects in tumor cells that can be exploited by traditional cytotoxic agents.

5.5 Future Directions

Early drug discovery efforts for alkylpolyamine analogs resulted in the erroneous conclusion that terminal alkyl groups larger than ethyl resulted in a dramatic loss of activity, and certainly this observation was true for the symmetrically substituted alkylpolyamines described in Section 5.2. However, the subsequent synthesis and evaluation of unsymmetrically substituted alkylpolyamines demonstrated that compounds with one large group could act as potent antitumor agents. Structure/function studies involving the unsymmetrically substituted alkylpolyamine analogs have revealed that considerable structural diversity is allowable among active analogs, and that small changes in chemical structure can lead to significant, and in some cases dramatic, changes in biological activity. Unfortunately, only a handful of alkylpolyamines have been advanced to comprehensive pre-clinical trials, and even fewer have been evaluated in human clinical trials. It is apparent that this trend is beginning to change,[40,101–104] and analogs such as **14** and **22** are now being progressed towards possible FDA approval. As the data from these trials are released, it will become apparent that drug discovery involving alkylpolyamine analogs is a fruitful undertaking. As such, the synthesis and evaluation of additional symmetrically and unsymmetrically substituted alkylpolyamine analogs are warranted.

References

1. R. J. Bergeron, J. S. McManis, W. R. Weimar, K. M. Schreier, F. Gao, Q. Wu, J. Ortiz-Ocasio, G. R. Luchetta, C. Porter and J. R. Vinson, *J. Med. Chem.*, 1995, **38**, 2278.
2. R. A. Casero and P. M. Woster, *J. Med. Chem.*, 2001, **44**, 1.
3. R. A. Casero, Jr and P. M. Woster, *J. Med. Chem.*, 2009, **52**, 4551.
4. R. A. Casero, Jr, B. Frydman, T. M. Stewart and P. M. Woster, *Proc. West. Pharmacol. Soc.*, 2005, **48**, 24.
5. T. Boncher, X. Bi, S. Varghese, R. A. Casero, Jr and P. M. Woster, *Biochem. Soc. Trans.*, 2007, **35**, 356.
6. R. A. Casero, Jr and L. J. Marton, *Nat. Rev. Drug Discov.*, 2007, **6**, 373.
7. O. Heby, L. Persson and M. Rentala, *Amino Acids*, 2007, **33**, 359.
8. H. M. Wallace, *Expert Opin. Pharmacother.*, 2007, **8**, 2109.
9. H. M. Wallace and K. Niiranen, *Amino Acids*, 2007, **33**, 261.
10. E. W. Gerner and F. L. Meyskens, Jr, *Nat. Rev. Cancer*, 2004, **4**, 781.
11. M. L. Edwards, N. J. Prakash, D. M. Stemerick, S. P. Sunkara, A. J. Bitonti, G. F. Davis, J. A. Dumont and P. Bey, *J. Med. Chem.*, 1990, **33**, 1369.
12. M. L. Edwards, R. D. Snyder and D. M. Stemerick, *J. Med. Chem.*, 1991, **34**, 2414.
13. A. J. Bitonti, P. P. McCann and A. Sjoerdsma, *Adv. Exp. Med. Biol.*, 1988, **250**, 717.

14. A. J. Bitonti, J. A. Dumont, T. L. Bush, M. L. Edwards, D. M. Stemerick, P. P. McCann and A. Sjoerdsma, *Proc. Natl. Acad. Sci. U. S. A.*, 1989, **86**, 651.

15. R. J. Baumann, W. L. Hanson, P. P. McCann, A. Sjoerdsma and A. J. Bitonti, *Antimicrob. Agents Chemother.*, 1990, **34**, 722.

16. A. J. Bitonti, J. A. Dumont, T. L. Bush, D. M. Stemerick, M. L. Edwards and P. P. McCann, *J. Biol. Chem.*, 1990, **265**, 382.

17. C. W. Porter, F. G. Berger, A. E. Pegg, B. Ganis and R. J. Bergeron, *Biochem. J.*, 1987, **242**, 433.

18. C. W. Porter, B. Ganis, P. R. Libby and R. J. Bergeron, *Cancer Res.*, 1991, **51**, 3715.

19. C. W. Porter and J. R. Sufrin, *Anticancer Res.*, 1986, **6**, 525.

20. R. J. Bergeron, A. H. Neims, J. S. McManis, T. R. Hawthorne, J. R. Vinson, R. Bortell and M. J. Ingeno, *J. Med. Chem.*, 1988, **31**, 1183.

21. R. A. Casero, Jr and A. E. Pegg, *FASEB J.*, 1993, **7**, 653.

22. R. J. Bergeron, Y. Feng, W. R. Weimar, J. S. McManis, H. Dimova, C. Porter, B. Raisler and O. Phanstiel, *J. Med. Chem.*, 1997, **40**, 1475.

23. R. J. Bergeron, J. S. McManis, C. Z. Liu, Y. Feng, W. R. Weimar, G. R. Luchetta, Q. Wu, J. Ortiz-Ocasio, J. R. Vinson and D. Kramer *et al.*, *J. Med. Chem.*, 1994, **37**, 3464.

24. L. Jeffers, D. Church, H. Basu, L. Marton and G. Wilding, *Cancer Chemother. Pharmacol.*, 1997, **40**, 172.

25. C. J. Bergeron, H. S. Basu, L. J. Marton, D. F. Deen, M. Pellarin and B. G. Feuerstein, *Cancer Chemother. Pharmacol.*, 1995, **36**, 411.

26. R. C. Roemmele and H. Rappoport, *J. Org. Chem.*, 1988, **53**, 2367.

27. H. Yajima, M. Takeyama, J. Kanaki, O. Nishimura and M. Fujino, *Chem. Pharm. Bull.*, 1978, **26**, 3752.

28. R. J. Bergeron, R. Muller, G. Huang, J. S. McManis, S. E. Algee, H. Yao, W. R. Weimar and J. Wiegand, *J. Med. Chem.*, 2001, **44**, 2451.

29. R. J. Bergeron, J. Wiegand and T. L. Fannin, *Dig. Dis. Sci.*, 2001, **46**, 2615.

30. R. J. Bergeron, J. Wiegand, J. S. McManis, W. R. Weimar, R. E. Smith, S. E. Algee, T. L. Fannin, M. A. Slusher and P. S. Snyder, *J. Med. Chem.*, 2001, **44**, 232.

31. A. C. Wolff, D. K. Armstrong, J. H. Fetting, M. K. Carducci, C. D. Riley, J. F. Bender, R. A. Casero, Jr and N. E. Davidson, *Clin. Cancer Res.*, 2003, **9**, 5922.

32. H. A. Hahm, D. S. Ettinger, K. Bowling, B. Hoker, T. L. Chen, Y. Zabelina and R. A. Casero, Jr, *Clin. Cancer Res.*, 2002, **8**, 684.

33. R. R. Streiff and J. F. Bender, *Invest. New Drugs*, 2001, **19**, 29.

34. R. J. Bergeron, G. W. Yao, H. Yao, W. R. Weimar, C. A. Sninsky, B. Raisler, Y. Feng, Q. Wu and F. Gao, *J. Med. Chem.*, 1996, **39**, 2461.

35. B. Frydman, A. V. Blokhin, S. Brummel, G. Wilding, Y. Maxuitenko, A. Sarkar, S. Bhattacharya, D. Church, V. K. Reddy, J. A. Kink, L. J. Marton, A. Valasinas and H. S. Basu, *J. Med. Chem.*, 2003, **46**, 4586.

36. B. Frydman, C. W. Porter, Y. Maxuitenko, A. Sarkar, S. Bhattacharya, A. Valasinas, V. K. Reddy, N. Kisiel, L. J. Marton and H. S. Basu, *Cancer Chemother. Pharmacol.*, 2003, **51**, 488.
37. A. Valasinas, A. Sarkar, V. K. Reddy, L. J. Marton, H. S. Basu and B. Frydman, *J. Med. Chem.*, 2001, **44**, 390.
38. V. K. Reddy, A. Valasinas, A. Sarkar, H. S. Basu, L. J. Marton and B. Frydman, *J. Med. Chem.*, 1998, **41**, 4723.
39. W. R. Waters, B. Frydman, L. J. Marton, A. Valasinas, V. K. Reddy, J. A. Harp, M. J. Wannemuehler and N. Yarlett, *Antimicrob. Agents Chemother.*, 2000, **44**, 2891.
40. J. S. Carew, S. T. Nawrocki, V. K. Reddy, D. Bush, J. E. Rehg, A. Goodwin, J. A. Houghton, R. A. Casero, Jr, L. J. Marton and J. L. Cleveland, *Cancer Res*, 2008, **68**, 4783.
41. V. K. Reddy, A. Sarkar, A. Valasinas, L. J. Marton, H. S. Basu and B. Frydman, *J. Med. Chem.*, 2001, **44**, 404.
42. A. Hacker, L. J. Marton, M. Sobolewski and R. A. Casero, Jr, *Cancer Chemother. Pharmacol.*, 2008, **63**, 45.
43. R. Lima e Silva, S. Kachi, H. Akiyama, S. JiKui, M. C. Hatara, S. Aslam, Y. Y. Gong, N. H. Khu, T. W. Lauer, S. F. Hackett, L. J. Marton and P. A. Campochiaro, *Exp. Eye Res.*, 2006, **83**, 1260.
44. A. Valasinas, V. K. Reddy, A. V. Blokhin, H. S. Basu, S. Bhattacharya, A. Sarkar, L. J. Marton and B. Frydman, *Bioorg. Med. Chem.*, 2003, **11**, 4121.
45. Y. Huang, E. R. Hager, D. L. Phillips, V. R. Dunn, A. Hacker, B. Frydman, J. A. Kink, A. L. Valasinas, V. K. Reddy, L. J. Marton, R. A. Casero, Jr and N. E. Davidson, *Clin. Cancer Res.*, 2003, **9**, 2769.
46. Y. Huang, J. C. Keen, E. Hager, R. Smith, A. Hacker, B. Frydman, A. L. Valasinas, V. K. Reddy, L. J. Marton, R. A. Casero, Jr and N. E. Davidson, *Mol. Cancer Res.*, 2004, **2**, 81.
47. Y. Huang, J. C. Keen, A. Pledgie, L. J. Marton, T. Zhu, S. Sukumar, B. H. Park, B. Blair, K. Brenner, R. A. Casero, Jr and N. E. Davidson, *J. Biol. Chem.*, 2006, **281**, 19055.
48. X. Feng, V. K. Reddy, H. Mayanja-Kizza, L. M. Weiss, L. J. Marton and S. Tzipori, *Antimicrob. Agents Chemother.*, 2009.
49. G. M. Rukunga and P. G. Waterman, *J. Nat. Prod.*, 1996, **59**, 850.
50. B. Frydman, S. Bhattacharya, A. Sarkar, K. Drandarov, S. Chesnov, A. Guggisberg, K. Popaj, S. Sergeyev, A. Yurdakul, M. Hesse, H. S. Basu and L. J. Marton, *J. Med. Chem.*, 2004, **47**, 1051.
51. N. H. Saab, E. E. West, N. C. Bieszk, C. V. Preuss, A. R. Mank, R. A. Casero, Jr and P. M. Woster, *J. Med. Chem.*, 1993, **36**, 2998.
52. R. J. Bergeron, N. J. Stolowich and C. W. Porter, *Synthesis*, 1982, **39**, 689.
53. R. J. Bergeron, J. R. Garlich and N. J. Stolowich, *J. Org. Chem.*, 1984, **49**, 2997.
54. R. J. Bergeron, T. R. Hawthorne, J. R. Vinson, D. E. Beck, Jr and M. J. Ingeno, *Cancer Res.*, 1989, **49**, 2959.

55. N. Seiler, J. G. Delcros, M. Vaultier, N. Le Roch, R. Havouis, F. Douaud and J. P. Moulinoux, *Cancer Res.*, 1996, **56**, 5624.
56. F. H. Bellevue III, M. Boahbedason, R. Wu, P. M. Woster, J. R. A. Casero, D. Rattendi, S. Lane and C. J. Bacchi, *Bioorg. Med. Chem. Lett.*, 1996, **6**, 2765.
57. M. L. Edwards, D. M. Stemerick, A. J. Bitonti, J. A. Dumont, P. P. McCann, P. Bey and A. Sjoerdsma, *J. Med. Chem.*, 1991, **34**, 569.
58. R. A. Casero, Jr, S. J. Ervin, P. Celano, S. B. Baylin and R. J. Bergeron, *Cancer Res.*, 1989, **49**, 639.
59. H. C. Ha, P. M. Woster, J. D. Yager and R. A. Casero, Jr, *Proc. Natl. Acad. Sci. U. S. A.*, 1997, **94**, 11557.
60. A. Pledgie, Y. Huang, A. Hacker, Z. Zhang, P. M. Woster, N. E. Davidson and R. A. Casero, Jr, *J. Biol. Chem.*, 2005.
61. Y. Wang, W. Devereux, P. M. Woster, T. M. Stewart, A. Hacker and R. A. Casero, Jr, *Cancer Res*, 2001, **61**, 5370.
62. Y. Wang, A. Hacker, T. Murray-Stewart, J. G. Fleischer, P. M. Woster and R. A. Casero, Jr, *Biochem. J.*, 2005, **386**, 543.
63. H. C. Ha, P. M. Woster and R. A. Casero, Jr, *Cancer Res.*, 1998, **58**, 2711.
64. H. K. Webb, Z. Wu, N. Sirisoma, H. C. Ha, R. A. Casero, Jr and P. M. Woster, *J. Med. Chem.*, 1999, **42**, 1415.
65. X. Bi, C. Lopez, C. J. Bacchi, D. Rattendi and P. M. Woster, *Bioorg. Med. Chem. Lett.*, 2006, **16**, 3229.
66. D. E. McCloskey, R. A. Casero, Jr, P. M. Woster and N. E. Davidson, *Cancer Res.*, 1995, **55**, 3233.
67. D. E. McCloskey, P. M. Woster, R. A. Casero, Jr and N. E. Davidson, *Clin. Cancer Res.*, 2000, **6**, 17.
68. D. E. McCloskey, J. Yang, P. M. Woster, N. E. Davidson and R. A. Casero, Jr, *Clin. Cancer Res.*, 1996, **2**, 441.
69. N. E. Davidson, H. A. Hahm, D. E. McCloskey, P. M. Woster and R. A. Casero, Jr, *Endocr. Relat. Cancer*, 1999, **6**, 69.
70. P. M. Kabra, H. K. Lee, W. P. Lubich and L. J. Marton, *J. Chromatogr.*, 1986, **380**, 19.
71. P. A. Marks, V. M. Richon, R. Breslow and R. A. Rifkind, *Curr. Opin. Oncol.*, 2001, **13**, 477.
72. K. Luger, A. W. Mader, R. K. Richmond, D. F. Sargent and T. J. Richmond, *Nature*, 1997, **389**, 251.
73. T. Jenuwein and C. D. Allis, *Science*, 2001, **293**, 1074.
74. R. W. Johnstone, *Nat. Rev. Drug Discov.*, 2002, **1**, 287.
75. M. Shogren-Knaak, H. Ishii, J.-M. Sun, M. J. Pazin, J. R. Davie and C. L. Peterson, *Science*, 2006, **311**, 844.
76. Y. Shi, F. Lan, C. Matson, P. Mulligan, J. R. Whetstine, P. A. Cole, R. A. Casero and Y. Shi, *Cell*, 2004, **119**, 941.
77. A. Karytinos, F. Forneris, A. Profumo, G. Ciossani, E. Battaglioli, C. Binda and A. Mattevi, *J. Biol. Chem.*, 2009, **284**, 17775.

78. Y. Tsukada and Y. Zhang, *Methods*, 2006, **40**, 318.
79. J. R. Whetstine, A. Nottke, F. Lan, M. Huarte, S. Smolikov, Z. Chen, E. Spooner, E. Li, G. Zhang, M. Colaiacovo and Y. Shi, *Cell*, 2006, **125**, 467.
80. M. Huarte, F. Lan, T. Kim, M. W. Vaughn, M. Zaratiegui, R. A. Martienssen, S. Buratowski and Y. Shi, *J. Biol. Chem.*, 2007.
81. G. Liang, R. J. Klose, K. E. Gardner and Y. Zhang, *Nat. Struct. Mol. Biol.*, 2007, **14**, 243.
82. J. Secombe, L. Li, L. Carlos and R. N. Eisenman, *Genes Dev.*, 2007, **21**, 537.
83. A. J. Bannister and T. Kouzarides, *Nature*, 2005, **436**, 1103.
84. T. Kouzarides, *Curr. Opin. Genet. Dev.*, 2002, **12**, 198.
85. C. Martin and Y. Zhang, *Nat. Rev. Mol. Cell. Biol.*, 2005, **6**, 838.
86. Y. Zhang and D. Reinberg, *Genes Dev.*, 2001, **15**, 2343.
87. G. Liang, J. C. Lin, V. Wei, C. Yoo, J. C. Cheng, C. T. Nguyen, D. J. Weisenberger, G. Egger, D. Takai, F. A. Gonzales and P. A. Jones, *Proc. Natl. Acad. Sci. U. S. A.*, 2004, **101**, 7357.
88. R. Schneider, A. J. Bannister, F. A. Myers, A. W. Thorne, C. Crane-Robinson and T. Kouzarides, *Nat. Cell Biol.*, 2004, **6**, 73.
89. M. G. Lee, C. Wynder, D. M. Schmidt, D. G. McCafferty and R. Shiekhattar, *Chem. Biol.*, 2006, **13**, 563.
90. Y. Huang, E. Greene, T. Murray Stewart, A. C. Goodwin, S. B. Baylin, P. M. Woster and R. A. Casero, Jr, *Proc. Natl. Acad. Sci. U. S. A.*, 2007, **104**, 8023.
91. P. Stavropoulos and A. Hoelz, *Expert Opin. Ther. Targets*, 2007, **11**, 809.
92. Y. Wang, T. Murray-Stewart, W. Devereux, A. Hacker, B. Frydman, P. M. Woster and R. A. Casero, Jr, *Biochem. Biophys. Res. Commun.*, 2003, **304**, 605.
93. J. Stranska, M. Sebela, P. Tarkowski, P. Rehulka, J. Chmelik, I. Popa and P. Pec, *Biochimie*, 2007, **89**, 135.
94. A. Cona, F. Manetti, R. Leone, F. Corelli, P. Tavladoraki, F. Polticelli and M. Botta, *Biochemistry*, 2004, **43**, 3426.
95. K. M. McGarvey, J. A. Fahrner, E. Greene, J. Martens, T. Jenuwein and S. B. Baylin, *Cancer Res.*, 2006, **66**, 3541.
96. J. A. Fahrner, S. Eguchi, J. G. Herman and S. B. Baylin, *Cancer Res.*, 2002, **62**, 7213.
97. Y. Akiyama, N. Watkins, H. Suzuki, K. W. Jair, M. van Engeland, M. Esteller, H. Sakai, C. Y. Ren, Y. Yuasa, J. G. Herman and S. B. Baylin, *Mol. Cell. Biol.*, 2003, **23**, 8429.
98. Y. Huang, T. Murray-Stewart, Y. Wu, S. B. Baylin, L. J. Marton, P. M. Woster and R. A. Casero, Jr., *Cancer Chemo. Pharmacol.*, 2009, **15**, 7217.
99. S. Flis, A. Gnyszka, K. Flis and J. Splawinski, *Eur. J. Pharmacol.*, 2010, **627**, 26.
100. M. Fouladi, *Cancer Invest.*, 2006, **24**, 521.

101. N. Babbar and E. W. Gerner, *Recent Results Cancer Res.*, 2011, **188**, 49.
102. A. S. Bachmann, *Hawaii Med. J.*, 2004, **63**, 371.
103. F. L. Meyskens, Jr and E. W. Gerner, *Cancer Prev. Res. (Philadelphia, PA)*, 2011, **4**, 628.
104. F. L. Meyskens, Jr, C. E. McLaren, D. Pelot, S. Fujikawa-Brooks, P. M. Carpenter, E. Hawk, G. Kelloff, M. J. Lawson, J. Kidao, J. McCracken, C. G. Albers, D. J. Ahnen, D. K. Turgeon, S. Goldschmid, P. Lance, C. H. Hagedorn, D. L. Gillen and E. W. Gerner, *Cancer Prev. Res. (Philadelphia, PA)*, 2008, **1**, 32.

CHAPTER 6

Targeting the Polyamine Catabolic Enzymes Spermine Oxidase, N^1-Acetylpolyamine Oxidase and Spermidine/Spermine N^1-Acetyltransferase

ANDREW C. GOODWIN,* TRACY R. MURRAY-STEWART AND ROBERT A. CASERO, JR

The Sidney Kimmel Comprehensive Cancer Center, The Johns Hopkins School of Medicine, Bunting/Blaustein Cancer Research Building, Room 551, 1650 Orleans Street, Baltimore, MD 21231, USA

6.1 Introduction

An important component of the regulation of polyamine homeostasis is the irreversibility of the biosynthetic enzyme reactions (discussed in Chapters 1, 2 and 4) and the presence of distinct pathways for the back-conversion of spermine to spermidine, and spermidine to putrescine (Figure 6.1). These polyamine catabolic reactions are mediated by three enzymes, spermidine/spermine N^1-acetyltransferase (SSAT), N^1-acetylpolyamine oxidase (APAO) and spermine oxidase (SMO). SMO catalyzes the oxidation of spermine, yielding spermidine, hydrogen peroxide (H_2O_2) and 3-aminopropanal. Alternatively, spermine can be acetylated by SSAT prior to oxidation by APAO to produce spermidine. Spermidine is recycled to putrescine by a similar two-step mechanism involving SSAT and APAO. In addition to generating putrescine from the higher polyamines, the acetylated polyamines formed by SSAT activity can be exported from the cell via

RSC Drug Discovery Series No. 17
Polyamine Drug Discovery
Edited by Patrick M. Woster and Robert A. Casero, Jr.
© Royal Society of Chemistry 2012
Published by the Royal Society of Chemistry, www.rsc.org

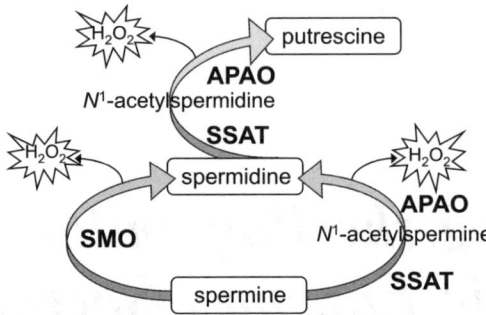

Figure 6.1   Illustration depicting back-conversion of the higher polyamines spermine and spermidine to putrescine, and production of the ROS byproduct, hydrogen peroxide (H_2O_2), by the mammalian polyamine catabolic enzymes (boldface). Spermine oxidase (SMO) catalyzes the oxidation of spermine, yielding spermidine, H_2O_2 and 3-aminopropanal. Alternatively, the N^1 nitrogen of spermine can be acetylated by spermidine/spermine N^1-acetyltransferase (SSAT) in the first of a two-step polyamine catabolic pathway. Acetylated spermine then serves as substrate for the N^1-acetylpolyamine oxidase (APAO), yielding spermidine, H_2O_2 and 3-acetamidopropanal. Spermidine is recycled to putrescine by a similar two-step mechanism involving SSAT and APAO. In addition, acetylated spermine and spermidine generated by SSAT activity can be exported from the cell via the polyamine transport system.

the polyamine transport system (see Chapter 7 for more details), reducing overall cellular polyamine pools.

The polyamine catabolic enzymes and their reaction byproducts have important functions in polyamine homeostasis, cellular response to pharmacological agents and disease etiology.[1-3] This multifunctional nature confers the opportunity for therapeutic benefit from drugs designed to either induce or inhibit polyamine catabolism, and examples of each strategy will be presented in this chapter. Section 6.2 will provide background on the characterization, structure, mechanisms of action and physiological functions of SSAT, APAO and SMO. Section 6.3 will describe approaches to selectively induce polyamine catabolic enzyme activity for therapeutic benefit. In these examples, the super-induction of polyamine catabolism, resulting in the production of cytotoxic reactive oxygen species (ROS) and aldehyde byproducts, is a marker of drug response. In contrast, moderate elevation of SSAT, APAO and SMO expression has been associated with numerous human disease states. Section 6.4 will summarize examples of situations in which it may be clinically beneficial to inhibit the activity of polyamine catabolic enzymes.

6.2   Structure, Mechanism of Action and Function of Polyamine Catabolic Enzymes

6.2.1   Spermidine/Spermine N^1-Acetyltransferase (SSAT)

Following the description of the irreversible polyamine biosynthetic enzymes, attention turned towards the degradation of the higher polyamines. Several

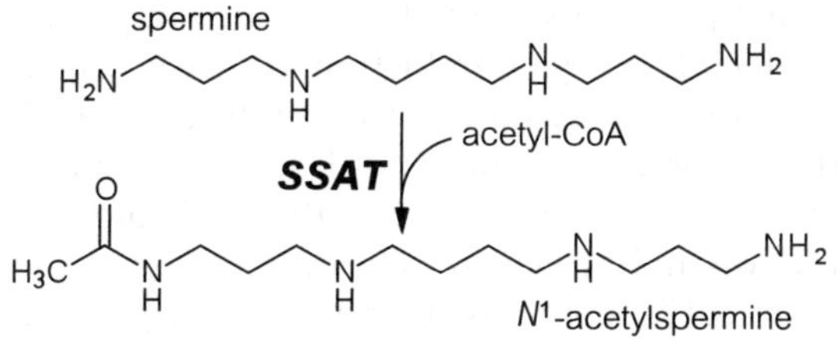

Scheme 6.1 SSAT enzyme reaction.

studies demonstrated the presence of acetylated polyamines in human urine and serum samples, leading to efforts to identify the enzyme(s) responsible for their formation.[4–7] Subsequent reports documented acetylation of spermine and spermidine in rat liver and kidney extracts that was dependent on acetyl-CoA and necessary for recycling of the higher polyamines to putrescine. These models led to the purification of the enzyme responsible, SSAT, and its designation as the rate-limiting enzyme in a two-step mechanism for polyamine back-conversion consisting of acetylation followed by oxidation.[8–11] SSAT catalyzes the acetylation of the N^1 nitrogen of spermine and spermidine (Scheme 6.1), as well as norspermine and norspermidine, but does not use putrescine, cadaverine, homospermine or histones as substrates.[12] While outside the scope of this chapter, spermidine can also be acetylated at the N^8 position by histone acetyltransferases (HATs), a result that has implications for the epigenetic effects of polyamine analog drugs targeting the chromatin-modifying enzymes discussed in Chapter 5.[13–19] Purification of SSAT and initial enzymatic studies were facilitated by the discovery that SSAT activity is highly inducible by a variety of agents, including hepatotoxic agents, growth hormone, folic acid, the natural polyamines and certain polyamine analogs.[10,20–28]

Using an expression library strategy based on protein sequence information, *SAT1*, the gene encoding SSAT, was first cloned from the H157 human non-small-cell lung-cancer cell line and its location mapped to Xp22.1.[29,30] The SSAT transcript consists of a 165 bp 5′ untranslated region (UTR), a 513 bp open reading frame (ORF) and a 382 bp 3′ UTR containing a poly-adenylation site. SSAT transcription is controlled by a TATA-less promoter, initiates 179 bp upstream of the translational start site and results in the splicing of six exons to encode a 171 amino acid (20 kDa) peptide. Highly conserved SSAT genes were subsequently cloned from the mouse and hamster genomes.[31] Transient transfection of SV40-transformed green monkey kidney cells with SSAT cDNA resulted in enzyme activity, decreased spermine and spermidine content, elevated putrescine concentrations and increased polyamine biosynthesis.[32] A second locus encoding an enzyme with 45% identity and 61% homology to SSAT was cloned and initially designated as SSAT2, but later studies demonstrated that this enzyme is not involved in polyamine metabolism and is now more appropriately referred to as thialysine *N*-ε-acetyltransferase.[33,34]

Analogous to the control of biosynthesis via modulation of ODC[35] and AdoMetDC[36] (see Chapter 4), SSAT is a highly regulated and primarily cytosolic enzyme that represents a rate-limiting step in polyamine catabolism.[1] SSAT is highly inducible; enzyme induction via treatment of the H157 non-small-cell lung-cancer cell line with cytotoxic polyamine analogs facilitated its cloning.[29,37] In the physiologic setting, SSAT activity is induced in the presence of high polyamine levels and is governed at the levels of transcription, mRNA processing, translation and protein stabilization.

The promoter of *SAT1* contains a polyamine-responsive element (PRE). Though the precise mechanism is not known, in the presence of high polyamine levels, the well-characterized transcription factor, Nuclear Factor-E2-related factor 2 (NRF2, recently reviewed in Osburn and Kensler[38]), interacts with polyamine-modulated factor 1 (PMF1), leading to binding at the *SAT1* PRE and increased gene expression.[39,40] In addition to increasing gene expression, elevated polyamines have been shown to reduce the occurrence of an alternative splicing event that yields an mRNA variant targeted for degradation rather than production of functional SSAT protein.[41] Polyamines also appear to reverse the inhibition of SSAT translation by an as-yet unidentified protein. Finally, interactions of polyamines with the SSAT protein prevent its ubiquitination and proteosomal degradation, resulting in increased SSAT protein half-life.[1,42] The combination of these regulatory mechanisms, in concert with controls on biosynthetic enzymes, ensures that polyamine pools are carefully maintained to meet cellular demands and avoid the toxicities associated with excess polyamine accumulation.[43]

In studies that have aided the understanding of SSAT enzymatic activity, several groups have reported high-resolution crystal structures of the mouse or human version of the enzyme in complex with several other molecules. SSAT consists of a β-sheet core surrounded by α-helices and exists as a homodimer via stabilizing interactions between amino acids 143 and 171 of the C-terminal arms. The surface of the holoenzyme contains two channels that incorporate the active site. The co-factor acetyl-CoA is able to bind in both channels, while the substrate spermine or spermidine binds in only one channel. The polyamine-binding site, which localizes the primary amine to be acetylated near the acyl group of acetyl-CoA, contains interspersed hydrophobic and acidic residues that accommodate the alternating basic and aliphatic nature of the polyamine backbone. The composition of the binding site corresponds to the previously demonstrated substrate specificity of SSAT, and mutation of these residues greatly decreases enzyme activity. Initial velocity studies indicate that SSAT has a sequential kinetic mechanism of action, in which both the polyamine and acetyl-CoA bind to the enzyme forming a ternary complex, at which point catalysis occurs. The SSAT mechanism, consisting of the formation of a tetrahedral intermediate, is similar to that of other GCN5-related *N*-acetyltransferase (GNAT) family members and relies on residues Y140 and E92 which serve as a general acid and base, respectively.[44–51]

In summary, SSAT catalyzes the first of a two-step polyamine catabolic pathway by acetylating the N^1 nitrogen of spermine or spermidine, using acetyl-CoA as the acetyl group donor. In cells with high SSAT activity, the enzyme can also utilize N^1-acetylspermine as a substrate, producing N^1,N^{12}-diacetylspermine.

These acetylated polyamines then can serve as substrates for the peroxisomal, FAD-dependent N^1-acetylpolyamine oxidase (APAO), yielding spermidine or putrescine, hydrogen peroxide (H_2O_2) and 3-acetamidopropanal.[42,52,53]

6.2.2 N^1-Acetylpolyamine Oxidase (APAO) and Spermine Oxidase (SMO)

Catabolism of the higher polyamines can achieve a reduction in overall poly-amine pools via excretion of acetylated polyamines or can serve as an alternative means to produce putrescine, such as when ornithine decarboxylase (ODC) is inhibited by α-difluoromethylornithine (DFMO) treatment.[23,54–57] Prior to the identification of the mammalian catabolic enzymes, studies demonstrated that spermine and spermidine are converted to spermidine and putrescine, respectively, in a process that involves oxidation.[58–63] Holtta was the first to demonstrate polyamine oxidase activity using purified APAO from rat liver.[64] This and subsequent studies of purified enzyme from pig liver and L1210 mouse leukemia cells established many of the basic features of APAO, including the confirmation of its role as the second step in polyamine catabolism following acetylation by SSAT.[65,66] Polyamine oxidation was later shown to occur in virtually all rodent tissue types examined.[67,68]

The ~60 kDa purified oxidase is dependent on the co-factor flavin-adenine dinucleotide (FAD), localizes to peroxisomes and functions optimally at pH 10.0. In contrast to SSAT, APAO preferentially utilizes N^1-acetylspermine and N^1-acetylspermidine as substrates versus the natural polyamines, is composed of a single subunit and is generally not inducible.[64–65,69–73] Several potent inhibitors of polyamine oxidation, particularly N^1,N^4-bis(2,3-butandienyl)-1,4-butanediamine (MDL 72527) (Figure 6.2), have been identified and have

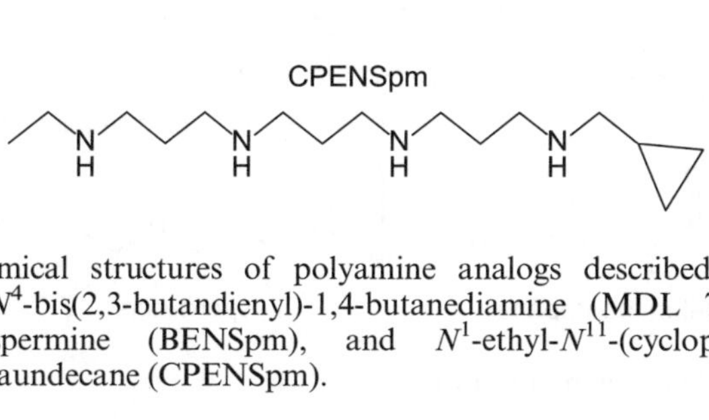

Figure 6.2 Chemical structures of polyamine analogs described in this chapter: N^1,N^4-bis(2,3-butandienyl)-1,4-butanediamine (MDL 72527), bis(ethyl)-norspermine (BENSpm), and N^1-ethyl-N^{11}-(cyclopropyl)methyl-4,8,-diazaundecane (CPENSpm).

become significant tools for biochemical studies and as preclinical therapeutic agents (see Section 6.4.1). These analogs have aided in the determination of the minimum structural requirements for APAO-substrate binding: an aliphatic α,ω-diamine with an alkyl substituent on one or both nitrogen atoms. There does not appear to be a stringent requirement for the length of the carbon chain between the nitrogen atoms, and substrate binding is mediated by side-chain hydrophobic interactions and electrostatic interactions between the protonated nitrogens and negatively charged residues.[74–76]

Efforts to clone the gene encoding the enzyme responsible for the above-described polyamine oxidation led to a clearer understanding of mammalian polyamine catabolism. As depicted in Scheme 6.2, APAO oxidizes N^1-acetylspermine and N^1-acetylspermidine generated by SSAT, producing spermidine or putrescine, H_2O_2 and 3-acetoaminopropanal. In contrast, a second polyamine oxidase, SMO, catalyzes the direct back-conversion of spermine to spermidine, producing H_2O_2 and 3-aminopropanal as byproducts (Scheme 6.3). Based on the reported sequence of the plant polyamine oxidase,[77] mammalian FAD-dependent polyamine oxidase genes were cloned from human cDNA libraries. The gene initially identified, now designated *SMOX* (originally named *PAOh1*), was found to encode a novel enzyme (SMO) that, unlike APAO, was inducible and acted on spermine with 3000-fold higher affinity versus spermidine or the acetylated polyamines.[78–80] Human *SMOX* consists of a 1894 bp mRNA containing a 1668 bp ORF encoding a 555 aa, 62 kDa protein with 23% identity and 50% similarity to the plant enzyme.[78] Multiple splice variants of human and mouse SMO have been reported, including two isoforms that have been demonstrated to be present and catalytically active in both the cytosol and nucleus.[81–84] The production of potentially mutagenic or cytotoxic ROS and 3-aminopropanal in the nucleus has significant implications for the role of SMO in drug response and disease etiology, as will be discussed in

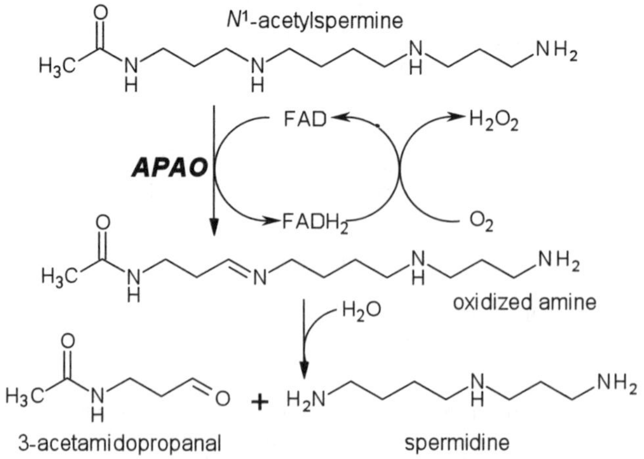

Scheme 6.2   APAO enzyme reaction.

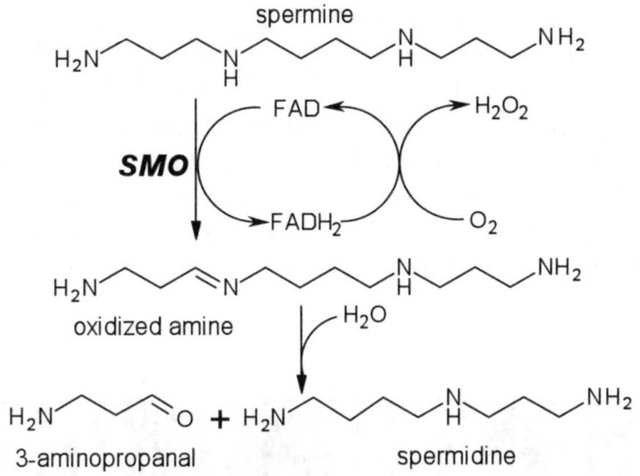

Scheme 6.3 SMO enzyme reaction.

Section 6.4. Ectopic expression of SMO in human cell lines resulted in reduced cellular levels of spermine but not N^1-acetylspermidine, and increased putrescine and spermidine pools.[78,79] Shortly after the discovery of SMO, *PAOX* (the gene encoding APAO) was cloned from both the mouse and human genomes.[85,86] Human *PAOX* consists of a 1850 bp mRNA yielding a 511 aa, 56 kDa protein that is highly similar (83% sequence identity) to the mouse version.

Confirming the studies on enzyme purified from animal tissue described above, APAO utilizes acetylated spermine and spermidine with equal affinity and only accommodates diacetylated spermine and spermine with low affinity. Further validating prior data, APAO contains a C-terminal peroxisomal transport signal, indicating that reactive products generated by APAO activity are unlikely to have general cellular effects unless acetylated polyamine levels or APAO activity is exceptionally high, or catalase activity is depressed. Recombinant APAO and SMO expressed in *E. coli* have high specific activities, indicating that post-translational modifications are unlikely to be consequential for enzyme activity. The polyamine analog MDL 72 527 was demonstrated to be a highly potent, irreversible competitive inhibitor of both recombinant SMO and APAO.[86–89]

As will be discussed throughout this chapter, the ability to break down the higher polyamines via either the SMO or SSAT-APAO pathways provides an added layer of complexity to cellular regulation of polyamine homeostasis. The intermediate generation of acetylated polyamines by SSAT, but not SMO, has a number of important implications. Due to the loss of a positive charge, the acetylated molecules are less potent than spermine and spermidine in their ability to carry out cellular functions. Further, acetylated polyamines, but not spermine or spermidine, can be exported from the cells in one component of cellular polyamine transport. Therefore, the catabolic pathways can affect either the recycling of higher polyamines to putrescine or the depletion of cellular polyamine pools via excretion.[52–53,90–93]

In addition to the mammalian APAO and SMO enzymes that are the focus of this chapter, conserved FAD-dependent polyamine oxidases have been described and cloned in higher plants,[77,94–105] yeast[106,107] and *Drosophila*.[108] Lysine-specific demethylase 1 (LSD1), a homologous FAD-dependent enzyme that is crucial in chromatin remodeling and regulation of gene expression (see Figure 6.3), is another important therapeutic target that does not utilize polyamines as substrates but can be modulated with specific polyamine analogs (see Chapter 5). While the structures of the mammalian polyamine catabolic

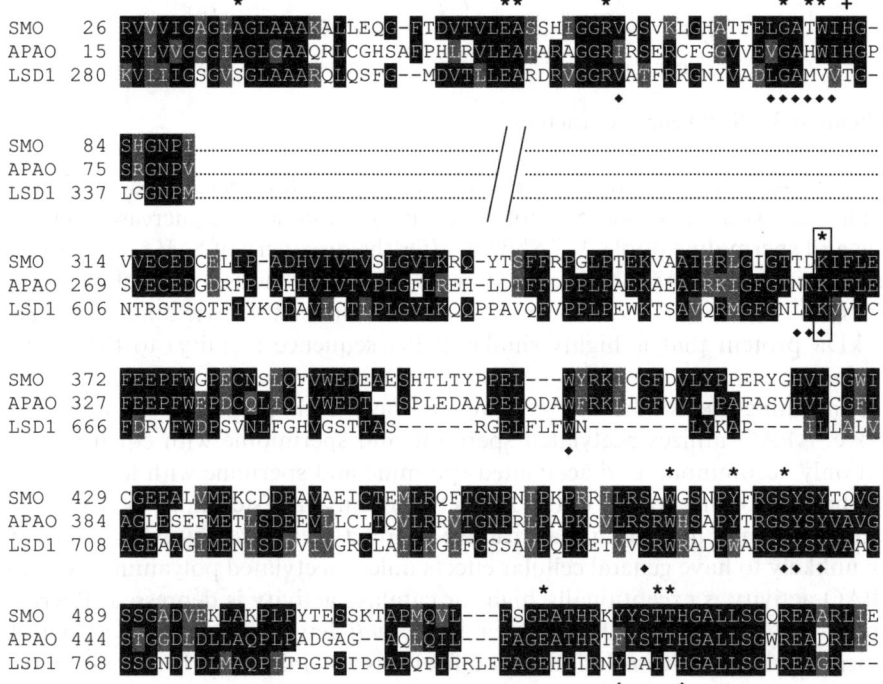

Figure 6.3   Amino acid homology alignment of the FAD-dependent amine oxidase domains of the human enzymes spermine oxidase (SMO), N^1-acetylpolyamine oxidase (APAO) and lysine-specific demethylase 1 (LSD1). Black shading indicates identical residues, and gray-shaded residues are conserved. The boxed lysine is a conserved residue shown to be essential for flavin oxidation and catalytic activity. Asterisks (*) identify other residues associated with FAD binding, as determined by homology to the maize PAO.[80,113] The histidine residue marked with a plus sign (+) corresponds to Glu62 of maize PAO and has been implicated in the differential pH requirements of plant *vs.* mammalian enzymes.[80] Diamonds (♦) identify residues located in the active site cavity of the LSD1 crystal structure that are required for positioning the FAD cofactor and substrate lysine, or that, when mutated, decrease catalytic activity.[118] Alignment was produced using ClustalW software with shading by BOXSHADE, both available at http://www.ch.embnet.org.

oxidases have unfortunately not been solved, crystal structures of yeast Fms1,[109] human LSD1[110,111] and plant polyamine oxidases[112–114] have contributed to the field's understanding of the structures and mechanisms of action of SMO and APAO. Mammalian and plant polyamine oxidases are quite similar in structure, despite the fact that APAO and SMO cleave their substrates at the exo side of the N^4 amino group, while plant PAOs act at the endo side of the N^4 amino group, leading to different reaction products.[77,85,112] A complete amino acid sequence was first obtained for the maize polyamine oxidase (MPAO), and comparison to other characterized enzymes revealed a sequence motif indicative of non-covalent FAD binding.[77] Binda and colleagues[113] further observed that the MPAO structure consisted of a 53 kDa monomer containing two domains formed from α-helices and β-strands, with the FAD-binding site located deep within the structure. At their interface, the FAD- and substrate binding domains form a U-shaped tunnel with the active site at its core. Analysis of mammalian SMO indicated that while general features of the active site were conserved with MPAO, this tunnel was wider than in the plant version. These studies also identified the Glu62 residue of MPAO as being an important factor in the pH sensitivity of the enzyme, and a conserved lysine residue (Lys300 in MPAO, Lys315 in mouse APAO, and Lys661 in LSD1) is linked to this flavin ring and appears to be critical for flavin oxidation and enzyme activity, possibly acting as an active site base.[80,114–118]

Several studies have analyzed the interaction of polyamine oxidases with numerous substrates and inhibitors to gain insight into substrate binding, kinetics and reaction mechanisms. Substrate or inhibitor molecules enter and adopt the conformation of the tunnel, displacing eight ordered water molecules and binding via a lock-and-key fit that does not induce any significant conformational changes of the oxidase. Substrate binding is mediated by van der Waals interactions and a series of unusual CH–O hydrogen bonds, and not via binding of positively charged amino groups to negatively charged residues in the enzyme tunnel. As a result, depending on the substrate molecule, carbon or nitrogen atoms can bind at the same location, offering an explanation for the existence of many closely related FAD-dependent oxidases across numerous species that have a wide variety of substrate specificities. The ability of a molecule to bind in the oxidase tunnel does not result in productive oxidation in all cases; the C–N bond to be attacked must be in the correct orientation and vicinity of the reactive flavin N^5 locus. In addition, the charge of the substrate amino groups is an important basis of substrate specificity for SMO and APAO. At physiological pH, spermine, N^1-acetylspermine and N^1-acetylspermidine carry four, three and two positively charged amino groups, respectively. However, based on the predicted pK_a and pH values at the active site, for catalysis to occur, SMO requires triprotonated spermine, while APAO only oxidizes monoprotonated substrates, a state that is favored in acetylated versus natural polyamines.[113–115,119] Takao and colleagues analyzed purified APAO in association with several polyamine analogs to predict that APAO catalysis requires the positive charge to be located three or four methylene chain lengths away from the cleaved C–N bond and that the active site must contain two anionic centers as well as a hydrophobic region.[120]

SMO and APAO enzyme kinetics appear to follow a ping-pong reaction mechanism. Following substrate binding, FAD is reduced and the amine oxidized via the direct transfer of a hydride from the substrate C–N bond to the flavin group. The oxidized polyamine is then released from the enzyme, at which time the slower second stage consists of the reduced flavin co-factor reacting with O_2 to produce H_2O_2 and restore SMO or APAO to the oxidized state. In contrast to the mechanism of SSAT described above, no intermediate product is produced during SMO or APAO catalysis.[87,119,121–124] The above studies have provided an improved insight into the structure and enzymology of the mammalian polyamine oxidases, though successful crystallization of SMO and APAO remains a critical challenge for the field. Numerous polyamine analogs have been evaluated for their ability to act as substrates or inhibitors of SMO and APAO, and understanding the substrate–enzyme interactions has and will continue to inform the design and synthesis of novel therapeutic agents.[88,89,120,122–126] Furthermore, a focused effort to develop potent and specific inhibitors of SMO and APAO will be crucial to expand on studies with the dual inhibitor, MDL 72 527, and continue efforts to exploit SMO and APAO as therapeutic targets.

6.3 Modulation of SMO, APAO and SSAT as a Therapeutic Strategy

6.3.1 Polyamine Catabolic Enzymes as Rational Drug Targets in Parasitic Diseases

Polyamine metabolic pathways have long been recognized as therapeutic targets in parasitic diseases, and this approach has yielded several clinical and commercial successes, as discussed in greater depth in Chapter 3. Polyamine oxidases have been found in some parasitic species but not others, thereby limiting the ability to target these enzymes. For example, spermidine/spermine N^1-acetyltransferases and polyamine oxidases were found in helminthes,[127] microsporidia[128,129] and *Trichomonas vaginalis*,[130] and a polyamine oxidase that acts on spermine and spermidine was discovered in the nematode *Ascaris*,[131] but *Trypanosoma cruzi* lacks polyamine oxidase activity.[132] While some early experiments demonstrating lethality of the polyamines or polyamine analogs to parasites were actually mediated by copper-containing serum amine oxidases, several successful examples targeting the induction of polyamine catabolism for antiparasitic activity have been documented.

Filarial worms *Ascaris ocum* and *Onchocerca volvulus* lack the polyamine biosynthetic enzyme ODC and are completely reliant on the host to provide polyamines. Berenil, which has no effect on mammalian polyamine catabolic enzymes, was shown to be extremely toxic to these helminthes in serum-free media via the inhibition of polyamine catabolism. Similar results were also seen when spermine was added to the cultures.[127] Interestingly, while the nematode PAO appears to be similar to the mammalian enzymes, MDL 72 527 had no inhibitory effect on the parasitic enzyme activity.[131] The polyamine analog with bis(ethyl)norspermine (BENSpm, see Figure 6.2) was found to be cytotoxic to *T. vaginalis* by inducing a large decrease in cellular polyamine levels. The analog inhibits parasitic SSAT (in contrast to the strong induction of the mammalian enzyme by BENSpm; see Section 6.3.2 below) and competes with spermine for uptake, and because the organism lacks polyamine biosynthetic enzymes, it cannot replenish polyamine levels as demonstrated in mammalian cells.[130,133] The polyamine catabolic enzymes of the protozoans *Encephalitozoon cuniculi* and *Cryptosporidum parvum* can be targeted based on differential inhibition or induction by polyamine analogs and differences in substrate specificity and pH optima as compared to the mammalian forms.[128,129] These strategies have been successfully employed to cure a murine model of *Cryptosporidum parvum* infection with polyamine analogs.[134,135]

The ability to utilize parasite polyamine catabolic enzymes as therapeutic targets is appealing for multiple reasons. First, selective induction of cytotoxic ROS production in parasite or parasite-infected host cells provides a mechanism of selectivity that is analogous to the efforts to target cancer cells with dysregulated polyamine metabolism, as described below in Section 6.3.2. Second, it appears as though additional selectivity for drug-induced cell death could be achieved based on enzymological differences between mammalian and parasitic catabolic enzymes. This approach has been successfully utilized with the polyamine biosynthetic enzyme ODC as a drug target in Trypanosomal diseases (see Chapters 3 and 4).

6.3.2 Cytotoxic Polyamine Analogs as Selective Chemotherapeutic Agents

The concept of polyamine analogs as candidate chemotherapeutic agents was first introduced in the 1980s.[24,136–140] Absolutely required for cell growth and differentiation, the highest concentrations of polyamines are found in cells undergoing rapid growth, including some tumor cells. Therefore, it was hypothesized that dysregulation of polyamine metabolism in cancer cells by these drugs would result in depletion of polyamine pools and cell-growth arrest. The first class of analogs synthesized and evaluated were symmetrically substituted bis(alkyl) polyamines. As expected, these and some other classes of analogs entered the cell via the polyamine uptake system, downregulated the activity of both ODC and AdoMetDC, and were cytotoxic in multiple tumor cell lines and xenografts.[140–150] An additional feature of specific polyamine analog-induced cell death was a significant (1000-fold or more in some cases) induction of SSAT and/or SMO.[26,151–163] While this chapter is focused on the interactions between polyamine analogs and polyamine catabolic enzymes, the evaluation of these compounds as candidate antitumor agents is covered more broadly in Chapter 5 and is the subject of recent review articles.[90,164]

The mechanism of action involved in these responses was first described in studies showing that the degree of SSAT induction was correlated with the sensitivity of human lung-cancer cell lines to bis(ethyl)polyamine analogs.[24,25,27,165–167] The increased activity of SSAT/APAO (or, as discussed below, SMO) generally results in polyamine depletion, production of cytotoxic levels of H_2O_2 and apoptosis. However, certain other polyamine analogs can induce polyamine catabolism without cell death or mediate cytotoxicity independent of any effect on SSAT or SMO.[168–171] In agreement with the initial findings, cells with elevated or reduced SSAT gene expression or activity exhibit a higher or lower sensitivity to polyamine analogs, respectively.[172–176] Furthermore, overexpression of SSAT genomic DNA in small-cell lung-cancer cells that are phenotypically insensitive to the effects of polyamine analog treatment rendered them capable of SSAT induction by BENSpm, and this induction correlated with a cytotoxic response.[177] The applicability of these findings to the clinic was supported by an immunohistochemistry study that showed strong SSAT staining in primary tumor tissue treated with the BENSpm.[178] SSAT induction by polyamine analogs was determined to occur via the same NRF2- and PMF1-dependent upregulation of transcription described above for increased *SAT1* expression in response to elevated polyamine levels. Importantly, a subset of lung-cancer cell lines and tumors has been shown to possess a mutant form of KEAP1, a negative regulator of NRF2, resulting in increased NRF2 activity.[179] Therefore, the use of polyamine analogs that cause SSAT-induced cytotoxicity mediated through NRF2 may be a therapeutic approach for this subset of lung cancers. In addition to increasing *SAT1* gene expression, polyamine analogs decrease the proteosomal degradation of SSAT.[51]

Following the cloning of *SMOX*, a key difference between SMO and APAO was determined to be that SMO, like SSAT, is highly inducible by BENSpm and other polyamine analogs.[78,156,180] Wang and colleagues used the induction of SMO in response to the polyamine analog N^1-ethyl-N^{11}-(cyclopropyl)methyl-4,8,diazaundecane (CPENSpm, see Figure 6.2) to characterize the mechanism of SMO regulation. Using actinomycin D and cycloheximide, inhibitors of *de novo* RNA and protein synthesis, respectively, they demonstrated that new mRNA and protein synthesis were required for increased SMO activity in response to CPENSpm. Increased *SMOX* promoter activity was accompanied by mRNA stabilization leading to an approximately twofold increase in mRNA half-life, analogous to observed effects of polyamines or analogs on *SAT1*. No increase in SMO protein half-life was detected, thus supporting the evidence that the major control of induction of SMO occurs at the mRNA level.[157] One study examining the toxicity of BENSpm in breast-cancer cell lines confirmed that H_2O_2 produced by SMO, not the SSAT/APAO pathway, was the primary contributor to cell death.[158]

Based on its simple chemical synthesis and the promising *in vitro* data described above, BENSpm was advanced into phase I human clinical trials as a candidate chemotherapeutic agent. Toxicities were significant and limited the use of BENSpm to once-daily dosing.[181] A single phase II study in advanced breast cancer was conducted but failed to demonstrate clear clinical benefit.[182] More recent studies have shown that combining polyamine analogs that induce polyamine catabolism with other chemotherapeutic agents results in synergistic activity, which may lead to future clinical evaluation.[183–190] Therefore, the potential of tumor-selective, analog-induced cell death mediated by polyamine catabolism remains a viable area for future investigation.

6.4 Inhibition of Polyamine Catabolism as a Therapeutic Approach

6.4.1 Role of SMO in Inflammation-Associated Tumorigenesis

It is estimated that chronic inflammation associated with microbial infection directly contributes to the etiology of about 20% of epithelial cancers. The chronic inflammatory microenvironment is characterized by immune dysregulation and elevated levels of ROS (including superoxide, hydrogen peroxide and singlet oxygen). These features result in activation of stress response pathways and oncogenes, downregulation of tumor suppressor genes, and damage to cells and tissues. They also contribute to tumor initiation, promotion and progression; however, the precise molecular links between inflammation and carcinogenesis remain to be clarified.[191–193]

Chu and colleagues directly evaluated the contribution of H_2O_2 produced by enteric bacteria to chronic inflammation and tumorigenesis using a mouse model in which the genes encoding the major H_2O_2 detoxification enzymes glutathione peroxidase 1 and 2 (*gpx1* and *gpx2*) were knocked out. No detectable phenotype was observed in mice with a single locus (*gpx1* or *gpx2*) deleted or when mice lacking both enzymes were housed in a germ-free facility, but double-knockout mice developed colitis (within 1 week) and intestinal tumors (by 6–9 months of age in 25% of animals) when housed under conventional conditions that permitted the establishment of commensal microflora.[194–196] These data strongly implicate oxidative damage by H_2O_2 as a contributor to inflammation-associated tumorigenesis, but fail to identify the cellular source of this ROS.

Several recent studies suggest that polyamine catabolism, and more specifically SMO, may represent one such molecular link between chronic inflammation, ROS production and carcinogenesis in multiple epithelial cell types. *Helicobacter pylori* is a bacterial pathogen that colonizes the stomach of more than 50% of the world's population. Infected individuals exhibit chronic gastritis, and a subset will progress to overt disease in the form of peptic ulcer disease, gastric adenocarcinoma or mucosa-associated lymphoid tissue (MALT) lymphoma.[197] A collaboration between the Casero and Wilson laboratories conclusively demonstrated that *H. pylori* infection rapidly induces SMO gene expression, protein levels and enzyme activity in gastric epithelial cells and human tissues from *H. pylori*-infected patients. Importantly, the approximately fivefold increase in SMO expression directly resulted in increased levels of ROS and 8-oxodeoxyguanosine (8-OHdG; a marker of oxidative DNA damage) that could be abrogated by catalase (detoxification enzyme for H_2O_2), MDL 72 527 (inhibitor of SMO and APAO) or siRNA directed against SMO. SMO also appears to mediate the induction of apoptosis in gastric macrophages, a mechanism utilized by *H. pylori* to evade the immune response and persistently infect the gastric epithelium.[198]

Additional studies were carried out to assess the ability of the cytokine tumor necrosis factor-alpha (TNF-α) to modulate SMO expression. TNF-α is widely produced by cells in response to injury or inflammation and represents a potent pro-inflammatory mediator. Exposure of Beas-2b human bronchial epithelial cells to TNF-α resulted in a rapid induction of SMO gene expression, protein and enzyme activity. This increased SMO activity was directly responsible for elevated cellular ROS levels and DNA damage, based on the elimination of these effects via pharmacological or siRNA-mediated inhibition of SMO.[199] Importantly, while SSAT is also induced by TNF-α,[200] these studies indicated that SMO, and not the SSAT-APAO pathway, was the source of the inflammation-associated ROS production and oxidative DNA damage. One reason for the greater pathological significance of toxic ROS and aldehyde production by SMO as compared to APAO may be the confinement of APAO activity to the peroxisome in the presence of catalase.[201]

Enterotoxigenic *Bacteroides fragilis* (ETBF) comprise a group of human pathogens that have been linked to acute diarrheal diseases in livestock and humans, inflammatory bowel diseases and colorectal cancer.[202–204] ETBF induces acute and persistent colitis *in vivo*, and ETBF-infected $APC^{+/-}$ (Min) mice develop numerous inflammation-associated colon tumors.[205–208] Therefore, this model was utilized to directly examine the role of SMO in inflammation-associated tumorigenesis. Recombinant ETBF toxin rapidly induces SMO in human intestinal epithelial cell lines, resulting in SMO-dependent DNA damage. ETBF-infected mice treated with MDL 72 527 (20 mg/kg three times per week *i.p.*) exhibited decreases in chronic inflammation, intestinal proliferation and cytokine induction. Further, MDL 72 527 treatment resulted in a 69% decrease in colon tumors in ETBF-infected Min mice.[246] These studies provide compelling evidence that induction of SMO-dependent ROS is a molecular link between chronic inflammatory stimuli and epithelial carcinogenesis.

Several additional studies provide support for the investigation of SMO as a chemopreventive therapeutic target. The polyamine oxidation inhibitor MDL 72 527 increased survival in the TRAMP model of prostate carcinogenesis.[209] In addition, descriptive studies have demonstrated an association between increased levels of SMO and *H. pylori*-associated gastritis, ulcerative colitis, prostatic intraepithelial neoplasia and prostate cancer.[210–212] While the body of evidence suggests that SMO, rather than APAO, is the source of inflammation-associated ROS contributing to tumorigenesis, conclusive determination awaits the development and evaluation of suitable specific pharmacologic inhibitors of each polyamine oxidase and genetic mouse models to aid in the evaluation of the role of the individual oxidases in the disease process.

6.4.2 Polyamine Catabolism in Ischemic Injuries

Ischemia–reperfusion injuries occur when a disruption in blood supply, associated with a period of limited oxygen and other metabolites, is followed by inflammation and oxidative damage upon resumption of blood flow. In models of brain ischemic injuries, studies demonstrated the elevation of SSAT mRNA levels, enzyme activity and N^1-acetylated polyamines.[213–215] The polyamine catabolic byproduct 3-aminopropanal was later identified as the cause of cell death in neurons and glial cells,[216] and administration of the polyamine oxidase inhibitor, MDL 72 527, or a drug that neutralizes this aldehyde reduced brain injury.[217–219] In plasma from patients with stroke or silent brain infarction, increased SMO and APAO enzyme activity and acrolein (produced from oxidase-generated aldehydes) were detected, indicating that they may be potential markers of disease.[220–222] Zahedi and colleagues further observed increased SSAT and SMO by northern blotting and immunohistochemistry in a rat model of traumatic brain injury.[223]

Increased SSAT is also associated with renal ischemia–reperfusion injury, a major cause of kidney failure. In animal models of this condition, elevated SSAT gene expression and protein levels were observed after reperfusion, while SSAT knockout mice exhibited significantly less tissue damage, a phenotype that was also observed in a model of endotoxin-induced acute kidney injury.[224–226] Further, overexpression of SSAT *in vitro* recapitulated many of the hallmarks of this disease, including adhesion defects, growth arrest, oxidative stress and DNA damage.[227,228] A similar association between polyamine catabolism and ischemia–reperfusion injuries has been observed in the heart[229–231] and liver,[225,232] indicating that these pathways are a common mediator of ischemic injury and therefore represent a compelling target for therapeutic intervention.

6.4.3 Association of Polyamine Catabolic Enzymes With Other Human Diseases

In addition to the examples detailed above, studies have documented an association between the polyamine catabolic enzymes and a number of other human medical conditions. A model of induced seizure suggested that polyamine catabolism potentiates neuronal damage in adult, but not juvenile rats.[233,234] SSAT polymorphisms or expression levels have also been associated with a number of other diseases with psychiatric components, including anxiety, depression, schizophrenia and suicide.[235–239] Based on these data and the role of SSAT, SMO and APAO in ischemic brain damage described in Section 6.4.2, further investigation into the modulation of polyamine catabolism as a therapeutic target in these neurological indications is warranted.

Elevated expression of polyamine catabolic enzymes has also been described in cervical intraepithelial neoplasia,[240] preeclampsia,[241] sepsis[242] and Keratosis follicularis spinulosa decalvans (KFSD or Siemens-1 syndrome).[243] Finally, a role for polyamine catabolism in cellular responses to radiation has been proposed.[244,245] All told, it appears certain that the polyamine catabolic enzymes play a crucial role in the course of numerous human disease conditions. While the individual etiologic mechanisms remain to be clarified, a major theme is the production of damaging ROS and aldehyde metabolites by the oxidases, SMO and APAO. In the therapeutic indications described in Sections 6.4.2 and 6.4.3, further basic understanding of the disease processes should facilitate drug-discovery efforts targeting the polyamine oxidases.

6.5 Conclusions

All told, the polyamine catabolic enzymes SSAT, APAO and SMO represent promising molecular targets for ongoing and future drug development. The polyamine oxidases SMO and APAO have been shown to generate cytotoxic levels of ROS in response to selective anti-tumor polyamine analogs. Each catabolic enzyme has been linked to the etiology, progression and cellular dysfunction associated with multiple human diseases. The potential to selectively activate or block the activity of these enzymes in particular cells or tissues is an exciting direction for continued research and development of novel therapeutic agents. To successfully exploit SSAT, APAO and SMO as drug targets, several challenges remain for the field. These include solving the crystal structures of human SMO and APAO, synthesizing specific and potent inhibitors of the polyamine oxidases, better understanding the molecular mechanisms underlying the induction of polyamine catabolism and clarifying the cellular pathways governing the uptake of polyamine-based drugs. Based on insights gained from three decades of research on polyamine catabolism, SSAT, APAO and SMO continue to be attractive therapeutic targets in a wide range of disease indications.

References

1. R. A. Casero and A. E. Pegg, *Biochem. J.*, 2009, **421**, 323–338.
2. N. Babbar, T. Murray-Stewart and R. A. Casero, *Biochem. Soc. Trans.*, 2007, **35**, 300–304.
3. Y. Wang and R. A. Casero, Jr, *J. Biochem. (Tokyo)*, 2006, **139**, 17–25.
4. M. M. Abdel-Monem, K. Ono, I. E. Fortuny and A. Theologides, *Lancet*, 1975, **2**, 1210.
5. M. Tsuji, T. Nakajima and I. Sano, *Clin. Chim. Acta*, 1975, **59**, 161–167.
6. R. G. Smith, D. Bartos, F. Bartos, D. P. Grettie, W. Frick, R. A. Campbell and G. D. J. Daves, *Biomed. Mass Spectrom.*, 1978, **5**, 515–517.
7. M. M. Abdel-Monem and K. Ohno, *J. Pharm. Sci.*, 1978, **67**, 1671–1673.
8. J. Blankenship and T. Walle, *Arch. Biochem. Biophys.*, 1977, **179**, 235–242.
9. N. Seiler, F. N. Bolkenius and B. Knodgen, *Biochim. Biophys. Acta*, 1980, **633**, 181–190.
10. I. Matsui and A. E. Pegg, *Biochem. Biophys. Res. Commun.*, 1980, **92**, 1009–1015.

11. F. D. Ragione and A. E. Pegg, *Biochemistry*, 1982, **21**, 6152–6158.
12. I. Matsui, L. Wiegand and A. E. Pegg, *J. Biol. Chem.*, 1981, **256**, 2454–2459.
13. P. R. Libby, *J. Biol. Chem.*, 1978, **253**, 233–237.
14. P. R. Libby, *Arch. Biochem. Biophys.*, 1980, **203**, 384–389.
15. B. G. Erwin, L. Persson and A. E. Pegg, *Biochemistry*, 1984, **23**, 4250–4255.
16. L. C. Wong, D. J. Sharpe and S. S. Wong, *Biochem. Genet.*, 1991, **29**, 461–475.
17. M. A. Desiderio, *Hepatology*, 1992, **15**, 928–933.
18. M. A. Desiderio, M. Weibel and P. S. Mamont, *Exp. Cell Res.*, 1992, **202**, 501–506.
19. K. Bandyopadhyay, J. L. Baneres, A. Martin, C. Blonski, J. Parello and R. A. Gjerset, *Cell Cycle*, 2009, **8**, 2779–2788.
20. I. Matsui and A. E. Pegg, *Biochim. Biophys. Acta*, 1980, **633**, 87–94.
21. I. Matsui and A. E. Pegg, *Cancer Res.*, 1982, **42**, 2990–2995.
22. I. Matsui and A. E. Pegg, *FEBS Lett.*, 1982, **139**, 205–208.
23. I. Matsui, H. Poso and A. E. Pegg, *Biochim. Biophys. Acta*, 1982, **719**, 199–207.
24. R. A. J. Casero, S. J. Ervin, P. Celano, S. B. Baylin and R. J. Bergeron, *Cancer Res.*, 1989, **49**, 639–643.
25. R. A. J. Casero, P. Celano, S. J. Ervin, C. W. Porter, R. J. Bergeron and P. R. Libby, *Cancer Res.*, 1989, **49**, 3829–3833.
26. P. R. Libby, R. J. Bergeron and C. W. Porter, *Biochem. Pharmacol.*, 1989, **38**, 1435–1442.
27. R. A. J. Casero, P. Celano, S. J. Ervin, L. Wiest and A. E. Pegg, *Biochem. J.*, 1990, **270**, 615–620.
28. N. W. Shappell, M. F. Fogel-Petrovic and C. W. Porter, *FEBS Lett.*, 1993, **321**, 179–183.
29. L. Xiao, P. Celano, A. R. Mank, A. E. Pegg and R. A. Casero, Jr, *Biochem. Biophys. Res. Commun.*, 1991, **179**, 407–415.
30. L. Xiao, P. Celano, A. R. Mank, C. Griffin, E. W. Jabs, A. L. Hawkins and R. A. Casero, Jr, *Biochem. Biophys. Res. Commun.*, 1992, **187**, 1493–1502.
31. M. Fogel-Petrovic, D. L. Kramer, B. Ganis, R. A. Casero, Jr and C. W. Porter, *Biochim. Biophys. Acta*, 1993, **1216**, 255–264.
32. C. Vargiu and L. Persson, *FEBS Lett.*, 1994, **355**, 163–165.
33. Y. Chen, S. Vujcic, P. Liang, P. Diegelman, D. L. Kramer and C. W. Porter, *Biochem. J.*, 2003, **373**, 661–667.
34. C. S. Coleman, B. A. Stanley, A. D. Jones and A. E. Pegg, *Biochem. J.*, 2004, **384**, 139–148.
35. A. E. Pegg, *J. Biol. Chem.*, 2006, **281**, 14529–14532.
36. A. E. Pegg, *Essays Biochem.*, 2009, **46**, 25–45.
37. R. A. J. Casero, P. Celano, S. J. Ervin, N. B. Applegren, L. Wiest and A. E. Pegg, *J. Biol. Chem.*, 1991, **266**, 810–814.
38. W. O. Osburn and T. W. Kensler, *Mutat. Res.*, 2008, **659**, 31–39.

39. Y. Wang, W. Devereux, T. M. Stewart and R. A. J. Casero, *J. Biol. Chem.*, 1999, **274**, 22095–22101.
40. Y. Wang, L. Xiao, A. Thiagalingam, B. D. Nelkin and R. A. Casero, Jr, *J. Biol. Chem.*, 1998, **273**, 34623–34630.
41. M. T. Hyvonen, A. Uimari, T. A. Keinanen, S. Heikkinen, R. Pellinen, T. Wahlfors, A. Korhonen, A. Narvanen, J. Wahlfors, L. Alhonen and J. Janne, *RNA*, 2006, **12**, 1569–1582.
42. A. E. Pegg, *Am. J. Physiol. Endocrinol. Metab.*, 2008, **294**, E995–E1010.
43. E. Agostinelli, G. Arancia, L. D. Vedova, F. Belli, M. Marra, M. Salvi and A. Toninello, *Amino Acids*, 2004, **27**, 347–358.
44. Y. Q. Zhu, D. Y. Zhu, L. Yin, Y. Zhang, C. Vonrhein and D. C. Wang, *Proteins*, 2006, **63**, 1127–1131.
45. M. C. Bewley, V. Graziano, J. Jiang, E. Matz, F. W. Studier, A. E. Pegg, C. S. Coleman and J. M. Flanagan, *Proc. Natl. Acad. Sci. U. S. A.*, 2006, **103**, 2063–2068.
46. S. S. Hegde, J. Chandler, M. W. Vetting, M. Yu and J. S. Blanchard, *Biochemistry*, 2007, **46**, 7187–7195.
47. E. J. Montemayor and D. W. Hoffman, *Biochemistry*, 2008, **47**, 9145–9153.
48. C. S. Coleman, H. Huang and A. E. Pegg, *Biochemistry*, 1995, **34**, 13423–13430.
49. C. S. Coleman, H. Huang and A. E. Pegg, *Biochem. J.*, 1996, **316**, 697–701.
50. C. S. Coleman and A. E. Pegg, *J. Biol. Chem.*, 1997, **272**, 12164–12169.
51. C. S. Coleman and A. E. Pegg, *Biochem. J.*, 2001, **358**, 137–145.
52. L. Persson, *Essays Biochem.*, 2009, **46**, 11–24.
53. C. Moinard, L. Cynober and J. P. de Bandt, *Clin. Nutr.*, 2005, **24**, 184–197.
54. A. E. Pegg, I. Matsui, J. E. Seely, M. L. Pritchard and H. Poso, *Med. Biol.*, 1981, **59**, 327–333.
55. N. Seiler, F. N. Bolkenius and O. M. Rennert, *Med. Biol.*, 1981, **59**, 334–346.
56. Y. Sato and K. Fujiwara, *J. Biochem.*, 1988, **104**, 98–101.
57. N. Seiler and F. N. Bolkenius, *Neurochem. Res.*, 1985, **10**, 529–544.
58. C. W. Tabor and S. M. Rosenthal, *J. Pharmacol. Exp. Ther.*, 1956, **116**, 139–155.
59. S. M. Rosenthal and C. W. Tabor, *J. Pharmacol. Exp. Ther.*, 1956, **116**, 131–138.
60. M. Siimes, *Acta Physiol. Scand. Suppl*, 1967, **298**, 1–66.
61. H. Tabor and C. W. Tabor, *Adv. Enzymol. Relat. Areas Mol. Biol.*, 1972, **36**, 203–268.
62. C. W. Tabor and H. Tabor, *Annu. Rev. Biochem.*, 1976, **45**, 285–306.
63. K. A. Flayeh and H. M. Wallace, *Biochem. Soc. Trans.*, 1990, **18**, 1225.
64. E. Holtta, *Biochemistry*, 1977, **16**, 91–100.
65. P. R. Libby and C. W. Porter, *Biochem. Biophys. Res. Commun.*, 1987, **144**, 528–535.
66. T. Tsukada, S. Furusako, S. Maekawa, H. Hibasami and K. Nakashima, *Int. J. Biochem.*, 1988, **20**, 695–702.

67. N. Seiler, F. N. Bolkenius, B. Knodgen and P. Mamont, *Biochim. Biophys. Acta*, 1980, **615**, 480–488.
68. V. Pavlov, I. Nikolov, D. Damjanov and O. Dimitrov, *Experientia*, 1991, **47**, 1209–1211.
69. P. S. Mamont, N. Seiler, M. Siat, A. M. Joder-Ohlenbusch and B. Knodgen, *Med. Biol.*, 1981, **59**, 347–353.
70. M. E. Beard, R. Baker, P. Conomos, D. Pugatch and E. Holtzman, *J. Histochem. Cytochem.*, 1985, **33**, 460–464.
71. H. Hayashi, H. Yoshida, F. Hashimoto and S. Okazeri, *Biochim. Biophys. Acta*, 1989, **991**, 310–316.
72. F. N. Bolkenius and N. Seiler, *Int. J. Biochem.*, 1981, **13**, 287–292.
73. Van den R. J. Munckhof, M. Denyn, W. Tigchelaar-Gutter, R. G. Schipper, A. A. Verhofstad, C. J. Van Noorden and W. M. Frederiks, *J. Histochem. Cytochem.*, 1995, **43**, 1155–1162.
74. P. Bey, F. N. Bolkenius, N. Seiler and P. Casara, *J. Med. Chem.*, 1985, **28**, 1–2.
75. F. N. Bolkenius, P. Bey and N. Seiler, *Biochim. Biophys. Acta*, 1985, **838**, 69–76.
76. F. N. Bolkenius and N. Seiler, *Biol. Chem. Hoppe Seyler*, 1989, **370**, 525–531.
77. P. Tavladoraki, M. E. Schinina, F. Cecconi, S. Di Agostino, F. Manera, G. Rea, P. Mariottini, R. Federico and R. Angelini, *FEBS Lett.*, 1998, **426**, 62–66.
78. Y. Wang, W. Devereux, P. M. Woster, T. M. Stewart, A. Hacker and R. A. Casero, Jr, *Cancer Res.*, 2001, **61**, 5370–5373.
79. S. Vujcic, P. Diegelman, C. J. Bacchi, D. L. Kramer and C. W. Porter, *Biochem. J.*, 2002, **367**, 665–675.
80. M. Cervelli, F. Polticelli, R. Federico and P. Mariottini, *J. Biol. Chem.*, 2003, **278**, 5271–5276.
81. T. Murray-Stewart, Y. Wang, W. Devereux and R. A. Casero, Jr, *Biochem. J.*, 2002, **368**, 673–677.
82. M. Cervelli, A. Bellini, M. Bianchi, L. Marcocci, S. Nocera, F. Polticelli, R. Federico, R. Amendola and P. Mariottini, *Eur. J. Biochem.*, 2004, **271**, 760–770.
83. M. Bianchi, R. Amendola, R. Federico, F. Polticelli and P. Mariottini, *Febs J.*, 2005, **272**, 3052–3059.
84. T. Murray-Stewart, Y. Wang, A. Goodwin, A. Hacker, A. Meeker and R. A. Casero Jr, *FEBS J.*, 2008, **275**, 2795–2806.
85. T. Wu, V. Yankovskaya and W. S. McIntire, *J. Biol. Chem.*, 2003, **278**, 20514–20525.
86. S. Vujcic, P. Liang, P. Diegelman, D. L. Kramer and C. W. Porter, *Biochem. J.*, 2003, **370**, 19–28.
87. A. Bellelli, S. Cavallo, L. Nicolini, M. Cervelli, M. Bianchi, P. Mariottini, M. Zelli and R. Federico, *Biochem. Biophys. Res. Commun.*, 2004, **322**, 1–8.
88. Y. Wang, T. Murray-Stewart, W. Devereux, A. Hacker, B. Frydman, P. M. Woster and R. A. Casero, Jr, *Biochem. Biophys. Res. Commun.*, 2003, **304**, 605–611.

89. Y. Wang, A. Hacker, T. Murray-Stewart, B. Frydman, A. Valasinas, A. V. Fraser, P. M. Woster and R. A. Casero, Jr, *Cancer Chemother. Pharmacol.*, 2005, **56**, 83–90.
90. R. A. Casero, Jr and L. J. Marton, *Nat. Rev. Drug Discov.*, 2007, **6**, 373–390.
91. T. Thomas and T. J. Thomas, *Cell. Mol. Life Sci.*, 2001, **58**, 244–258.
92. H. M. Wallace, *Essays Biochem.*, 2009, **46**, 1–9.
93. N. Seiler, *Curr. Drug Targets*, 2003, **4**, 565–585.
94. T. A. Smith, *Biochem. Biophys. Res. Commun.*, 1970, **41**, 1452–1456.
95. T. A. Smith, W. L. Haygarth and J. F. Williams, *Biochem. Soc. Trans.*, 1976, **4**, 74–77.
96. R. Kaur-Sawhney, H. E. Flores and A. W. Galston, *Plant Physiol.*, 1981, **68**, 494–498.
97. H. Yanagisawa, A. Kato, S. Hoshiai, A. Kamiya and N. Torii, *Plant Physiol.*, 1987, **85**, 906–909.
98. R. Federico, C. Alisi, A. Cona and R. Angelini, *Adv. Exp. Med. Biol.*, 1988, **250**, 617–623.
99. R. Federico, A. Cona, R. Angelini, M. E. Schinina and A. Giartosio, *Phytochemistry*, 1990, **29**, 2411–2414.
100. P. N. Moschou, M. Sanmartin, A. H. Andriopoulou, E. Rojo, J. J. Sanchez-Serrano and K. A. Roubelakis-Angelakis, *Plant Physiol.*, 2008, **147**, 1845–1857.
101. T. Kamada-Nobusada, M. Hayashi, M. Fukazawa, H. Sakakibara and M. Nishimura, *Plant Cell Physiol.*, 2008, **49**, 1272–1282.
102. Y. Takahashi, R. Cong, G. H. Sagor, M. Niitsu, T. Berberich and T. Kusano, *Plant Cell Rep.*, 2010.
103. A. Radova, M. Sebela, P. Galuszka, I. Frebort, S. Jacobsen, H. G. Faulhammer and P. Pec, *Phytochem. Anal.*, 2001, **12**, 166–173.
104. P. Tavladoraki, M. N. Rossi, G. Saccuti, M. A. Perez-Amador, F. Polticelli, R. Angelini and R. Federico, *Plant Physiol.*, 2006, **141**, 1519–1532.
105. M. Cervelli, M. Bianchi, A. Cona, C. Crosatti, M. Stanca, R. Angelini, R. Federico and P. Mariottini, *FEBS J.*, 2006, **273**, 3990–4002.
106. M. Nishikawa, T. Hagishita, H. Yurimoto, N. Kato, Y. Sakai and T. Hatanaka, *FEBS Lett.*, 2000, **476**, 150–154.
107. J. Landry and R. Sternglanz, *Biochem. Biophys. Res. Commun.*, 2003, **303**, 771–776.
108. A. Tatarenkov, A. G. Saez and F. J. Ayala, *Gene*, 1999, **231**, 111–120.
109. Q. Huang, Q. Liu and Q. Hao, *J. Mol. Biol.*, 2005, **348**, 951–959.
110. M. Yang, C. B. Gocke, X. Luo, D. Borek, D. R. Tomchick, M. Machius, Z. Otwinowski and H. Yu, *Mol. Cell*, 2006, **23**, 377–387.
111. Y. Chen, Y. Yang, F. Wang, K. Wan, K. Yamane, Y. Zhang and M. Lei, *Proc. Natl. Acad. Sci. U. S. A.*, 2006, **103**, 13956–13961.
112. C. Binda, A. Coda, R. Angelini, R. Federico, P. Ascenzi and A. Mattevi, *Acta Crystallogr. D Biol Crystallogr*, 1998, **54**, 1429–1431.
113. C. Binda, A. Coda, R. Angelini, R. Federico, P. Ascenzi and A. Mattevi, *Structure*, 1999, **7**, 265–276.

114. C. Binda, R. Angelini, R. Federico, P. Ascenzi and A. Mattevi, *Biochemistry*, 2001, **40**, 2766–2776.
115. Henderson M. Pozzi, V. Gawandi and P. F. Fitzpatrick, *Biochemistry*, 2009, **48**, 1508–1516.
116. Henderson M. Pozzi and P. F. Fitzpatrick, *Arch. Biochem. Biophys.* 2010, **498**, 83–88.
117. F. Polticelli, J. Basran, C. Faso, A. Cona, G. Minervini, R. Angelini, R. Federico, N. S. Scrutton and P. Tavladoraki, *Biochemistry*, 2005, **44**, 16108–16120.
118. P. Stavropoulos, G. Blobel and A. Hoelz, *Nat. Struct. Mol. Biol.*, 2006, **13**, 626–632.
119. M. S. Adachi, P. R. Juarez and P. F. Fitzpatrick, *Biochemistry*, 2010, **49**, 386–392.
120. K. Takao, S. Shibata, T. Ozawa, M. Wada, Y. Sugitia, K. Samejima and A. Shirahata, *Amino Acids*, 2009, **37**, 401–405.
121. A. Bellelli, R. Angelini, M. Laurenzi and R. Federico, *Arch. Biochem. Biophys.*, 1997, **343**, 146–148.
122. A. Cona, F. Manetti, R. Leone, F. Corelli, P. Tavladoraki, F. Polticelli and M. Botta, *Biochemistry*, 2004, **43**, 3426–3435.
123. M. Royo and P. F. Fitzpatrick, *Biochemistry*, 2005, **44**, 7079–7084.
124. M. H. Pozzi, V. Gawandi and P. F. Fitzpatrick, *Biochemistry*, 2009, **48**, 12305–12313.
125. T. Wu, K. Q. Ling, L. M. Sayre and W. S. McIntire, *Biochem. Biophys. Res. Commun.*, 2005, **326**, 483–490.
126. M. Bianchi, F. Polticelli, P. Ascenzi, M. Botta, R. Federico, P. Mariottini and A. Cona, *FEBS J.*, 2006, **273**, 1115–1123.
127. S. Muller, R. M. Wittich and R. D. Walter, *Adv. Exp. Med. Biol.*, 1988, **250**, 737–743.
128. C. J. Bacchi, D. Rattendi, E. Faciane, N. Yarlett, L. M. Weiss, B. Frydman, P. Woster, B. Wei, L. J. Marton and M. Wittner, *Microbiology*, 2004, **150**, 1215–1224.
129. N. Yarlett, G. Wu, W. R. Waters, J. A. Harp, M. J. Wannemuehler, M. Morada, D. Athanasopoulos, M. P. Martinez, S. J. Upton, L. J. Marton and B. J. Frydman, *Mol. Biochem. Parasitol.*, 2007, **152**, 170–180.
130. N. Yarlett, M. P. Martinez, B. Goldberg, D. L. Kramer and C. W. Porter, *Microbiology*, 2000, **146**, 2715–2722.
131. S. Muller and R. D. Walter, *Biochem. J.*, 1992, **283**, 75–80.
132. S. Majumder and F. Kierszenbaum, *Mol. Biochem. Parasitol.*, 1993, **60**, 231–239.
133. N. Yarlett and C. J. Bacchi, *Biochem. Soc. Trans.*, 1994, **22**, 875–879.
134. W. R. Waters, B. Frydman, L. J. Marton, A. Valasinas, V. K. Reddy, J. A. Harp, M. J. Wannemuehler and N. Yarlett, *Antimicrob. Agents Chemother.*, 2000, **44**, 2891–2894.
135. N. Yarlett, W. R. Waters, J. A. Harp, M. J. Wannemuehler, M. Morada, J. Bellcastro, S. J. Upton, L. J. Marton and B. J. Frydman, *Antimicrob. Agents Chemother.*, 2007, **51**, 1234–1239.

136. A. E. Pegg, *Cancer Res.*, 1988, **48**, 759–774.
137. R. A. J. Casero, B. Go, H. W. Theiss, J. Smith, S. B. Baylin and G. D. Luk, *Cancer Res.*, 1987, **47**, 3964–3967.
138. C. W. Porter and J. R. Sufrin, *Anticancer Res.*, 1986, **6**, 525–542.
139. C. W. Porter and R. J. Bergeron, *Adv. Enzyme Regul.*, 1988, **27**, 57–79.
140. R. J. Bergeron, A. H. Neims, J. S. McManis, T. R. Hawthorne, J. R. Vinson, R. Bortell and M. J. Ingeno, *J. Med. Chem.*, 1988, **31**, 1183–1190.
141. C. W. Porter, J. McManis, R. A. Casero and R. J. Bergeron, *Cancer Res.*, 1987, **47**, 2821–2825.
142. C. W. Porter, P. F. J. Cavanaugh, N. Stolowich, B. Ganis, E. Kelly and R. J. Bergeron, *Cancer Res.*, 1985, **45**, 2050–2057.
143. M. L. Edwards, N. J. Prakash, D. M. Stemerick, S. P. Sunkara, A. J. Bitonti, G. F. Davis, J. A. Dumont and P. Bey, *J. Med. Chem.*, 1990, **33**, 1369–1375.
144. T. L. Bowlin, N. J. Prakash, M. L. Edwards and A. Sjoerdsma, *Cancer Res.*, 1991, **51**, 62–66.
145. N. E. Davidson, A. R. Mank, L. J. Prestigiacomo, R. J. Bergeron and R. A. Casero, Jr, *Cancer Res.*, 1993, **53**, 2071–2075.
146. C. W. Porter, R. J. Bernacki, J. Miller and R. J. Bergeron, *Cancer Res.*, 1993, **53**, 581–586.
147. M. E. Dolan, M. J. Fleig, B. G. Feuerstein, H. S. Basu, G. D. Luk, R. A. J. Casero and L. J. Marton, *Cancer Res.*, 1994, **54**, 4698–4702.
148. D. E. McCloskey, R. A. Casero, Jr, P. M. Woster and N. E. Davidson, *Cancer Res.*, 1995, **55**, 3233–3236.
149. M. Huber and R. Poulin, *Cancer Res.*, 1995, **55**, 934–943.
150. C. Hegardt, O. T. Johannsson and S. M. Oredsson, *Eur. J. Biochem.*, 2002, **269**, 1033–1039.
151. A. E. Pegg, B. G. Erwin and L. Persson, *Biochim. Biophys. Acta*, 1985, **842**, 111–118.
152. B. G. Erwin and A. E. Pegg, *Biochem. J.*, 1986, **238**, 581–587.
153. A. E. Pegg, R. Pakala and R. J. Bergeron, *Biochem. J.*, 1990, **267**, 331–338.
154. N. H. Saab, E. E. West, N. C. Bieszk, C. V. Preuss, A. R. Mank, R. A. J. Casero and P. M. Woster, *J. Med. Chem.*, 1993, **36**, 2998–3004.
155. R. A. J. Casero, A. R. Mank, N. H. Saab, R. Wu, W. J. Dyer and P. M. Woster, *Cancer Chemother. Pharmacol.*, 1995, **36**, 69–74.
156. W. Devereux, Y. Wang, T. M. Stewart, A. Hacker, R. Smith, B. Frydman, A. L. Valasinas, V. K. Reddy, L. J. Marton, T. D. Ward, P. M. Woster and R. A. Casero, *Cancer Chemother. Pharmacol.*, 2003, **52**, 383–390.
157. Y. Wang, A. Hacker, T. Murray-Stewart, J. G. Fleischer, P. M. Woster and R. A. Casero, Jr, *Biochem. J.*, 2005, **386**, 543–547.
158. A. Pledgie, Y. Huang, A. Hacker, Z. Zhang, P. M. Woster, N. E. Davidson and R. A. J. Casero, *J. Biol. Chem.*, 2005, **280**, 39843–39851.
159. R. Jiang, W. Choi, A. Khan, K. Hess, E. W. Gerner, R. A. J. Casero, W. K. Yung, S. R. Hamilton and W. Zhang, *Int. J. Oncol.*, 2007, **31**, 431–440.

160. A. Hacker, L. J. Marton, M. Sobolewski and R. A. J. Casero, *Cancer Chemother. Pharmacol.*, 2008, **63**, 45–53.
161. Y. Huang, E. R. Hager, D. L. Phillips, V. R. Dunn, A. Hacker, B. Frydman, J. A. Kink, A. L. Valasinas, V. K. Reddy, L. J. Marton, R. A. J. Casero and N. E. Davidson, *Clin Cancer Res.*, 2003, **9**, 2769–2777.
162. E. Gabrielson, E. Tully, A. Hacker, A. E. Pegg, N. E. Davidson and R. A. J. Casero, *Cancer Chemother. Pharmacol.*, 2004, **54**, 122–126.
163. R. F. Minchin, S. Knight, A. Arulpragasam and M. Fogel-Petrovic, *Int. J. Cancer*, 2006, **118**, 509–512.
164. R. A. Casero and P. M. Woster, *J. Med. Chem.* 2009.
165. R. A. J. Casero, A. R. Mank, L. Xiao, J. Smith, R. J. Bergeron and P. Celano, *Cancer Res.*, 1992, **52**, 5359–5363.
166. C. W. Porter, B. Ganis, P. R. Libby and R. J. Bergeron, *Cancer Res.*, 1991, **51**, 3715–3720.
167. N. W. Shappell, J. T. Miller, R. J. Bergeron and C. W. Porter, *Anticancer Res.*, 1992, **12**, 1083–1089.
168. J. Yang, L. Xiao, K. A. Berkey, P. A. Tamez, J. K. Coward and R. A. J. Casero, *J. Cell. Physiol.*, 1995, **165**, 71–76.
169. H. C. Ha, P. M. Woster, J. D. Yager and R. A. J. Casero, *Proc. Natl. Acad. Sci. U. S. A.*, 1997, **94**, 11557–11562.
170. Y. Chen, D. L. Kramer, P. Diegelman, S. Vujcic and C. W. Porter, *Cancer Res.*, 2001, **61**, 6437–6444.
171. Stanic', I., A. Facchini, R. M. Borzi, C. Stefanelli and F. Flamigni, *J. Cell. Physiol.*, 2009, **219**, 109–116.
172. K. Niiranen, M. Pietila, T. J. Pirttila, A. Jarvinen, M. Halmekyto, V. P. Korhonen, T. A. Keinanen, L. Alhonen and J. Janne, *J. Biol. Chem.*, 2002, **277**, 25323–25328.
173. L. Alhonen, A. Karppinen, M. Uusi-Oukari, S. Vujcic, V. P. Korhonen, M. Halmekyto, D. L. Kramer, R. Hines, J. Janne and C. W. Porter, *J. Biol. Chem.*, 1998, **273**, 1964–1969.
174. G. Marverti, S. Bettuzzi, S. Astancolle, C. Pinna, M. G. Monti and M. S. Moruzzi, *Eur. J. Cancer*, 2001, **37**, 281–289.
175. D. E. McCloskey and A. E. Pegg, *J. Biol. Chem.*, 2003, **278**, 13881–13887.
176. Y. Chen, D. L. Kramer, J. Jell, S. Vujcic and C. W. Porter, *Mol. Pharmacol*, 2003, **64**, 1153–1159.
177. T. Murray-Stewart, N. B. Applegren, W. Devereux, A. Hacker, R. Smith, Y. Wang and R. A. J. Casero, *Biochem. J.*, 2003, **373**, 629–634.
178. E. W. Gabrielson, A. E. Pegg and R. A. J. Casero, *Clin Cancer Res.*, 1999, **5**, 1638–1641.
179. A. Singh, V. Misra, R. K. Thimmulappa, H. Lee, S. Ames, M. O. Hoque, J. G. Herman, S. B. Baylin, D. Sidransky, E. Gabrielson, M. V. Brock and S. Biswal, *PLoS Med.*, 2006, **3**, e420.
180. R. A. Casero, Jr, Y. Wang, T. M. Stewart, W. Devereux, A. Hacker, R. Smith and P. M. Woster, *Biochem. Soc. Trans.*, 2003, **31**, 361–365.
181. H. A. Hahm, D. S. Ettinger, K. Bowling, B. Hoker, T. L. Chen, Y. Zabelina and R. A. J. Casero, *Clin Cancer Res.*, 2002, **8**, 684–690.

182. A. C. Wolff, D. K. Armstrong, J. H. Fetting, M. K. Carducci, C. D. Riley, J. F. Bender, R. A. J. Casero and N. E. Davidson, *Clin Cancer Res.*, 2003, **9**, 5922–5928.

183. S. Hector, C. W. Porter, D. L. Kramer, K. Clark, J. Prey, N. Kisiel, P. Diegelman, Y. Chen and L. Pendyala, *Mol. Cancer Ther.*, 2004, **3**, 813–822.

184. S. Hector, R. Tummala, N. D. Kisiel, P. Diegelman, S. Vujcic, K. Clark, M. Fakih, D. L. Kramer, C. W. Porter and L. Pendyala, *Cancer Chemother. Pharmacol.*, 2008, **62**, 517–527.

185. K. Dredge, J. A. Kink, R. M. Johnson, I. Bytheway and L. J. Marton, *Cancer Chemother. Pharmacol.*, 2009.

186. R. Tummala, P. Diegelman, S. M. Fiuza, L. A. Batista de Carvalho, M. P. Marques, D. L. Kramer, K. Clark, S. Vujcic, C. W. Porter and L. Pendyala, *Oncol. Rep.*, 2010, **24**, 15–24.

187. A. Pledgie-Tracy, M. Billam, A. Hacker, M. D. Sobolewski, P. M. Woster, Z. Zhang, R. A. Casero and N. E. Davidson, *Cancer Chemother. Pharmacol.*, 2010, **65**, 1067–1081.

188. S. S. Qutob, D. Proulx, F. M. Mesak and C. E. Ng, *Radiat. Res.*, 2005, **163**, 357–363.

189. W. L. Allen, E. G. McLean, J. Boyer, A. McCulla, P. M. Wilson, V. Coyle, D. B. Longley, R. A. J. Casero and P. G. Johnston, *Mol. Cancer Ther.*, 2007, **6**, 128–137.

190. R. Tummala, P. Diegelman, S. Hector, D. L. Kramer, K. Clark, P. Zagst, G. Fetterly, C. W. Porter and L. Pendyala, *Cancer Chemother. Pharmacol.*, 2010.

191. L. M. Coussens and Z. Werb, *Nature*, 2002, **420**, 860–867.

192. B. B. Aggarwal, S. Shishodia, S. K. Sandur, M. K. Pandey and G. Sethi, *Biochem. Pharmacol.*, 2006, **72**, 1605–1621.

193. R. Franco, O. Schoneveld, A. G. Georgakilas and M. I. Panayiotidis, *Cancer Lett.*, 2008, **266**, 6–11.

194. R. S. Esworthy, R. Aranda, M. G. Martin, J. H. Doroshow, S. W. Binder and F. F. Chu, *Am. J. Physiol. Gastrointest. Liver Physiol*, 2001, **281**, G848–G855.

195. R. S. Esworthy, S. W. Binder, J. H. Doroshow and F. F. Chu, *Biol. Chem.*, 2003, **384**, 597–607.

196. F. F. Chu, R. S. Esworthy, P. G. Chu, J. A. Longmate, M. M. Huycke, S. Wilczynski and J. H. Doroshow, *Cancer Res.*, 2004, **64**, 962–968.

197. J. G. Kusters, A. H. van Vliet and E. J. Kuipers, *Clin. Microbiol. Rev.*, 2006, **19**, 449–490.

198. R. Chaturvedi, Y. Cheng, M. Asim, F. I. Bussiere, H. Xu, A. P. Gobert, A. Hacker, R. A. J. Casero and K. T. Wilson, *J. Biol. Chem.*, 2004, **279**, 40161–40173.

199. N. Babbar and R. A. J. Casero, *Cancer Res.*, 2006, **66**, 11125–11130.

200. N. Babbar, A. Hacker, Y. Huang and R. A. J. Casero, *J. Biol. Chem.*, 2006, **281**, 24182–24192.

201. A. E. Pegg, *IUBMB Life*, 2009, **61**, 880–894.

202. C. L. Sears, *Clin. Microbiol. Rev.*, 2009, **22**, 349–369.
203. H. M. Wexler, *Clin. Microbiol. Rev.*, 2007, **20**, 593–621.
204. E. C. Wick and C. L. Sears, *Curr. Opin. Infect. Dis.*, 2010, **23**, 470–474.
205. V. Nakano, D. A. Gomes, R. M. Arantes, J. R. Nicoli and M. J. Avila-Campos, *Curr. Microbiol.*, 2006, **53**, 113–117.
206. S. Rabizadeh, K. J. Rhee, S. Wu, D. Huso, C. M. Gan, J. E. Golub, X. Wu, M. Zhang and C. L. Sears, *Inflamm. Bowel Dis.*, 2007, **13**, 1475–1483.
207. K. J. Rhee, S. Wu, X. Wu, D. L. Huso, B. Karim, A. A. Franco, S. Rabizadeh, J. E. Golub, L. E. Mathews, J. Shin, R. B. Sartor, D. Golenbock, A. R. Hamad, C. M. Gan, F. Housseau and C. L. Sears, *Infect. Immun.*, 2009, **77**, 1708–1718.
208. S. Wu, K. J. Rhee, E. Albesiano, S. Rabizadeh, X. Wu, H. R. Yen, D. L. Huso, F. L. Brancati, E. Wick, F. McAllister, F. Housseau, D. M. Pardoll and C. L. Sears, *Nat. Med.*, 2009.
209. H. S. Basu, T. A. Thompson, D. R. Church, C. C. Clower, F. Mehraein-Ghomi, C. A. Amlong, C. T. Martin, P. M. Woster, M. J. Lindstrom and G. Wilding, *Cancer Res.*, 2009, **69**, 7689–7695.
210. H. Xu, R. Chaturvedi, Y. Cheng, F. I. Bussiere, M. Asim, M. D. Yao, D. Potosky, S. J. Meltzer, J. G. Rhee, S. S. Kim, S. F. Moss, A. Hacker, Y. Wang, R. A. Casero, Jr and K. T. Wilson, *Cancer Res.*, 2004, **64**, 8521–8525.
211. S. K. Hong, R. Chaturvedi, M. B. Piazuelo, L. A. Coburn, C. S. Williams, A. G. Delgado, R. A. J. Casero, D. A. Schwartz and K. T. Wilson, *Inflamm. Bowel Dis.*, 2010, **16**, 1557–1566.
212. A. C. Goodwin, S. Jadallah, A. Toubaji, K. Lecksell, J. L. Hicks, J. Kowalski, G. S. Bova, A. M. De Marzo, G. J. Netto and R. A. J. Casero, *Prostate*, 2008, **68**, 766–772.
213. M. Zoli, P. Pedrazzi, I. Zini and L. F. Agnati, *Brain Res. Mol. Brain Res.*, 1996, **38**, 122–134.
214. A. M. Rao, J. F. Hatcher, A. Dogan and R. J. Dempsey, *J. Neurochem.*, 2000, **74**, 1106–1111.
215. Nagesh G. Babu, K. A. Sailor, D. Sun and R. J. Dempsey, *Neurosci. Lett.*, 2001, **300**, 17–20.
216. S. Ivanova, G. I. Botchkina, Y. Al-Abed, M. r. Meistrell, F. Batliwalla, J. M. Dubinsky, C. Iadecola, H. Wang, P. K. Gregersen, J. W. Eaton and K. J. Tracey, *J. Exp. Med.*, 1998, **188**, 327–340.
217. A. Dogan, A. M. Rao, J. Hatcher, V. L. Rao, M. K. Baskaya and R. J. Dempsey, *J. Neurochem.*, 1999, **72**, 765–770.
218. S. Ivanova, F. Batliwalla, J. Mocco, S. Kiss, J. Huang, W. Mack, A. Coon, J. W. Eaton, Y. Al-Abed, P. K. Gregersen, E. Shohami, E. S. J. Connolly and K. J. Tracey, *Proc. Natl. Acad. Sci. U. S. A.*, 2002, **99**, 5579–5584.
219. P. L. Wood, M. A. Khan, J. R. Moskal, K. G. Todd, V. A. Tanay and G. Baker, *Brain Res.*, 2006, **1122**, 184–190.
220. H. Tomitori, T. Usui, N. Saeki, S. Ueda, H. Kase, K. Nishimura, K. Kashiwagi and K. Igarashi, *Stroke*, 2005, **36**, 2609–2613.

221. M. Yoshida, H. Tomitori, Y. Machi, D. Katagiri, S. Ueda, K. Horiguchi, E. Kobayashi, N. Saeki, K. Nishimura, I. Ishii, K. Kashiwagi and K. Igarashi, *Atherosclerosis*, 2009, **203**, 557–562.

222. G. H. Kim, R. J. Komotar, M. E. McCullough-Hicks, M. L. Otten, R. M. Starke, C. P. Kellner, M. C. Garrett, M. B. Merkow, M. Rynkowski, K. A. Dash and S. Connolly, *Can. J. Neurol. Sci.*, 2009, **36**, 14–19.

223. K. Zahedi, F. Huttinger, R. Morrison, T. Murray-Stewart, R. A. Casero and K. I. Strauss, *J. Neurotrauma*, 2010, **27**, 515–525.

224. K. Zahedi, Z. Wang, S. Barone, A. E. Prada, C. N. Kelly, R. A. Casero, N. Yokota, C. W. Porter, H. Rabb and M. Soleimani, *Am. J. Physiol. Renal Physiol.*, 2003, **284**, F1046–F1055.

225. K. Zahedi, A. B. Lentsch, T. Okaya, S. Barone, N. Sakai, D. P. Witte, L. J. Arend, L. Alhonen, J. Jell, J. Janne, C. W. Porter and M. Soleimani, *Am. J. Physiol. Gastrointest. Liver Physiol.*, 2009, **296**, G899–G909.

226. K. Zahedi, S. Barone, D. L. Kramer, H. Amlal, L. Alhonen, J. Janne, C. W. Porter and M. Soleimani, *Am. J. Physiol. Cell Physiol.*, 2010, **299**, C164–C174.

227. Z. Wang, K. Zahedi, S. Barone, K. Tehrani, H. Rabb, K. Matlin, R. A. Casero and M. Soleimani, *J. Am. Soc. Nephrol.*, 2004, **15**, 1844–1852.

228. K. Zahedi, J. J. Bissler, Z. Wang, A. Josyula, L. Lu, P. Diegelman, N. Kisiel, C. W. Porter and M. Soleimani, *Am. J. Physiol. Cell Physiol.*, 2007, **292**, C1204–C1215.

229. Y. J. Zhao, C. Q. Xu, W. H. Zhang, L. Zhang, S. L. Bian, Q. Huang, H. L. Sun, Q. F. Li, Y. Q. Zhang, Y. Tian, R. Wang, B. F. Yang and W. M. Li, *Eur. J. Pharmacol.*, 2007, **562**, 236–246.

230. J. H. Ryu, Y. S. Cho, Y. S. Chun and J. W. Park, *Biochem. Biophys. Res. Commun.*, 2008, **366**, 438–444.

231. L. Han, C. Xu, Y. Guo, H. Li, C. Jiang and Y. Zhao, *Int. J. Cardiol.*, 2009, **132**, 142–144.

232. S. Barone, T. Okaya, S. Rudich, S. Petrovic, K. Tenrani, Z. Wang, K. Zahedi, R. A. Casero, A. B. Lentsch and M. Soleimani, *Am. J. Physiol. Cell Physiol.*, 2005, **289**, C826–35.

233. M. Baudry and I. Najm, *Neurosci. Lett.*, 1994, **171**, 151–154.

234. Y. Hayashi and M. Baudry, *Brain Res. Dev. Brain Res.*, 1995, **87**, 96–99.

235. A. Sequeira, F. G. Gwadry, J. M. Ffrench-Mullen, L. Canetti, Y. Gingras, R. A. J. Casero, G. Rouleau, C. Benkelfat and G. Turecki, *Arch. Gen. Psychiatry*, 2006, **63**, 35–48.

236. T. A. Klempan, D. Rujescu, C. Merette, C. Himmelman, A. Sequeira, L. Canetti, L. M. Fiori, B. Schneider, A. Bureau and G. Turecki, *Am. J. Med. Genet. B Neuropsychiatr Genet*, 2009, **150B**, 934–943.

237. C. R. Bermudo-Soriano, C. Vaquero-Lorenzo, M. Diaz-Hernandez, M. M. Perez-Rodriguez, J. Fernandez-Piqueras, J. Saiz-Ruiz and E. Baca-Garcia, *Prog. Neuropsychopharmacol. Biol. Psychiatry*, 2009, **33**, 345–348.

238. M. Guipponi, S. Deutsch, K. Kohler, N. Perroud, F. Le Gal, M. Vessaz, T. Laforge, B. Petit, F. Jollant, S. Guillaume, P. Baud, P. Courtet, R. La

Harpe and A. Malafosse, *Am. J. Med. Genet. B Neuropsychiatr. Genet,* 2009, **150B**, 799–807.

239. C. Vaquero-Lorenzo, C. Riaza Bermudo-Soriano, M. M. Perez-Rodriguez, M. Diaz-Hernandez, J. Lopez-Castroman, J. Fernandez-Piqueras, J. Saiz-Ruiz and E. Baca-Garcia, *Am. J. Med. Genet. B Neuropsychiatr. Genet.,* 2009, **150B**, 515–519.

240. M. S. Rogers, S. F. Yim, K. C. Li, C. C. Wang and M. Arumanayagam, *Gynecol. Oncol.,* 2002, **84**, 383–387.

241. G. S. Lee, Y. S. Joe, S. J. Kim and J. C. Shin, *Arch. Gynecol. Obstet.,* 2009.

242. A. Celik, S. Okada, S. Tsujinaka, K. Soda and F. Konishi, *Int. Surg.,* 2010, **95**, 1–7.

243. G. Gimelli, S. Giglio, O. Zuffardi, L. Alhonen, S. Suppola, R. Cusano, C. Lo Nigro, R. Gatti, R. Ravazzolo and M. Seri, *Hum. Genet.,* 2002, **111**, 235–241.

244. K. Mita, K. Fukuchi, K. Hamana, S. Ichimura and M. Nenoi, *Int. J. Radiat. Biol.,* 2004, **80**, 369–375.

245. M. Bianchi, A. Bellini, M. Cervelli, P. Degan, L. Marcocci, F. Martini, M. Scatteia, P. Mariottini and R. Amendola, *Biochim. Biophys. Acta,* 2007, **1773**, 774–783.

246. A. Goodwin, S. Wu, D. Huso, X. Wu, J. Hicks, C. Destefano Shields, A. Hacker-Prietz, S. Rabizadeh, A. De Marzo, C. Sears and R. Casero, Jr, *Cancer Prev. Res.,* 2010, **3**, A56.

CHAPTER 7

Design of Polyamine Transport Inhibitors as Therapeutics

OTTO PHANSTIEL IV* AND JENNIFER JULIAN ARCHER

College of Medicine, Health Sciences Campus at Lake Nona, University of Central Florida, 6850 Lake Nona Blvd., Orlando, FL 32827-7408, USA

7.1 Introduction

High polyamine transport activity and upregulated polyamine biosynthesis are hallmarks of aggressive cancers. There is a direct relationship between high intracellular polyamine content and cell proliferation.[1,2] Indeed, the gene coding for the key polyamine biosynthetic enzyme, ornthine decarboxylase (ODC), is an oncogene.[3] Efforts to control proliferation by lowering intracellular polyamine levels have met with limited success. For example, α-difluoromethylornithine (DFMO) was developed as an irreversible inhibitor of ODC and was shown to reduce intracellular polyamine levels and inhibit cell growth, and was well tolerated in humans at optimized dosages.[4–6] However, an unforeseen compensatory mechanism arose during DFMO therapy. Cells were shown to compensate *via* upregulation of polyamine transport. As a result, therapies involving DFMO alone were disappointing in the clinic.[7,8]

Therefore, the development of polyamine transport inhibitors (PTIs) is primarily an attempt to block this compensatory mechanism and allow DFMO to exert its maximal effect and induce sustained polyamine depletion. Indeed, a combination therapy involving DFMO and a PTI is warranted based upon DFMO's high tolerability in humans. Potent PTIs, which are selective for the

RSC Drug Discovery Series No. 17
Polyamine Drug Discovery
Edited by Patrick M. Woster and Robert A. Casero, Jr.
© Royal Society of Chemistry 2012
Published by the Royal Society of Chemistry, www.rsc.org

polyamine transporter (PAT), are required for this approach to be successful. This is a particular challenge to researchers, as the genes and proteins associated with the polyamine transport system in mammals were, until recently, virtually unknown. This review will summarize the attempts to date to develop these novel inhibitors.

7.2   Models of Polyamine Transport

What is known about the mammalian PAT has been reviewed in detail.[9] In brief, PAT is an energy-dependent, carrier-mediated, active transport system, which is phenotypically observable. This process is not limited to mammals, and most of our information on polyamine transport is gleaned from studies in other organisms (*E. coli, D. melanogaster, C. elegans*).[10–12] In terms of mammalian polyamine transport, two models have been proposed by the Belting[13,14] and Poulin[15] groups, respectively.

Belting's model has been supported by other investigators[16,17] and involves a multi-step endocytic process.[14] In this system, polyamines first bind to high-affinity sites on heparan sulfate (HS) proteoglycans in caveolae, where glypican-1 has been identified as a vehicle. The caveolae then become endocytosed, where processing of the HS chains occurs by heparanases.[14] Further processing by NO released from SNO groups on glypican-1 induces cleavage at GlcNH3$^+$ sites liberating the polyamines into the aqueous portion of the compartment.[14] Cell-sorting events then direct the continued trafficking of polyamines (Figure 7.1) *via* unknown pathways.

In contrast, Poulin's group suggested that polyamines enter the cell *via* a simple active transport system and are then sequestered by polyamine sequestering vesicles (PSVs).[15] These PSVs were found to colocalize with acidic vesicles of the late endocytic compartment and the *trans* Golgi. The data indicated that polyamine accumulation is primarily driven by the activity of a vesicular polyamine carrier, which facilitates escape from the PSV.[15]

One caveat is that this latter model is predicated upon evidence gained from fluorescent polyamine probes, which may traffic differently than native polyamines. For example, in our laboratory, an appended fluorophore converted a regularly transported polyamine, *i.e.* spermine, into a poorly transported substrate, *i.e.* N^1-anthracenylmethyl-spermine (Ant-343), which was later shown to be a potent transport inhibitor.[18] This is an important nuance, as

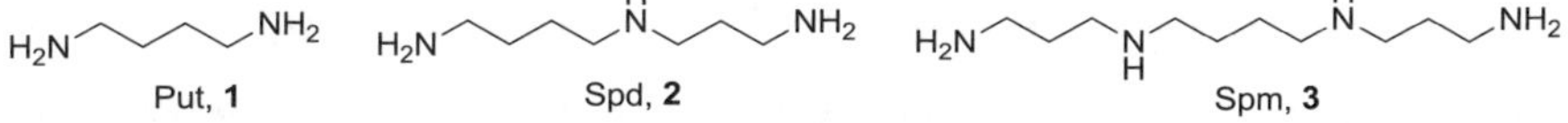

Figure 7.1   Native polyamines **1–3**. Mammalian cells can synthesize putrescine (Put) from ornithine via ODC. The higher polyamines, spermidine (Spd) and spermine (Spm), are made by successive aminopropylation steps using spermidine synthase and spermine synthase, respectively.

Table 7.1 Differential affinities and rates of transport of the native polyamines.

Cell Line	Kinetics	Put	Spd	Spm
DRO	K_m (μM)	1.7 ± 0.3	0.5 ± 0.1	0.4 ± 0.1
	V_{max} (μM)	28 ± 7	6 ± 2	10 ± 3
ZR-75-1	K_m (μM)	3.7 ± 0.4	0.49 ± 0.15	0.2 ± 0.06
	V_{max} (μM)	ND	ND	ND
T47-D	K_m (μM)	7.4 ± 1.1	4.4 ± 0.9	2.7 ± 0.4
	V_{max} (μM)			
MDA-MB-231	K_m (μM)	1.28 ± 0.03	0.113 ± 0.03	0.28 ± 0.06
	V_{max} (μM)			
L1210	K_m (μM)	8.5 ± 0.9	2.2 ± 0.3	1.5 ± 0.4
	V_{max} (μM)	27.3 ± 10.8	140.5 ± 47.7	65.6 ± 16.4

other fluorescent polyamines have found a use in screening cancer cells for their respective PAT activity.[19]

It is important to realize that the Belting and Poulin models may not be mutually exclusive and may work cooperatively or as dual import systems. Indeed, claims of multiple PATs in cells by several groups have been reported. However, the data are inconclusive. For example, Aziz suggested that there are two PATs operative in bovine pulmonary artery smooth muscle cells (PASMCs), a selective PAT (which allows only Spd and Spm entry) and a non-selective transporter, which permits all native polyamines entry (Put, Spd and Spm).[20] Interpretation of transport behavior is often confounded by the fact that typically putrescine has a higher K_m value than either spermidine or spermine for PAT. As shown in Table 7.1, the higher polyamines (Spd and Spm) typically have lower K_m values than putrescine, which indicates a higher binding affinity for the putative cell surface receptor. Thus, kinetically it may appear that there are two separate transporters, when in fact the differences in transport can often be explained simply by the different binding affinities of the three native polyamines for PAT.

DFMO was shown to enhance polyamine transport by increasing V_{max} without having a dramatic effect on K_m in Ehrlich tumor cells.[21] This suggests that polyamine transport can be modulated without changing the overall affinity of PAT for its ligands. This is consistent with Belting's observation where a combination of enhanced HS processing activity and increased expression of the high-spermine-affinity binding sites on HS would account for these parameters.[22,23]

7.2.1 Homeostasis, Antizyme and Polyamine Transport

Any discussion of polyamine transport must consider its role in contributing to polyamine homeostasis, which is accomplished by balancing polyamine bio-synthesis, transport and polyamine degradation. This is important, as there are intracellular controls which may help or hinder the action of a transport inhibitor. For example, antizyme (AZ) plays a key role in regulating polyamine transport. As polyamine levels rise within cells, AZ is induced, and both polyamine biosynthesis *via* ODC and import *via* PAT are inhibited.[24,25] Using glutamine synthetase-Chinese hamster ovary (GS-CHO) cells and an inducible AZ system, Mitchell *et al.* showed that spermidine uptake kinetics were altered after induction of unmodified full length AZ. Specifically, there was a 90% decrease in V_{max} and a doubling of the K_m from 0.84 to 1.7 µM upon AZ induction.[24] Polyamine compounds which activate AZ will limit their own import into cells.[24] An important corollary of these findings is that AZ-inducing agents may make excellent PTI agents.

Degradation and consumption of polyamines by the cell can play a role in regulating transport activity as well. Polyamine consumption refers to metabolic alterations to the polyamine architecture, which modify its covalent structure and remove it from the available polyamine pools. This includes N-acetylation of Spd and Spm by spermidine/spermine acetyl transferase (SSAT)[26,27] followed by oxidation and/or cellular export as well as the catabolism of spermidine by deoxyhypusine synthase (DOHS) to form hypusine.[28,29] These net removal steps can each theoretically effect polyamine import, which could offset their loss and help maintain charge balance.

In summary, cells use polyamine transport to maintain polyamine homeostasis. PTIs offer new chemical tools to block import and thereby disrupt homeostasis. This property is key to the success of combination therapies involving a PTI and DFMO, which aim to induce sustained polyamine depletion.

7.3   Transportons and Anti-Transportons, New Words and Definitions

In writing this review, it became apparent that there was no suitable word in the literature that described a molecule that was transported into a cell *via* a transport system. Moreover, there was no direct way to refer to a substance that bound and did not enter a cell versus one that successfully navigated into the cell. Most literature refers to the transported entity *via* an adjective such as the *transported* solute. The terms 'ligand' and 'substrate' did not define the transportability of the substance. Moreover, no formal enzymatic product is generated from the 'substrate' as a result of transport. The term 'transport agonist' was entertained but discarded, as it is unknown whether polyamines and their analogs signal their own uptake. Note that there is some intriguing evidence suggesting that indeed bound polyamines may induce signaling, as will be discussed later in this review.

In attempt to clarify this issue, we propose the following new terms:

- Transporton: any substance that is transported from one location to another *via* a transport system. Note that in biological systems, transportons typically induce a cellular response when transported.
- Anti-transporton: any substance that interacts with the transport system but shows no net transport *via* that transport system. For example, a

polyamine may bind to an extracellular receptor like heparan sulfate proteoglycans (HSPGs), become internalized *via* endocytosis but then be recycled back out to the cell surface. This example will result in zero net import of the polyamine. A polyamine with transport behavior fitting this description would be deemed an anti-transporton.

A caveat in this nomenclature is that most PTI molecules will likely lie along a continuum of activities between these absolute definitions. For example, a PTI may be partially transported into the cell, but the bulk of it remains stuck to the cell surface and vice versa. This is further confounded because the current tools cannot differentiate between substances, which are tight-surface binders versus those that enter the cell. Tight binders that cannot be washed off the cell surface would be counted as imported materials in the current HPLC methods. Indeed, a microscopy study of anthrylmethyl-polyamines showed that an *N*-(anthrylmethyl)-tetraamine conjugate remained stuck to the cell surface after numerous washings and was only moderated imported, whereas the related anthrylmethyl-triamine was effectively washed off the cell surface and showed enhanced trafficking into the cell.[18]

Since both transportons and anti-transportons will be competitive inhibitors of native polyamine import, the term 'transport inhibitor' was generally insufficient to describe the differences in their modes of action. As our knowledge of the trafficking of these agents increases, these new terms will help to define their modes of action.

7.4   Role of Polyamine Transportons

Ironically, PTIs evolved from early studies of polyamine mimicry and poly-amine metabolism. Polyamine metabolism was once a poorly understood regulatory system in mammals. This prompted chemists to begin synthesizing polyamine mimics as putative transportons of the PAT and as potential modulators of polyamine metabolism. The problem was that none of the structural tolerances of the transport system were known at the time. Over time, tedious structure–activity relationships were developed from studies of homologous series of polyamine analogs which provided essential insights into the molecular design of efficient PAT transportons. The synthesis of these homologous series also required the development of new synthetic approaches to polyamine backbones.[30]

In the early 1980s, Porter and Bergeron used a homologous series of *N*-alkylated polyamines to correlate the role of charge and tether length of polyamine derivatives with their ability to target the PAT (Figure 7.2).[31] Certain polyamine carbon sequences correlated with increased toxicity in L1210 cells, presumably due to the differential import of specific polyamine sequences.[31] From this group came a series of diethylated conjugates, DENSPM **4** and DESPM **5**. These molecules were recognized by the polyamine synthetic, catabolic and transport machinery. They, however, could not substitute for the biological function of the native polyamines and caused a cytotoxic response. Downregulation in ODC and *S*-adenosylmethionine decarboxylase (Ado-MetDC) activity and the induction of SSAT activity were observed. In general, imported polyamine analogs can act as antimetabolites, which deplete the intracellular polyamines, or as mimics which displace the natural polyamines from their binding sites, but do not provide a growth-promoting function.[32] These outcomes and similar findings with **6** have been described in more detail.[33]

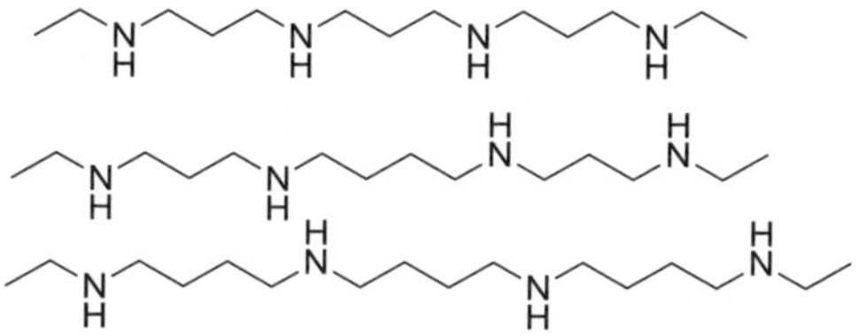

Figure 7.2 *N*-Ethylated polyamines **4–6**.

These intriguing findings were rationalized as follows. First, import of polyamine mimics was interpreted by the cell as an influx of native polyamines. The cells responded by attempting to re-establish polyamine homeostasis to address the increased polyamine content. The result was a reduction in poly-amine biosynthesis, increased catabolism (*via* SSAT and polyamine oxidase, PAO) and downregulation of PAT.[32] As a result, cell death occurred due to complete polyamine depletion, including spermine. These N-ethylated poly-amine mimics were also able to compete with native polyamines for the transporter, which confirmed that they utilized the PAT for cellular entry.[34] These were considered to be competitive PTIs. Here, we further classify these molecules as PAT transportons because they readily enter cells *via* PAT.

While polyamine transportons are clearly important and may have direct clinical applications, they are not the subject of this review. They are mentioned briefly here, as they played an important role in developing and optimizing the following polyamine anti-transportons.

7.5 Anti-Transportons

As a result of these efforts to understand the structural preferences of the PAT, several anti-transportons were discovered. Therefore, discussions of how PTIs were developed often include the mutual story of selective PAT transportons and anti-transportons.

7.5.1 Polypyridinium Quaternary Salts

The first evidence of anti-transportons is a group of quaternary polypyridinium salts developed by Minchin *et al.* in 1989, which do not technically contain polyamines.[35] This study stemmed from the observation that an increased import of polyamines was accompanied by an increase in the accumulation of the anticancer agent, MGBG **10**.[35] This suggested that non-polyamine structures could target the PAT. Unlike primary and secondary amines found in linear polyamine analogs, where the protonation state and charge are pH-dependent, pyridinium quaternary salts are fully charged, and their charges are virtually independent of environmental conditions. The planar nature of these analogs causes them to be more rigid than the native polyamines, and they represent designs with fixed charges at discrete spaces. As such, these structures provided a unique opportunity to evaluate the effect of charge on binding to PAT.

The K_i values of polypyridium molecules **7–9** in B-16 murine cells were 54.4 ± 10.6, 0.82 ± 0.09 and 0.15 ± 0.02 μM, respectively (Figure 7.3). These polypyridinium salts were able to competitively block [^{14}C] radiolabeled putrescine from entering PAT-active B16 murine melanoma cells in the presence of 1 mM DFMO. B16 cells are known to have low PAT activity on their own, but in the presence of DFMO, PAT activity is highly upregulated.[35] In agreement with Bergeron's findings,[36] these authors suggested that linearly spaced charges, rather than charged polyamines *per se*, are the requirement for binding to PAT.[35] As such, electrostatic interactions, rather than hydrogen bonding, are the key to binding to PAT. Upon binding of the diquaternary salt **7** to PAT, an increased apparent K_m for putrescine was observed, but there was no significant effect on the V_{max}. As such, putrescine uptake was strictly competitive.

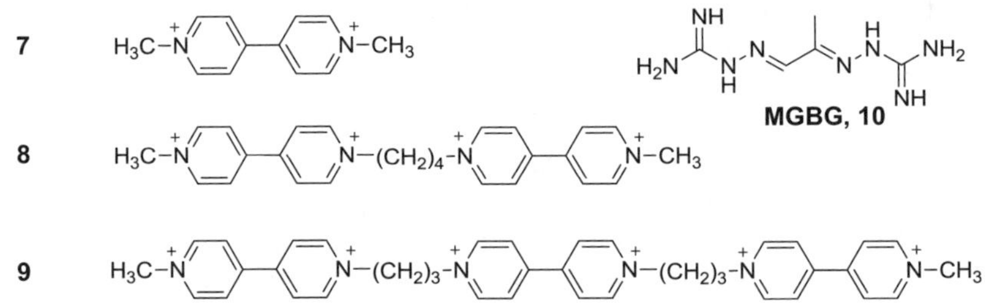

Figure 7.3 Polypyridinium quaternary salts **7–9** and methylglyoxal-bis(guanylhydrazone) (MGBG, **10**).

The series of di-, tetra- and hexapyridinium quaternary salts allowed for the evaluation of increasing linear charges as a function of binding to PAT. As charge increased, the binding affinity increased as evident from the decrease in K_i values obtained going from di- to a tetra- to a hexapyridinium quaternary salts (**7–9**, Figure 7.3). Thus, the apparent trend was that K_i is inversely proportional to charge when it comes to binding to PAT.

In addition to the number of charges, spacing between the charges also plays an important role in PAT binding. Using a series of diaminoalkanes, these authors showed two peaks of activity corresponding to inter-nitrogen distances of 0.6–0.7 nm and 1.0–1.1 nm. These are consistent with earlier findings by Porter *et al.*, who demonstrated the role of charge and distance in PAT binding, with the optimal distance being similar to that separating the N^1 and N^8 atoms of spermidine.[31]

It is still unclear, however, if **7–9** can utilize PAT. The authors stated that there is 'no evidence to suggest that these compounds are substrates for uptake.'[35] Evidence that they do not utilize PAT to gain cellular entry comes from the fact that these molecules did not become more toxic in the presence of DFMO. DFMO classically increases the toxicity of PAT-internalized substrates as PAT activity increases in response to polyamine depletion.[37] However, these constructs do appear to become internalized *via* an alternative route and were shown to be highly toxic by themselves, and so it is not clear if their intrinsic toxicity masks the effect of DFMO treatment. Assuming the authors are correct, then these are the first reported PAT anti-transportons.

7.5.2 Irreversible and Sulfur-Containing PAT Inhibitors

The concept of utilizing PTIs as a means of solving the DFMO efficacy problem evolved from early attempts to modulate PAT activity. In one of the earliest reported examples of a suicide inhibitor of polyamine transport, Poulin *et al.* demonstrated that 1-ethyl-3-(3-dimethylaminopropyl)carbodi-imide (**EDC, 11**, Figure 7.4) irreversibly inhibited ^{3}H-spermidine transport with an EC_{50} value of 96 ± 16 µM.[38] The native polyamines were all effective (EC_{50} values <10 µM) in protecting against EDC-mediated inactivation of radiolabeled putrescine or spermidine transport.

This group also synthesized MESC **12**, which was designed to be conjugated to a reporter in order to capture and isolate proteins involved in polyamine transport. To the best of our knowledge, the use of MESC and its derivatives for this purpose was never published, but it was discovered that MESC, its synthetic precursor **13** and a side product DEASC **14** were able to inhibit polyamine transport competitively at varying degrees with K_i values in the micromolar range.[39]

MESC had the highest K_i value with all of the polyamines tested and therefore was the least potent PTI. The disulfides **13** and **14** were both more potent than MESC **12**, and only **13** had a lower K_i value (1.6 ± 0.5 µM) than the K_m of putrescine (3.7 ± 0.4 µM). Interestingly, **14** was shown to be a noncompetitive inhibitor of putrescine in addition to being a competitive inhibitor.

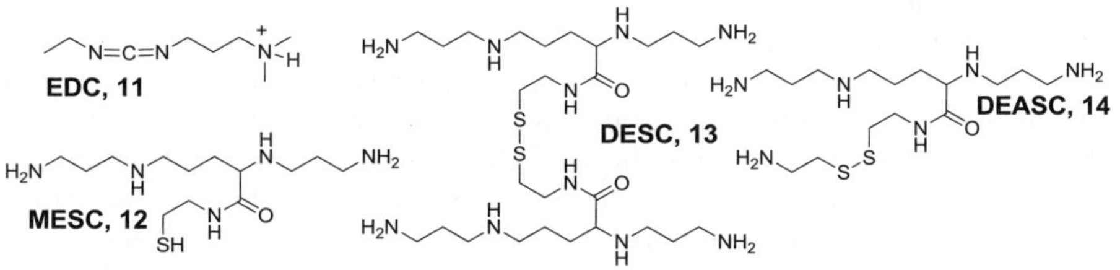

Figure 7.4 Irreversible PTI **11** and sulfur-containing PTI designs **12–14**.

This was the first example of a mixed-mode PTI. In the following year, the transporton MGBG **10** (Figure 7.3) was shown to have mixed inhibition kinetics as well.[40]

These PTIs appear to be true anti-transportons, where only trace amounts of MESC **12** and DESC **13** were found to be internalized at high concentrations (200 µM) as determined by HPLC. Adding cycloheximide (CHX), which is known to upregulate PAT activity *via* repression of AZ, enhanced the uptake of spermidine but did not enhance internalization of either conjugate, further emphasizing that **12** and **13** are true anti-transportons that do not internalize *via* PAT.

In order to be therapeutically relevant, PTIs need to prevent growth of DFMO-treated cancer cells in the presence of native polyamines. In ZR-75-1 cells, DFMO reduced growth to approximately 50% of the control, and as little as 300 nM of spermidine was sufficient to fully restore growth of these cells. The most potent antagonist, DESC **13**, however, was not able to prevent spermidine from restoring growth of these cells, even at a 500-fold molar ratio of DESC to spermidine.

Retrospectively, this was likely due to the fact that **13** did not completely inhibit the uptake of polyamines and instead simply reduces the rate in which they accumulate. DESC **13** (at 200 µM) caused a decrease in the rate of spermidine uptake by 50% after 6 h and still permitted accumulation of the polyamine within the cell. With the addition of CHX, uptake was further reduced to 80–85% of the control. This result is not surprising because the K_m of spermidine is more than 10-fold lower than the respective K_i values of any of the reported conjugates.[39] As will be demonstrated later in this review, a low K_i is necessary, but not sufficient for a PTI alone to be able to reduce cell growth.

These early agents were important as they demonstrated the need to balance PTI toxicity and potency. Indeed, insights gained from the study of these compounds led to a new series of branched polyamine dimers, which attempted to address these issues.

7.5.3 Dimeric Branched Polyamine Motifs

DESC **13** was not very potent as a PTI and was unstable in media containing other thiols, which limited its use in biological systems. Due to the growing interest in PTIs, Poulin's group later modified **13** in the hopes of improving the design. It was thought that because **13** was more potent than **12**, tethering two polyamine chains *via* their internal amine centers may produce more potent inhibitors (see **15** in Figure 7.5). As such, Poulin's group synthesized a series of spermidine and sym-norspermidine dimers connected *via* their respective central nitrogens. The tether was varied (from $R = CH_2CH_2$, ethene, ethyne and phenyl linkages) to assess the effect of the spacer unit on PAT inhibition (Figure 7.5). The K_i values of these inhibitors and K_m values of the native polyamines were determined in T-47D and ZR-75-1 breast-cancer cells.[41]

All of the new constructs synthesized had a better potency than DESC **13** except for those with a 2-butynyl tether.[41] This demonstrated that tethering *via*

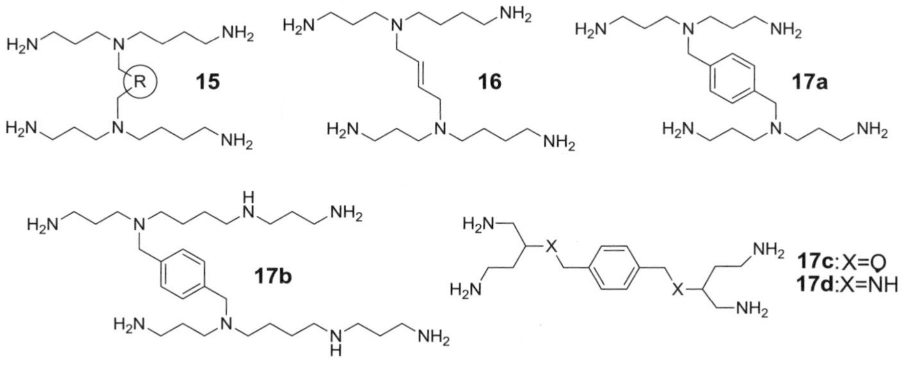

Figure 7.5 Disubstituted systems with branched polyamine PTI designs 15–17.

the secondary amine was not detrimental to recognition by the PAT receptor. The spermidine and sym-norspermidine disubstituted motifs gave similar results. In all cases, the inhibitors were better at preventing putrescine uptake over spermidine and spermine uptake, and gave K_i values similar to K_m values of the higher polyamines (Spd and Spm). The selective inhibition of Put was not surprising because the K_m of Put in this T-47D cell line was higher than either the K_m of Spd or Spm, which were similar in value. Thus, these agents were more effective inhibitors of Put import (than Spd and Spm) because Put has a lower affinity for PAT in this cell line. This trend has been seen in many other cell lines as well.

The high potency observed with the spermidine and sym-norspermidine dimeric designs with the phenyl and ethene crosslinkers should be evaluated further, as only kinetics were measured in this paper, and their influence on growth inhibition was not determined.

A follow-up paper in 2003 determined the K_i values xylylated dimers of putrescine and polyamines (**17a–d** in Figure 7.5) in T-47D human breast cancer cells.[42] Interestingly, **17a** and **17b** had similar K_i values (1.5 ± 0.1 and 1.1 ± 0.1 μM, respectively). Compounds **17c** and **17d** gave similar K_i values (12.8 ± 0.6 and 9.3 ± 1.1 μM, respectively), whereas the K_i of Put was > 500 μM. The authors suggested that the aminopropyl moiety confers an optimal ability to dimeric xylylated polyamines for competing against Spd uptake.[42]

7.5.4 Acridinyl Linear Polyamine Conjugates

While the branched polyamine chains were effective in Poulin's system, Delcros and Blagbrough published their account of a series of linear acridine–polyamine conjugates **18** and **19** and evaluated these agents in CHO and CHO-MG cell lines.[43] In general, the amino-linked analogs **19** were more potent than the corresponding amide-linked agents **18** due to differences in cell uptake. The mono-substituted analogs (Figure 7.6) showed moderate targeting of the PAT-active CHO line over the PAT-defective CHO-MG cell line. Native polyamines

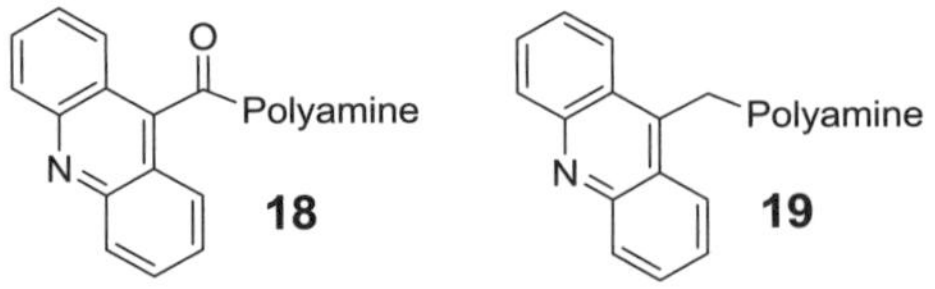

Figure 7.6 Linear polyamine acridine conjugates.

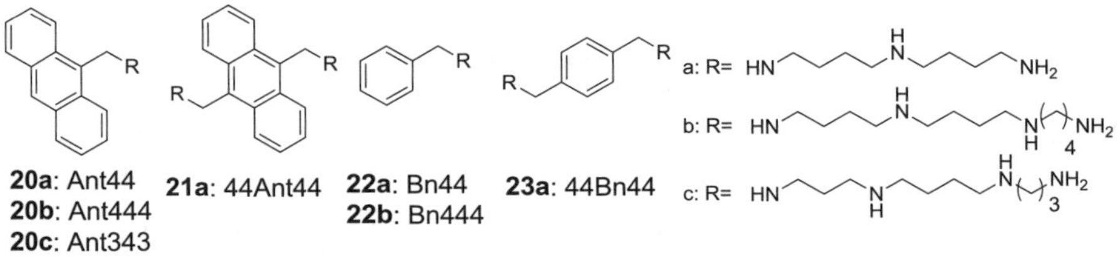

Figure 7.7 Arylmethyl-polyamine derivatives **20–23**.

were shown to rescue cells treated with these analogs, indicating that the analogs used PAT for import. DFMO had a synergistic effect on the potency of these acridine analogs. Despite the high affinity of these agents for the PAT, the authors suggested that particular acridine–tetraamine analogs inhibited their own import and were relatively poor substrates for entering into cells *via* PAT. In this regard, the acridinyl tetraamine conjugates behaved largely as anti-transportons and demonstrated that linear tetraamine motifs were effective PAT binding motifs.

7.5.5 Aryl-Based Anti-Transportons

Our group attempted to design toxic PAT-selective transportons that would permit selective delivery into PAT expressing cells (cancers) preferentially over non-PAT-active cells (normal tissue). In brief, a series of conjugates consisting of triamines, tetraamines and a diamine control appended to an anthracenyl methyl group were synthesized (Figure 7.7). These were then evaluated and ranked *via* CHO-MG/CHO IC$_{50}$ ratio determinations. Their K_i values were determined in L1210 cells for comparison to the historical analogs of Bergeron. This work was recently reviewed.[44] In developing our extensive structure–activity relationships, potent PTIs were also discovered.

By maintaining the anthrylmethyl substituent and varying only the polyamine message, many insights were gained in terms of molecular design. Like the native polyamines, the binding affinity increased (K_i decreased) as the number of charges increased. The K_i values in L1210 indicated the expected trend where diamines (K_i in 10 μM range) bind less tightly than triamines (K_i in low micromolar concentrations, *e.g.* **20a**) and tetraamines (K_i in the low nanomolar

range, *e.g.* **20b**), respectively. All of the anthracenylmethyl–polyamine conjugates were able to bind to the PAT receptor but had varying affinities as
demonstrated by dramatic differences in K_i values in L1210 cells. These K_i trends
are also observed in the corresponding unsubstituted polyamines, suggesting
that the spacing between the charges from the amine nitrogens plays an
important role in PAT receptor recognition. The butylene spacer was shown to
be optimal. This supported the previous findings of Bergeron[31] and Minchon[35]
in their respective studies of polyamine chain length.

As the tetraamine message was varied, the affinity for the transporter was
also modulated, suggesting that charge-spacing optimization could quadruple
the affinity for the transport receptor (**20c**: 202 nM *vs.* **20b**: 51 nM). In a
deconvolution microscopy experiment, the tetraamine Ant-444 **20b** was mostly
observed on the cell surface with a minimal amount internalized after 3 min. In
contrast, Ant-44 **20a**, which was shown to be a potent transporton, was found
to be mostly internalized after 3 min. This study suggested that N^1-anthrace-
nylmethyl-tetraamines poorly entered the cell *via* the PAT and behaved like
true anti-transportons.[18]

In this case, a high affinity for the PAT receptor (*i.e.* K_i in the nanomolar
range) hindered net internalization. In contrast, productive transport was
observed with triamine compounds having a moderate binding affinity (*i.e.* K_i
in the micromolar range), which presumably facilitated eventual release from
the receptor. These results were consistent with those found by Delcros and
Blagbrough in their study of acridine-tetraamines.[43] In short, L1210 K_i values
in the micromolar range were a characteristic of effective transportons, whereas
values in the nanomolar range were associated with anti-transporton
properties.

To further test the hypothesis that the Ant-tetraamines like **20b** were potent
PAT anti-transportons, CHO cells were given the transporton **20a** at its IC_{50}
dose and then competed against a series of nine anthryl-tetraamines (*e.g.* **20b**
and **20c**) dosed at non-toxic concentrations. All of the tetraamines could successfully block the toxicity of **20a** confirming their ability to behave as potent
PTIs (unpublished results). Unfortunately, these anthracenylmethyl–
tetraamine conjugates were very toxic *in vivo* and therefore have limited
application as therapeutics. Derivatives **20b** and **20c**, for example, were found
to be potent antagonists of recombinant *N*-methyl-D-aspartate (NMDA)
receptors and toxic in mice, which limited their therapeutic utility.[45] Nevertheless, these agents suggested that tetraamines are more optimal PTIs and
helped to further define the parameters necessary for successful PTI design.

Recently, the potent transporton **20a** was further improved by appending an
additional homospermidine message in the 10-position of the anthracene ring,
creating a disubstituted system, 44-Ant-44 (**21a**), with a dramatic improvement
in PAT selectivity and transporton delivery as seen in a CHO/CHOMG
screen.[46] This disubstituted molecule demonstrated over a 2200-fold preference
for entering and killing the PAT-active CHO line over its CHOMG PAT-
deficient mutant.[46] A similar effect was seen with benzyl platforms, wherein the
disubstituted xylyl system (44Bn44, **23a**) was superior to the mono-substituted

system (Bn44, **22a**) in terms of PAT targeting and transporton activity.[46] In contrast to Poulin's results with branched disubstituted xylyl derivatives, these *linear* disubstituted motifs provided potent transportons and not anti-transportons. This clearly indicates that the polyamine message can dictate the PTI's mode of action.

7.5.6 Trimeric Polyamine Scaffolds

Efforts to further improve upon this design led to the corresponding tri-substituted benzene derivative **24** (Figure 7.8).[46] Surprisingly, appending a third polyamine message to an aromatic core appeared to inhibit its entry. As such, the trimeric design **24** behaved as a potent anti-transporton. This was later confirmed in CHO cells. The trimer-44 anti-transporton showed effective blockade of the entry of the Ant44 transporton **20a**.[46] In addition, **24** was able to prevent native polyamines from rescuing growth of DFMO-treated CHO cells (unpublished results).

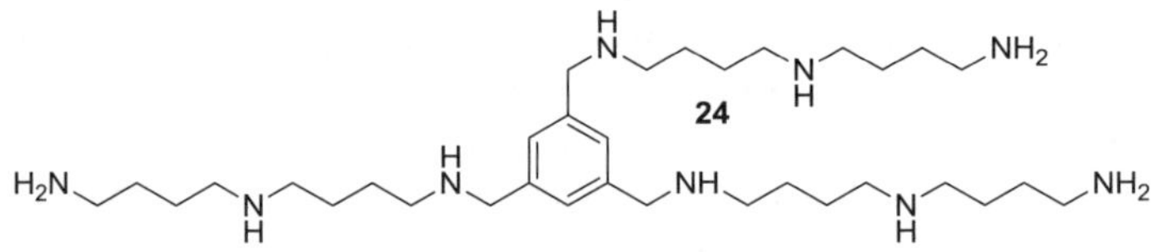

Figure 7.8   Trisubstituted aryl PTI design **24**.

7.5.7 Polyamine–Glutaraldehyde Polymers

Intentional design of PAT anti-transportons was reported by Aziz *et al.* in 1995.[47] High levels of polyamines were found during normal lung development and in damaged lung tissue. In contrast to hyperoxic lung injuries, where ODC activity was found to be upregulated, chronic hypoxia induced low ODC activity, and increased import of putrescine. As such, this group made long linear polymeric chains of putrescine, spermidine and spermine in order to inhibit uptake of polyamines into cultured bovine PASMCs.[47]

Reduced polyamine–glutaraldehyde polymers were synthesized *via* the condensation of native polyamines as their HCl salts with a 3% solution of glutaraldehyde in PBS. NaBH$_4$ was used to reduce the imines to form the final product which were then dialyzed to obtain an appropriate molecular-weight range (12.5–15 kDa). Interestingly, these conjugates are end-hydroxylated, which may provide protection from serum amine oxidases (see Figure 7.9).[47–49] Indeed, the toxicity of poly-Spm **25** was not altered by the addition of a serum amine oxidase inhibitor, aminoguanidine (AG).[48]

Poly-Put was only highly effective at inhibiting radiolabeled putrescine from entering the PASMCs, while poly-Spd was capable of inhibiting Put to a high degree and Spd moderately, and was partially effective against Spm. Poly-Spm

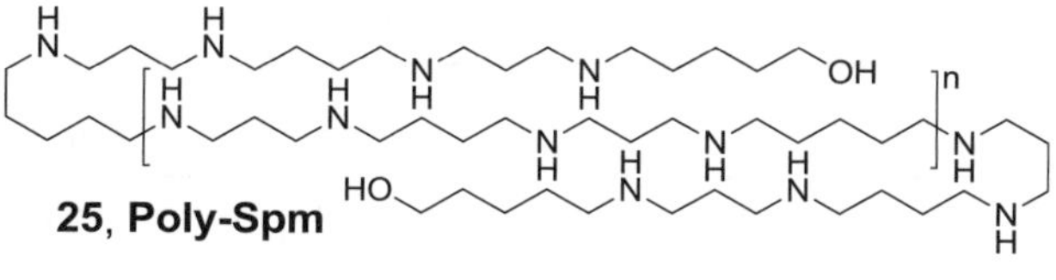

Figure 7.9 Poly-spermine anti-transporton **25**.

25 significantly inhibited all three native polyamines from entering these cells.[47] Furthermore, these conjugates were specific at inhibiting polyamines and not other charged molecules. This was demonstrated by their inability to block cellular entry of the positively charged amino acid [^{14}C]-Lysine.

These conjugates are one of the first examples of PAT anti-transportons predicated upon a polyamine architecture. Radiolabeled versions of these polyamine–glutaraldehyde polymers yielded the same levels of radiolabel in cell lysates regardless of whether the cells were incubated at 4°C or 37°C. Since polyamine transport is an active process, intracellular levels of a PAT transporton should always be significantly higher at 37°C than at 4°C. In contrast, the levels should be the same for an anti-transporton at either temperature because there is no net transport into the cell.

Similar results were found in other cell lines including the human anaplastic carcinoma cell line DRO90-1(DRO).[50] After replacing the media for 72 h in DRO cells treated with poly-Spm, polyamine uptake was not restored. The authors admit that it still may be possible that poly-Spm entered these cells by other means. Interestingly, Poly-Spm **25** alone was cytotoxic and caused a reduction in Spd and Spm levels by 66% and 77%, respectively, despite its inability to be internalized. This suggests that binding to the PAT receptor alone is sufficient to signal degradation of endogenous polyamines. This is in contrast to a PAT transporton like DENSPM **4** (Figure 7.2), which was assumed to induce SSAT and polyamine degradation upon internalization.

Depletion of endogenous spermine is typically cytotoxic to cells. In contrast, depletion of intracellular Put and Spd levels, as demonstrated using DFMO, is typically only cytostatic. When poly-Spm **25** was treated in conjunction with DFMO, the combination appeared to work cooperatively in reducing polyamine levels (Spd and Spm were reduced by 90% and 58%, respectively) further than either agent alone. Interestingly, the toxicity of **25** in DRO cells was not reversed by the addition of spermidine to the media.[50] However, other workers showed that spermine could rescue MES-SA and Dx5 cells from the cyotoxicity of **25**.[48]

Poly-Spm **25** had a significant effect on the metabolic activity of the enzymes involved in polyamine homeostasis in DRO cells as demonstrated in Table 7.2. The mechanism in which polyamines are depleted by poly-Spm in DRO cells appears to be *via* reduced synthesis and increased polyamine degradation. Interestingly, these effects mimic those induced by the native polyamine spermine, which is a natural substrate of PAT. If, indeed, poly-Spm is an anti-transporton and does not enter the cell, then these results suggest that binding to the PAT receptor is sufficient to induce a putative signaling event that leads

Table 7.2   Relative enzyme activities in DRO cells.

Activities/treatment	ODC	SAM-DC	SSAT	Antizyme	Transport
Poly-Spm	Decrease[a]	Decrease	Increase	Increase	Decrease
Spm	Decrease	Decrease	Increase	Increase	ND
DFMO	Decrease	Increase	No change	No change	Increase
Poly-Spm + DFMO	Decrease (faster)	ND	ND	ND	Decrease

ND: no data (not determined).
[a]ODC is actually upregulated by poly-Spm for the first 10 min, but then a severe decrease in ODC was observed.

to the observed changes in PA metabolic enzymatic activities. We speculate that the cell may be preparing for an influx of Spm, but because poly-Spm does not enter the cell, it results in polyamine depletion instead.

The observation that both internalized and non-internalized spermine derivatives induce polyamine depletion suggests that the signaling cascade could be initiated upon binding to the extracellular receptor, an interaction that both molecules have in common. This feature is worthy of further study and could play a role in future PTI designs.

Mice with DRO-subcutaneous xenographs had tumors growing 12-fold in volume over 4 weeks. At lower i.p. doses of **25** alone (100 µg), tumor growth was inhibited 70%. Higher doses (300 µg or more) actually enhanced tumor growth to more than 140% of the controls.[50] Interestingly, minced tumors showed a dose-dependent inhibition of polyamine uptake. However, the highest dose (1000 µg) only gave a maximum inhibition of 25% compared to the controls. It is important to note that the tumors grew at this high concentration, whereas at lower doses, poly-Spm showed efficacy.[50] We speculate that at the higher doses, the poly-Spm conjugate may be metabolized into polyamine products, which the cells could utilize in place of native polyamines. Since this was not seen in cell culture, it apparently required metabolism by the whole animal.

Experiments designed to test the ability of poly-Spm **25** to inhibit polyamine transport revealed different transport kinetics in multi-drug-resistant (MDR) cell lines compared to their parental counterparts. Dx5 cells and K562/R7 cells were used and are MDR variants of MES-SA (human uterine sarcoma) and K562 (human myelogenous leukemia), respectively. Intriguingly, the parental lines had a higher rate of polyamine import than their MDR counterparts, as shown by several-fold higher V_{max} values.[48] The K_m values were essentially the same, which implied that these cancer cells can intentionally limit transport without changing the ligand affinity of the receptor, much like DFMO increases V_{max} without changing K_m. A compensatory increase in polyamine biosynthesis was not observed, since ODC activity was found to be the same in both MES-SA and Dx5. The reduced rate of polyamine import (lower V_{max}) in the MDR lines may explain why poly-Spm was less effective in blocking import of polyamines to this cell line.

Poly-Spm **25** was previously shown to be cytotoxic and induced polyamine depletion, particularly spermine, within all cell lines tested. The IC_{50} values of poly-Spm were much higher (about five times more resistant to the drug) in the

MDR cell lines than in the parental cancer lines. Indeed, the overall trend observed was that cells with a higher polyamine transport activity were more sensitive to poly-Spm. Similar results in toxicity and growth inhibition were obtained in ARO81 (ARO) and BHT 101 cell lines confirming that poly-Spm **25** behaves the same universally in anaplastic thyroid carcinoma lines.[50]

In short, poly-Spm **25** is an example of a toxic anti-transporton, which was unsuccessful *in vivo* due to unanticipated metabolic processing found in intact animals.

7.5.8 Linear Spermine–Amide Dimers

While polymeric polyamine scaffolds had unique properties, other groups have explored smaller, well-defined architectures. For example, Burns *et al.* explored linear spermine dimers conjugated *via* N^1-amides to platforms of varying chain lengths (Figure 7.10).[51] These derivatives were shown to have very low K_i values in the nanomolar range. Interestingly, varying the chain length of the spacer unit in **26** did not significantly alter binding affinity. However, differences in spacing (the n value in **26**) did affect toxicity. When the carbon spacer was greater than seven, the conjugates showed toxicity on their own and were able to inhibit cell growth in the absence of DFMO. Derivatives with shorter tethers had a minimal effect on cell growth in the absence of DFMO.

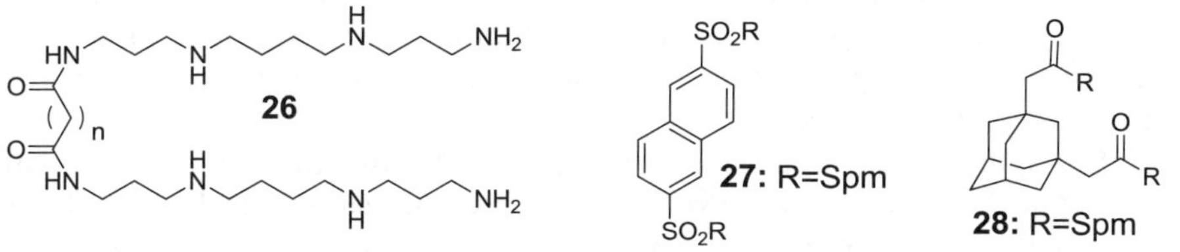

Figure 7.10 Spermine-based diamides **26–28**.

Derivatives of **26** were able to inhibit growth in the low micromolar range in MDA-MB-231 cells, when combined with DFMO in the presence of spermidine. In this case, growth inhibition corresponded well with PAT inhibition. When changing the central linker to 2,6-naphthalenedisulfonic acid **27**, or 1,3-adamantanedicarboxylic **28**, both were equally potent at inhibiting PAT compared to the aliphatic linkers. The naphthyl system was poor at inhibiting growth, and the adamantyl system was not capable of inhibiting growth at all up to 300 μM tested.[51] This suggests that these conjugates may be PAT transportons, which internalize and they themselves or their metabolites can substitute for native polyamines. Clearly, these promising compounds should be further evaluated *in vivo*.

7.5.9 Amino Acid–Spermine Conjugates

While the linear spermine–amide dimers showed promise, Burns' group also created a PAT anti-transporton predicated upon a simple lysine–spermine conjugate, *e.g.* Lys-Spm **29** (Figure 7.11). This molecule showed a high efficacy both in multiple cell culture systems and in mouse models. By conjugating lysine to spermine *via* an amide bond, the Burns group generated a relatively non-toxic anti-transporton (*i.e.* **29**), which could be used in combination with DFMO.[52,53]

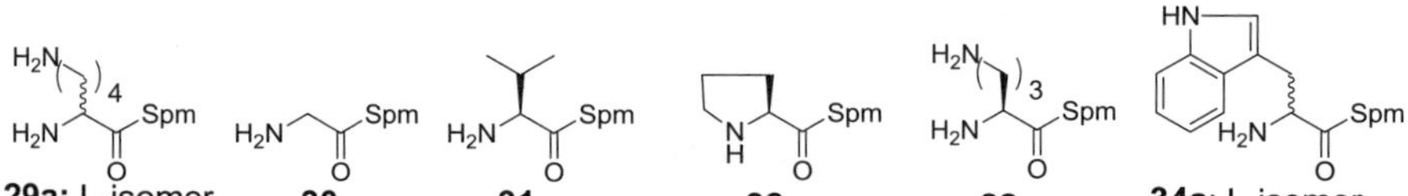

29a: L-isomer
29b: D-isomer
30
31
32
33
34a: L-isomer
34b: D-isomer

Figure 7.11   Spermine amino-acid derivatives **29–34**.

The L-stereoisomer of Lys-Spm (**29a**) was able to block the uptake of all the native polyamines in MDA-MB-231 (MDA) human breast-carcinoma cells. As little as 1 μM of **29a** was able to block 72% of [^{3}H] spermidine from entering the MDA cells and was shown to be non-toxic at concentrations up to 1 mM (>90% viability compared to the control). Transport inhibition was essentially complete at 30 μM (>98% inhibition). Interestingly, intracellular putrescine levels were doubled, but this did not affect growth. Although speculative, it is possible that that the aggressive cell line requires exogenous polyamines for growth (in addition to endogenously synthesized pools) and may attempt to increase putrescine production to compensate for the blockade of the exogenous polyamine source.

L-Lys-Spm **29a** was tested to see if it inhibited growth in combination with DFMO in the presence of 1 μM Spd *via* the MTS assay. The MTS assay was optimized for these experimental conditions and performed over 6 days in the presence of AG. When either DFMO or **29a** was tested alone in the presence of 1 μM Spd, there was no significant effect on growth. The combination therapy, however, revealed dose-dependent growth inhibition, and **29a** (30 μM) with DFMO (230 μM) gave significant growth inhibition (*i.e.* 40–45% compared to the control). Unlike other polyamine derivatives, L-Lys-Spm did not rescue DFMO-treated cells and thus does not likely substitute for the native polyamines within the cell. Additionally, there was no apparent combined toxicity seen with this combination drug treatment.

In order to compare the efficacy of PTIs in terms of growth to each other, and in different cell lines, a customized EC_{50} parameter was established by Burns. DFMO gives different reductions in viability depending upon the cell line and the time of incubation. In order to compare between cell lines, the authors first determined the maximum inhibition of growth obtained with DFMO.

A hypothetical example is provided to clarify this EC_{50} parameter. Assume a dose of 5 mM DFMO resulted in 50% inhibition of growth of a particular cell line. Next, Spd was titrated into these DFMO-treated cells to determine the minimum concentration of Spd needed to rescue the DFMO-treated cells back to 100% viability (MC_{100}). Lastly, the PTI was titrated into DFMO + Spd

(MC$_{100}$) treated cells. At some concentration, the PTI would block the import of Spd, and the cells would revert to the 50% viability seen with DFMO only. In this example, the Burns group would use the IC$_{25}$ value (75% viability) as the EC$_{50}$ value. The EC$_{50}$ value is halfway between the 50% viability observed with DFMO only and the 100% viability observed with DFMO + Spd. The EC$_{50}$ values of **29a** in breast-, prostate-, bladder-, melanoma- and lung-cancer cell lines demonstrated that the combination therapy was effective in multiple cancers.

L-Lys-Spm **29a** was also shown to be a reversible PAT inhibitor. When the drug was removed from the media, polyamine transport was recovered over time. While neither L-Lys-Spm nor DFMO was toxic in the presence of Spd, the combination of these agents (DFMO + **29a**) showed a significant decrease in cell growth over a 3-week time period. Interestingly, Put levels were lowered in short-term studies with little effect on Spd levels. However, the combination therapy reduced Spd by 95% compared to the control over the 3-week period. Spermine levels were not affected, demonstrating that this combination therapy is cytostatic and not cytotoxic. This example demonstrates that polyamine depletion is time-dependent; a fact that will surely influence the clinical applications of this technology.

Due to the success of Lys-Spm, other amino acid–spermine conjugates were synthesized and compared to **29a** in MDA-MB-231 breast-cancer cells. With the exception of ornithine analog **33**, none of the other derivatives rivaled the potency of the Lys-Spm derivative (Figure 7.11).[52] It is possible that changes in the stereochemistry of the transport inhibitor can cause a non-optimal binding orientation, if a chiral fit is involved. When comparing the tryptophan series, L-Trp-Spm **34a** and D-Trp-Spm **34b** gave different K_i values (119 and 15 nM, respectively) and EC$_{50}$ values (63 and 2.5 µM, respectively). These results suggested that a chiral target is involved in the binding of these ligands.

There is good evidence to suggest that some of the amino acid–spermine analogs are transportons of PAT. Conjugates whose K_i values did not correlate well with EC$_{50}$ values were tested for their ability to rescue the growth inhibition of DFMO with no Spd added. Both the glycine **30** and proline **32** derivatives, respectively, were able to rescue growth and are thus suggested to be PAT transportons that can substitute for the native polyamines or are degraded into metabolites that can be utilized as native polyamines (Figure 7.11).

Besides the appended N-substituent, the polyamine message itself plays a key role in PAT recognition. The N^1-acetylspermine (K_i) and spermine (K_m) gave similar values (177 nM and 113 nM, respectively). Interestingly, when the polyamine message was changed from spermine (3,4,3) to norspermine (3,3,3-tetraamine), the K_i value rose to > 1000 µM for N^1-acetylnorspermine. N^1-acetylspermine has three amine groups and gave a similar K_i to the K_m of spermidine, which suggested that PAT recognizes N^1-acetylspermine as a triamine. Thus, as long as the proper charge and spacer units are in place, amides can be utilized to link polyamines to other cargoes without disrupting the polyamine 'reading frame'.[43]

In terms of polyamine depletion as a strategy, a reduction in tumor growth may require a significant reduction in Put levels as well as Spd levels. Inhibitors of spermine synthetase have been used to suggest that spermine plays a minor role in cell growth and has a low turnover.[53] This difference in the contribution to growth by each of the native polyamines provides an opportunity to limit toxicity of PTI agents. Although cell-line-dependent, typically when intracellular spermine levels are reduced along with putrescine and spermidine, cellular cytotoxicity is seen.[53] Since the native polyamine can be inter-converted metabolically, anti-transportons which block import of all three native polyamines may be the most effective in combination with DFMO. However, the work by the Burns group suggests that there is an opportunity to moderate the cytotoxicity of PTI + DFMO therapy and instead induce a cytostatic response. Presumably, this outcome could be accomplished *via* anti-transportons, which are effective primarily against Put and Spd.

A subcutaneous xenograph mouse model was developed using the MDA-MB-231 cells. In this model, mice were given L-Lys-Spm alone (45 mg of **29a** kg^{-1} $dose^{-1}$, i.p., three times a day, 5 days a week), DFMO alone (1% in drinking-water available *ad libitum* for the 14-day duration of the study) or a combination of both. L-Lys-Spm, DFMO and the combination therapy provided inhibition of tumor growth by 33%, 30% and 50%, respectively. Overall, the *in vivo* results suggest a synergetic effect between DFMO and the PTI.

In a nude mouse model using PC-3 cells, DFMO or inhibitor alone had no significant growth-inhibitory effect in contrast to what was observed in MDA-MB-231 cells. The combination of DFMO and the inhibitor provided a maximum of 61% growth inhibition with the highest concentrations of DFMO and inhibitor used.[54] Clearly, synergy must operative in this case. Further evidence of this is shown by measuring the polyamine content in the treated tumors compared to the untreated control. Giving mice DFMO only caused a 33% reduction in Put in the tumor (with no effect on growth). The combination therapy caused a 64% and 23% reduction in Put and Spd in the tumor, respectively. Although modest, the 23% reduction in Spd was not observed with either agent alone.

In a follow-up paper, the isomeric D-Lys-Spm **29b** was shown to be potent in combination with DFMO in a mouse model of squamous cell carcinoma (SCC).[2] The SCC cells themselves contained abnormally high levels of putrescine, which took days to reduce with DFMO. In contrast, the combination therapy took only 24 h to significantly reduce Put levels. In the combination therapy trial involving a 4-week treatment period followed by 6 weeks off-treatment, there were several complete cures (50%) as indicated by no return of tumors after 10 weeks (versus DFMO only, 8%). Tumor regression was also observed in animals without complete cures. In this mouse model, DFMO-only therapy reduced proliferation, whereas **29b** alone did not. However, the combination therapy (DFMO + **29b**) caused increased apoptosis, which was not observed with either agent alone.

This group also evaluated ODC expression and polyamine content in the tumors. When mice were treated with DFMO only or DFMO + **29b**, ODC activity was suppressed in the SCCs by more than 90%, even though these tumors had a high ODC activity at the outset of the experiment. D-Lys-Spm **29b** did not impact ODC activity.[2] Overall, the SCC model appears to be a valid and useful model to explore the effects of combination PTI and DFMO treatments.

7.5.10 Heparin Sulfate (HS)-Binding Agents

As previously mentioned, polyamines bind to heparan-sulfate-containing proteoglycans (HSPGs) or other highly sulfated glycosaminoglycans, GAGs. This conclusion was supported upon the following observations. First, CHO cells deficient in PG synthesis show reduced PA uptake. Second, HS-lyase, an enzyme responsible for degrading cell surface HS, was able to cause a 61% reduction in spermine uptake in wt-cells.[13] As such, agents which share a binding site on HSPGs should also be inhibitors of PAT.

7.5.10.1 HIV-Tat

HIV-Tat is a cationic peptide (*i.e.* GRKKRRQRRRPPQC) which can bind to HS-PGs with high affinity and is a known protein transduction agent. In one study, HIV-Tat was shown to competitively inhibit the uptake of polyamines and significantly prevented polyamine uptake in human bladder carcinoma T24 cells treated with DFMO.[55] These results suggest that HIV-Tat and polyamines may share and compete for the same cell-surface receptor. However, separate reports suggest that HIV-Tat has a higher affinity for highly sulfated-HS than polyamines (K_d of 37 µM for spermine[23] and 1.5 µM for HIV-Tat).[56] Interestingly, uptake of AlexaFluor-647-labeled Tat peptide and exogenous polyamines were both enhanced by DFMO treatment.

Growth inhibition was observed when HIV-Tat was treated in combination with DFMO in the presence of Spd, but not with HIV-Tat alone. Thus, since HIV-Tat enters cells and competes with native polyamines, it is a PAT transporton *in vitro*. In the mouse, HIV-Tat, in combination with DFMO, was able to inhibit T24 carcinoma-cell tumor growth, demonstrating its efficacy as a PTI. Treatment with DFMO or HIV-Tat alone in the *in vivo* model had no appreciable effect. However, the combination therapy showed a dramatic efficacy *in vivo*. When comparing weights of the tumors, the combination therapy provided a 92% reduction in tumor mass compared to the controls. No serious side effects were noted. As such, the Tat-peptide in combination with DFMO is a potent inhibitor of tumor growth *in vivo*. The demonstrated ability of small cationic peptides which function as PTIs could lead to the design of novel peptide-containing transport inhibitors. Indeed, the Tat protein and a simple conjugate of amino acids and polyamines have already shown efficacy in this regard (*e.g.* **29**).[2,55]

7.5.10.2 HS-Specific Antibodies

Peptide-based PTIs, which are specific for HS, were realized in the form of antibodies. The Belting group prepared single-chain variable fragment anti-HS antibodies and demonstrated their specificity for different regions of HS.[57] Mutant CHO cells which had low overall PG expression ($\approx$ 5%) had low antibody binding ($\approx$ 6%). In addition, mutants with completely defective HS synthesis did not accumulate any antibody, as shown by flow cytometry. Of the series of antibodies tested, the highest inhibition (60%) was realized with the RB4EA12 antibody. As expected, polyamine uptake was shown to correspond with the ability of the polyamine to bind to the cell surface, and inhibition was shown to be dose-dependent.

Interestingly, HIV-Tat was inhibited by both RBEA12 and an antibody that poorly inhibited PA uptake. This suggests that PAs and HIV-Tat have over-lapping yet different recognition elements for HS. PA uptake was inhibited by RBEA12 in multiple cell lines including CHO, A549 and HeLa cells. This suggested that RBEA12 may have general application as a PTI.

RBEA12 was also shown to be a potent inhibitor of exogenous PA uptake and growth in the presence of DFMO. Interestingly, DFMO increased the affinity of the REBEA12 antibody for HS twofold. The native polyamines showed a threefold increase in affinity for HS under the same conditions. In addition, the RBEA12 antibody (presumably because it blocks PA uptake) also increased ODC levels several-fold, and the antibody alone had a growth-inhibitory effect. Since there was no difference in cell growth for RBEA12 and the control when serum-free media were used, it was suggested that the anti-body also inhibited HS from binding growth factors in the serum which could activate cell growth. ODC knockouts CHO cells (CHO-ODC-) that require exogenous polyamines supplemented in the media to grow were significantly inhibited by RBEA12 alone. The reduced cell number was shown to be a cytostatic and not a cytotoxic effect. While there are a few exceptions, anti-bodies in general do not penetrate cell membranes.[58] As such, these PAT antibodies are likely anti-transportons.

7.5.11   Lipophilic Polyamine Conjugates

Burns' lead compound, Lys-Spm, was recently modified further by attachment of lipophilic carbon chains to increase the oral delivery profile of this PTI class.[59] As shown in Table 7.3, significant increases in growth inhibition potency were observed. Introduction of the lipophilic component also resulted in lower K_i values presumably due to a higher affinity for the cell surface receptor (Table 7.3).

The authors propose that Schwyzer's theory may explain how the lipid tail augmented the potency of the parent Lys-Spm.[59] Presumably, the lipophilic portion of the aliphatic Lys-Spm derivative first embeds itself into the cell membrane, and then the polyamine portion finds its receptor *via* a two-dimensional search of the membrane surface, rather than a three-dimensional search *via* diffusion. Thus, targeting both the PAT receptor and the membrane *via* lipopolyamine constructs may enhance PAT receptor binding.

Table 7.3 Biological properties of **29** and **35–37**.

Construct	Isomer	EC_{50} in MDA-231 (μM)	K_i values (spd) (nM)	IC_{50} in MDA-231 (μM)
Lys-Spm **29**	L	7.0	32	>1000
	D	2.7	28	
C_{16}-acyl **35**	L	0.06	7.5	62
	D	0.076	10.5	57
C_{16}-alkyl **36**	L	0.019		
	D	–		
C_{20}-acyl **37**	L	0.010		
	D	–		

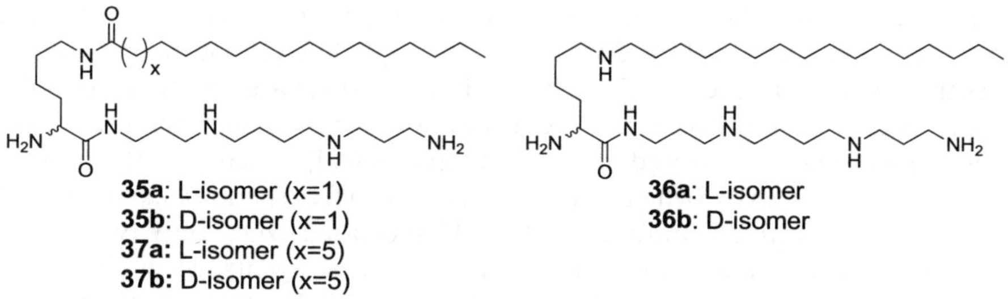

35a: L-isomer (x=1)
35b: D-isomer (x=1)
37a: L-isomer (x=5)
37b: D-isomer (x=5)

36a: L-isomer
36b: D-isomer

Figure 7.12 Lipopolyamines predicated upon amino acids **35–37**.

Unlike tryptophan **34**, changes in stereochemistry did not show a significant difference in relative activities. Interestingly, the original Lys-Spm conjugate **29** was essentially non-toxic with an IC_{50} in MDA-231 breast cancer cells of > 1000 μM. However, the L and D C_{16}-acyl conjugates **35** had IC_{50} values of 62 and 57 μM, respectively (Figure 7.12).[59] Although not severely toxic, there was clearly an increased toxicity associated with introduction of the lipid chain and enhanced potency. However, the therapeutic window was actually increased due to enhanced affinity imparted by the lipid tail, which facilitated further animal testing.

Based upon the promising results with the C_{16}-acyl derivatives, a series of N^{ε}-alkyl and N^{ε}-acyl Lys-Spm derivatives were synthesized with different chain lengths. The EC_{50} values of these series were determined in MDA-231 (breast), PC-3 (prostate), A375 (melanoma) SK-OV-3 (ovarian) and gave similar results in all cell lines. The C_{20}-acyl derivative **37** had the best growth-inhibitory activity of the acyl series, and C_{16}-alkyl **36** had the best growth-inhibitory activity of the alkyl series (Figure 7.12). These had the longest chain lengths of the respective series tested and suggested that longer chain lengths lower the EC_{50} value.[59]

Additional evidence that longer chain lengths enhance growth-inhibitor potency comes from a study using oral hamster keratinocyte cells (HCPC cells). The parent Lys-Spm, C_8-acyl and C_{16}-acyl derivatives gave EC_{50} values of 20.5,

2.2 and 0.15 µM, respectively. Thus, a 10-fold increase in potency was observed with each successive eight-carbon addition to the lipid tail.[59]

The authors also demonstrated that these agents alone did not affect intracellular polyamine levels in SK-Mel cells. Select members did, however, affect enzyme activity. For example, the Lys-D-Spm **29b** and C_{16}-L-acyl **35** derivatives were able to reduce ODC activity (40 and 46%, respectively), and all were able to decrease SSAT (40% reduction compared to the control).[59] The errors are not reported for these experiments; thus the significance of these observed reductions compared to the control is not clear. The authors claim from preliminary cell-based results that these conjugates do not cross the cell membrane and presumably work on the polyamine transport cell surface receptor (*i.e.* they are anti-transportons).

There are curious findings in this field, where signal transduction upon anti-transporton binding to the PAT receptor can initiate a change in polyamine metabolic enzymes. This theory has yet to be tested but is seen with poly-Spm **25**,[50] lipopolyamines such as **35**[59] and Belting's RB4EA12 antibody,[57] which presumably do not penetrate the cell membrane and yet are able to affect polyamine metabolic enzymes. These agents did not give the same responses, however. Compound **35** reduced both ODC and SSAT, whereas Poly-Spm **25** increased SSAT activity (which is also observed with free spermine), but reduced ODC activity. Belting's antibody RB4EA12 increased ODC activity (two- to threefold).[57] It is important to note that none of these agents, when given with DFMO, was able to restore cellular growth. Thus, neither they nor their metabolites can provide a functional substitute for native polyamines inside the cell.

Due to their enhanced potency in the cell culture, compounds C_{16}-D-acyl (**35**) and C_{16}-L-alkyl Lys Spm (**36**) were selected for *in vivo* mouse studies. Both compounds gave a very positive therapeutic response when given in combination with DFMO. Interestingly, **35** + DFMO was 70-fold more potent *in vivo* than Lys-Spm + DFMO in the SCC mouse model. The C_{16}-D-acyl conjugate (**35** at 0.5 mg/kg i.p. twice daily) gave significant responses (71%) in the group tested with the combination therapy. Looking at the long-term performance, nine out of 17 had a complete response with one SCC returning after a 6-week rest period. In the second smaller trial to evaluate oral dosing, mice were given **35** orally (3 mg/kg/day), and all of the SCCs responded, including some complete responses over 6 weeks.[59] Thus, this promising drug seems to work both orally and by i.p. injection.

While the Burns group showed a clear trend towards increasing potency with increasing lipid chain length, not all lipids may have this same ability, however.[60] In terms of lipid design, it is not yet clear whether Burns' one tail design is superior to lipid motifs containing more than one lipid tail. Two reports suggest there may be some advantages to further exploration of the lipid component.

7.5.11.1 Dual-Tailed Lipopolyamines

In a study of 3,4-(bis-alkenyloxy)-benzyl-polyamines (*e.g.* **38**), twin oleoyl (C_{18}) tails were maintained as part of the lipo-polyamine design, and the polyamine

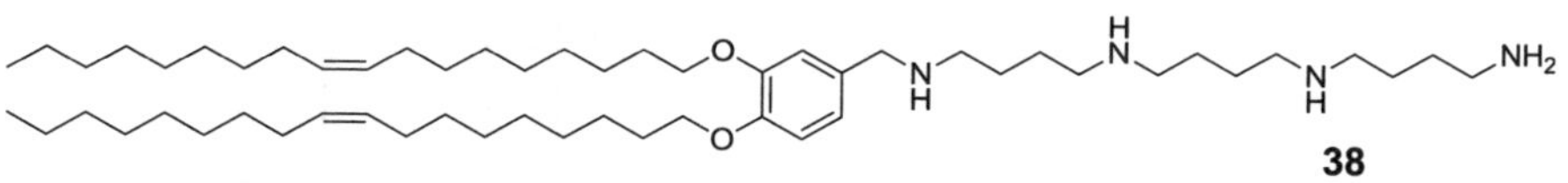

Figure 7.13 Dual-tailed lipopolyamine construct **38**.

message was incrementally varied (Figure 7.13). A clear trend was apparent in terms of cytotoxicity, solubility and transfection ability. The longer the polyamine message, the greater the water solubility, and the more efficient the transfection agent. However, longer polyamine messages also imparted toxicity. In this regard, there is likely an optimal balance between the number of polyamine charges, lipid length and cytotoxicity.

While a head-to-head comparison between the C_{16}-D-acyl derivative **35** and **38** was not performed, both compounds were shown to compete with spermine for cellular entry by different investigators.[59,60] This suggests that further attention to the lipid portion of this design could yield more optimal PTIs because the effectiveness of delivery *via* membrane integration is influenced by the lipid component of these designs.

Indeed, Blagbrough *et al.* in a related study of lipopolyamine transfection agents demonstrated that very long lipid chains appended to the spermine backbone can enhance transfection efficiency (Table 7.4).[61] There was an interesting observation that the arachidoyl side chain ($R = CO(CH_2)_{18}CH_3$; **39**) and closely related arachidonoyl ($R = CO(CH_2)_3(CH = CHCH_2)_4(CH_2)_3CH_3$; **40**) side chains imparted significantly different transfection efficiencies and toxicities in the HeLa-derived HtTA cancer cell line over the FEK4 primary skin cell line.[61] A direct comparison of transfection ability is difficult to glean, as the compounds were dosed at their optimum concentrations for transfection and not on an equimolar basis *per se*. Nevertheless, when dosed at their respective optimal transfection doses, the arachidoyl derivative **39** was much less efficient in terms of transfection than the corresponding unsaturated arachidonoyl derivative **40** (6% *vs.* 80% in HtTA, respectively). However, the cytotoxicity profiles were the opposite of this trend (**39**: 75% and **40**: 17% viability in HtTA cells, when dosed at the same N/P ratio of 4). While the transfection trends could be explained by the higher dosing of **40** (9.4 µg) *vs.* **39** (4.8 µg), the viability data are nearly equimolar (N/P = 4). In this regard, subtle chains in the lipid tail resulted in dramatic differences in cytotoxicity.

As shown in Table 7.4, Blagbrough *et al.* in a related siRNA study showed that alterations of the lipid message strongly influenced the cytotoxicity and siRNA delivery efficiency of the lipo-polyamine.[61] Clearly, the extended lipids, *i.e.* the erucoyl tails, imparted good efficacy and lower toxicity.

In summary, both the lipid and polyamine components need further optimization in this class of lipo-polyamines. While the goal of lipo-polyamine transfection agents and lipo-polyamine PTIs may be different, their ability to bind to the extracellular polyamine receptor is paramount for their success. The current literature provides particular suggestions for improved PTI designs.

Table 7.4 Cytotoxicity and delivery fficiencies of long-chain lipo-polyamines.

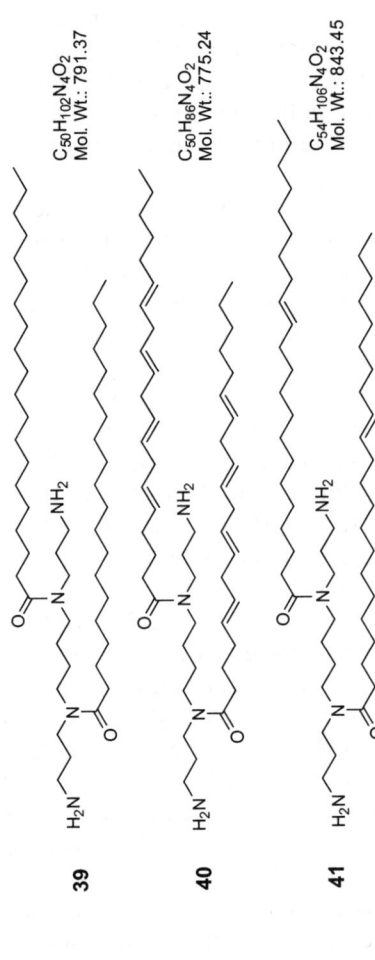

39 C₅₀H₁₀₂N₄O₂
Mol. Wt.: 791.37

40 C₅₀H₈₆N₄O₂
Mol. Wt.: 775.24

41 C₅₄H₁₀₆N₄O₂
Mol. Wt.: 843.45

N^4,N^9-Spermine derivative	Viability (%) in FEK4	Viability (%) in HtTA	Transfection (%) in HtTA	Transfection (%) in FEK4
Diarachidoyl **39**	90	80	41	35
Diarachidonoyl **40**	55	18	90	85
Dierucoyl **41**	82	77	90	85

Cytotoxicity and lipofection affect siRNA (12.5 pmol) complexed with TransIT-TKO (4 µl), arachidoyl (N^4,N^9-diarachidoyl spermine **39**, 6.0 µg), arachidonoyl (N^4,N^9-diarachidonoyl spermine **40**, 6.0 µg) or erucoyl (N^4,N^9-dierucoyl spermine **41**; 6.0 µg) on the primary skin cell line FEK4 and on the HeLa-derived cancer cell line HtTA.

Specifically, extension of the polyamine chain will likely impart greater toxicity to lipo-polyamines, whereas the extension of the lipid portion may provide compounds which are less toxic and more efficacious.

7.6 Conclusions

Using both trial and error, numerous groups have begun to define the parameters for successful inhibition of polyamine transport in humans. It appears that a combination therapy involving a non-toxic PTI combined with DFMO may be a successful anticancer strategy. Although, to date, no studies have been performed in humans, there was good efficacy in several mouse models using **35**. As such, a successful combination therapy (DFMO + PTI) in humans may not be far in the future.

Clinical success in this area would usher in a new paradigm for polyamine-based chemotherapeutics and may shift the field's focus from PAT transportons to anti-transportons. The fact that aggressive cancers seem to be more sensitive to this approach than normal cells holds additional promise. If one can target cells predicated upon their polyamine transport activity, then a cell-selective targeting outcome may be observed. This possibility provides additional promise for the further development of these drugs.

As the secrets to successful polyamine transport inhibition become unveiled, it is clear that investigations outside the polyamine community will play important contributions. Research efforts in protein transduction, endosomal trafficking and transfection[62] will likely contribute to the optimal designs of PTIs which may come in the form of cationic proteins, lipo-polyamines or PAT-selective antibodies. Indeed, the examples provided in this review suggest that many of these areas are directly related. In conclusion, this area is ripe for new discovery and will require an optimal balance between binding affinity, metabolic stability and cytotoxicity.

References

1. E. W. Gerner, F. L. Meyskens, Jr, S. Goldschmid, P. Lance and D. Pelot, *Amino Acids*, 2007, **33**, 189–195.
2. Y. Chen, R. S. Weeks, M. R. Burns, D. W. Boorman, A. Klein-Szanto and T. G. O'Brien, *Int. J. Cancer*, 2006, **118**, 2344–2349.
3. K. Ravanko, K. Jarvinen, A. Paasinen-Sohns and E. Hölttä, *Cancer Res.*, 2000, **60**, 5244–5253.
4. F. L. Meyskens, Jr and E. W. Gerner, *Clin. Cancer Res.*, 1999, **5**, 945–951.
5. M. D. Abeloff, M. Slavik, G. D. Luk, C. A. Griffin, J. Hermann, O. Blanc, A. Sjoerdsma and S. B. Baylin, *J. Clin. Oncol.*, 1984, **2**, 124–130.
6. C. J. Fabian, B. F. Kimler, D. A. Brady, M. S. Mayo, C. H. Chang, J. A. Ferraro, C. M. Zalles, A. L. Stanton, S. Masood, W. E. Grizzle, N. F. Boyd, D. W. Arneson and K. A. Johnson, *Clin. Cancer Res.*, 2002, **8**, 3105–3117.

7. N. Seiler, *Curr. Drug Targets*, 2003, **4**, 537–564.
8. E. W. Gerner and F. L. Meyskens, Jr, *Nat. Rev. Cancer*, 2004, **4**, 781–792.
9. N. Seiler, J. G. Delcros and J. P. Moulinoux, *Int. J. Biochem. Cell Biol.*, 1996, **28**, 843–861.
10. K. Igarashi and K. Kashiwagi, *Biochem. J.*, 1999, **344**, 633–642.
11. C. Tsen, M. Iltis, N. Kaur, C. Bayer, J.-G. Delcros, L. von Kalm and O. Phanstiel IV, *J. Med. Chem.*, 2008, **51**, 324–330.
12. A. Heinick, K. Urban, S. Roth, D. Spies, F. Nunes, O. Phanstiel, IV, E. Liebau and K. Lüersen, *FASEB J.*, 2010, **24**, 206–217.
13. M. Belting, S. Persson and L.-Å. Fransson, *Biochem. J.*, 1999, **338**, 317–323.
14. M. Belting, K. Mani and M. Jönsson, *J. Biol. Chem.*, 2003, **278**, 47181–47189.
15. D. Soulet, B. Gagnon, S. Rivest, M. Audette and R. Poulin, *J. Biol. Chem.*, 2004, **279**, 49355–49366.
16. T. Uemura, D. E. Stringer, K. A. Blohm-Mangone and E. W. Gerner, *Am. J. Physiol. Gastrointest Liver Physiol.*, 2010, **299**, G517–G522.
17. U. K. Basu Roy, N. S. Rial, K. L. Kachel and E. W. Gerner, *Mol. Carcinogenesis*, 2008, **47**, 538–553.
18. C. Wang, J.-G. Delcros, J. Biggerstaff and O. Phanstiel IV, *J. Med. Chem.*, 2003, **46**, 2672–2682.
19. J.-P. Annereau, V. Briel, C. Dumontet, Y. Guminski, T. Imbert, M. Broussas, S. Vispé, S. Bréand, N. Guilbaud, J.-M. Barret and C. Bailly, *Leukemia Res.*, 2010, **34**, 1383–1389.
20. S. M. Aziz, J. W. Olson and M. N. Gillespie, *Am. J. Respir. Cell Mol. Biol.*, 1994, **10**, 160–166.
21. J. C. Paz, F. Sánchez-Jiménez and M. A. Medina, *Amino Acids*, 2001, **21**, 271–279.
22. K. Ding, S. Sandgren, K. Mani, M. Belting and L. Å. Fransson, *J. Biol. Chem.*, 2001, **276**, 46779–46791.
23. M. Belting, B. Havsmark, M. Jönsson, S. Persson and L.-Å. Fransson, *Glycobiology*, 1996, **6**, 121–129.
24. J. L. Mitchell, C. L. Simkus, T. K. Thane, P. Tokarz, M. M. Bonar, B. Frydman, A. L. Valasinas, V. K. Reddy and L. J. Marton, *Biochem. J.*, 2004, **384**, 271–279.
25. J. L. Mitchell, R. R. Diveley, Jr and A. Bareyal-Leyser, *Biochem. Biophys. Res. Commun.*, 1992, **186**, 81–88.
26. B. G. Erwin and A. E. Pegg, *Biochem. J.*, 1986, **238**, 581–587.
27. R. A. Casero Jr and A. E. Pegg, *FASEB J.*, 1993, **7**, 653–661.
28. M. H. Park and E. C. Wolff, *J. Biol. Chem.*, 1988, **263**, 15264–15269.
29. E. C. Wolff, M. H. Park and J. E. Folk, *J. Biol. Chem.*, 1990, **265**, 4793–4799.
30. V. Kuksa, R. Buchan and P. K. T. Lin, *Synthesis*, 2000, 1189–1207.
31. C. W. Porter, J. Miller and R. J. Bergeron, *Cancer Res.*, 1984, **44**, 126–128.
32. Y. Huang, A. Pledgie, R. A. Casero Jr and N. E. Davidson, *Anticancer Drugs*, 2005, **16**, 229–241.

33. R. J. Bergeron, W. R. Weimar, G. R. Luchetta, C. A. Sninsky, R. R. Streiff and J. Wiegand, *Drug Met. Dispos.*, 1996, **24**, 334–343.
34. R. J. Bergeron, Y. Feng, W. R. Weimar, J. S. McManis, H. Dimova, C. Porter, B. Raisler and O. Phanstiel, *J. Med. Chem.*, 1997, **40**, 1475–1494.
35. R. F. Minchin, R. L. Martin, L. A. Summers and K. F. Ilett, *Biochem. J.*, 1989, **262**, 391–395.
36. C. W. Porter, R. J. Bergeron and N. J. Stolowich, *Cancer Res.*, 1982, **42**, 4072–4078.
37. C. A. Rinehart and K. Y. Chen, *J. Biol. Chem.*, 1984, **259**, 4750–4756.
38. K. Torrossian, M. Audette and R. Poulin, *Biochem. J.*, 1996, **319**, 21–26.
39. M. Huber, J. G. Pelletier, K. Torossian, P. Dionne, I. Gamache, R. Charest-Gaudreault, M. Audette and R. Poulin, *J. Biol. Chem.*, 1996, **271**, 27556–27563.
40. P. Brachet, J. E. Long and E. R. Seidel, *Biochem. Pharmacol.*, 1998, **56**, 517–526.
41. L. Covassin, M. Desjardins, R. Charest-Gaudreault, M. Audette, M.-J. Bonneau and R. Poulin, *Bioorg. Med. Chem. Lett.*, 1999, **9**, 1709–1714.
42. L. Covassin, M. Desjardins, D. Soulet, R. Charest-Gaudreault, M. Audette and R. Poulin, *Bioorg. Med. Chem. Lett.*, 2003, **13**, 3267–3271.
43. J-G. Delcros, S. Tomasi, S. Carrington, B. Martin, J. Renault, I. S. Blagbrough and P. Uriac, *J. Med. Chem.*, 2002, **45**, 5098–5111.
44. O. Phanstiel IV, N. Kaur and J-G. Delcros, *Amino Acids*, 2007, **33**, 305–313.
45. L. Jin, H. Sugiyama, M. Takigawa, D. Katagiri, H. Tomitori, K. Nishimura, N. Kaur, O. Phanstiel IV, M. Kitajima, H. Takayama, T. Okawara, K. Williams, K. Kashiwagi and K. Igarashi, *J. Pharmacol. Exp. Ther.*, 2007, **320**, 47–55.
46. N. Kaur, J-G. Delcros and O. Phanstiel IV, *J. Med. Chem.*, 2008, **51**, 1393–1401.
47. S. M. Aziz, M. P. Gosland, P. A. Crooks, J. W. Olson and M. N. Gillespie, *J. Pharmacol. Exp. Ther.*, 1995, **274**, 181–186.
48. S. M. Aziz and M. N Gillespie, *J. Pharmacol. Exp. Ther.*, 1996, **278**, 185–192.
49. R. J. Bergeron, R. Müller, J. Bussenius, J. S. McManis, R. L. Merriman, R. E. Smith, H. Yao and W. R. Weimar, *J. Med. Chem.*, 2000, **43**, 224–235.
50. M. Yatin, G. M. Venkataraman, R. Marcinek and K. B. Ain, *Thyroid*, 1999, **9**, 805–14.
51. G. F. Graminski, C. L. Carlson, J. R. Ziemer, F. Cai, N. M. Vermeulen, S. M. Vanderwerf and M. R. Burns, *Bioorg. Med. Chem. Lett.*, 2002, **12**, 35–40.
52. M. R. Burns, C. L. Carlson, S. M. Vanderwerf, J. R. Ziemer, R. S. Weeks, F. Cai, H. K. Webb and G. F. Graminski, *J. Med. Chem.*, 2001, **44**, 3632–3644.
53. L. J. Martin and A. E. Pegg, *Ann. Rev. Pharmacol. Toxicol.*, 1995, **35**, 55–91.

54. B. H. Devens, R. S. Weeks, M. R. Burns, C. L. Carlson and M. K. Brawer, *Prostate Cancer Prostatic Dis.*, 2000, **3**, 275–279.
55. K. Mani, S. Sandgren, J. Lilja, F. Cheng, K. Svensson, L. Persson and M. Belting, *Mol. Cancer Ther.*, 2007, **6**, 782–778.
56. A. Ziegler, P. Nervi, M. Dürrenberger and J. Seelig, *Biochemistry*, 2005, **44**, 138–148.
57. J. E. Welch, P. Bengtson, K. Svensson, A. Wittrup, G. J. Jenniskens, G. B. Ten Dam, T. H. van Kuppevelt and M. Belting, *Int. J. Oncol.*, 2008, **3**, 749–756.
58. K. Yanase and M. P. Madaio, *J. Autoimmun.*, 2005, **24**, 145–151.
59. M. R. Burns, G. F. Graminski, R. S. Weeks, Y. Chen and T. G. O'Brien, *J. Med. Chem.*, 2009, **52**, 1983–1993.
60. R. A. Gardner, M. Belting, K. Svensson and O. Phanstiel IV, *J. Med. Chem.*, 2007, **50**, 308–318.
61. H. M. Ghonaim, S. Li and I. S. Blagbrough, *Pharmaceut. Res.*, 2009, **26**, 19–31.
62. J. J. Green, R. Langer and D. G. Anderson, *Acc. Chem. Res.*, 2008, **41**, 749–759.

Non-Covalent Polynuclear Platinum Compounds as Polyamine Analogs

YUN QU,[a] JOSEPH J. MONIODIS,[b] AMANDA L. HARRIS,[a] XIAOHONG YANG,[a] ALEX HEGMANS,[a] LAWRENCE F. POVIRK,[c] SUSAN J. BERNERS-PRICE[b,d] AND NICHOLAS P. FARRELL*[a,d]

[a] Department of Chemistry, Virginia Commonwealth University, Richmond, VA 23284-2006, Australia; [b] Chemistry M313, School of Biomedical, Biomolecular & Chemical Sciences, University of Western Australia, Crawley, WA 6009, Australia; [c] Department of Pharmacology, 380A Goodwin Laboratory, Massey Cancer Center, Virginia Commonwealth University, 401 College St., Richmond, VA 23298-0035, USA; [d] Institute of Glycomics, Griffith University, Gold Coast Campus, Qld. 4222, Australia

8.1 Introduction

DNA binding is the mechanistic paradigm by which cytotoxic platinum complexes are believed to exert their antitumour activity.[1,2] In general, the associated processes of DNA function such as transcription and repair may be affected by DNA-targeted drugs.[3,4] The development of polynuclear platinum complexes (PPCs) represents an approach to systematically alter the cellular response induced by cisplatin by changing the nature and structure of the DNA lesion induced.[5] The development of BBR3464, the only platinum compound

RSC Drug Discovery Series No. 17
Polyamine Drug Discovery
Edited by Patrick M. Woster and Robert A. Casero, Jr.
© Royal Society of Chemistry 2012
Published by the Royal Society of Chemistry, www.rsc.org

not based on the mononuclear cisplatin chemotype to have entered human clinical trials, validated this approach.[5,6] The poly (di/tri)nuclear) chemotype is very diverse, and its structural essence is summarized by the presence of at least two platinum coordination units linked by flexible diamine chains $H_2N–R–NH_2$.[7,8] Dinuclear compounds with straight-chain diamines and linear bridging polyamines based on spermidine and spermine as well as *cis* and *trans* geometric isomers are encompassed. This chapter reviews recent developments on the chemistry and biology of polynuclear platinum drugs and especially the recognition that 'non-covalent' agents based on this motif represent a further challenge to the structure–activity paradigms for platinum antitumor agents by demonstrating that Pt–DNA bond formation is not a strict requirement for manifestation of cytotoxicity and antitumor activity.

8.2   Covalently Binding Polynuclear Platinum Complexes

The clinical development and DNA-binding profile of BBR3464 have been well documented.[5,6] The presence of the high charge (4+) is a unique feature contributing to the potency of the drug—the charge results in a greater affinity for DNA, higher cellular accumulation and more potent cytotoxicity compared to compounds with 'simple' straight-chain diamines, as exemplified by the prototype dinuclear compound $[\{trans\text{-}PtCl(NH_3)_2\}_2H_2N(CH_2)_6NH_2]^{2+}$.[7,8] The non-optimal pharmacokinetics with associated plasma decomposition of BR3464 has spurred the search for more suitable compounds.[9] The general profile of the trinuclear BBR3464 can be replicated using charged polyamines in spermidine or spermine-linked dinuclear platinum complexes, and these compounds represent viable 'second-generation' complexes for drug development (Figure 8.1). Of specific interest are spermine-linked dinuclear compounds based on BBR3610 (Figure 8.2). This bridging polyamine was originally designed to have the same length and charge distribution as the central unit in BBR3464 and was equally potent in tissue-culture and animal-tumour models.[10–12] Variation of the leaving group, by replacing chloride with carboxylate, or the substitution-inert group, by replacing NH_3 with 1,2-diaminocyclohexane (dach), has produced compounds with reduced metabolic deactivation compared to BBR3464. These newer analogs represent promising agents for further studies.[13–15] Polyamines linked to DNA-interacting moieties have been used in attempts to target cytotoxics to cancer cells.[16] Notably a chlorambucil–polyamine combination was significantly more cytotoxic than the free alkylating agent, reminiscent of the situation with BBR3610.[17] In this sense, the platinum-polyamines described here represent a class of 'chimeric' molecules with potential for dual biological activity—the polyamine conjugate does not affect the final outcome of the target reaction, in this case the formation of {Pt,Pt} interstrand crosslinks.[18,19] Polyamine catabolism is linked to antiproliferative activity and apoptosis, and has been a target of polyamine drug development.[20] It is of considerable interest to note that both oxaliplatin

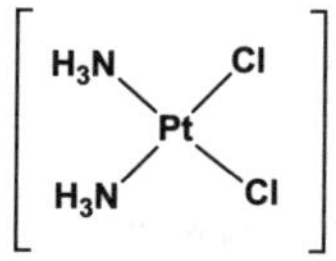

cis-DDP

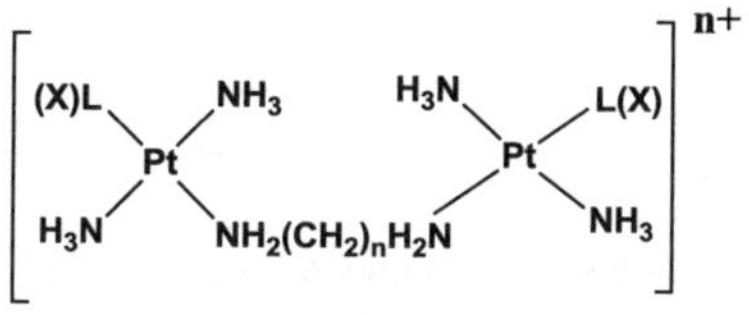

X = Cl, 1,1/t,t
L = NH$_3$, (0,0)

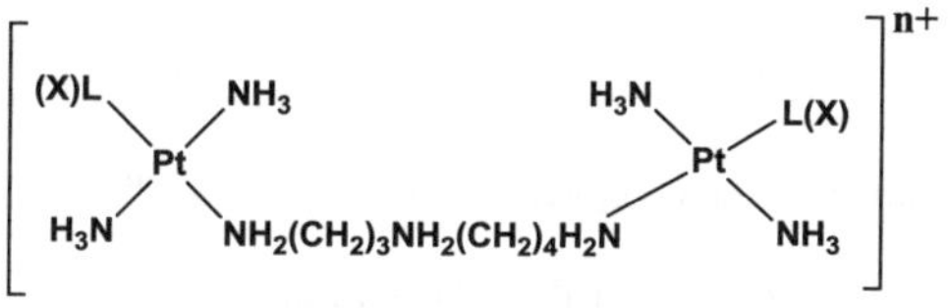

X = Cl, BBR3571
L = NH$_3$, (0,0 spm)
L = NH$_2$(CH$_2$)$_6$NH$_3$$^+$ (0,0 spmda)

X = Cl, BBR 3464
L = NH$_3$, (**1**)
L = NH$_2$(CH$_2$)$_6$NH$_3$$^+$ (**2**, TriplatinNC)

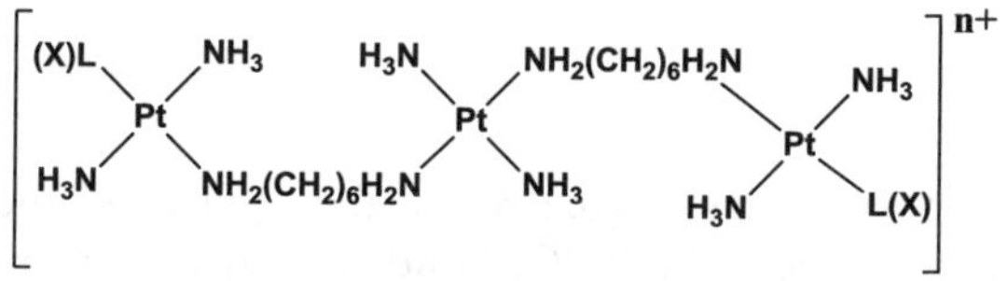

Figure 8.1 Structures of polynuclear platinum complexes showing the main structural types. 1,1/t,t refers to monofunctional Pt coordination spheres with one substitution-labile group (Cl$^-$) in mutually *trans* positions. (0,0) refers to 'non-covalent' platinum spheres with no substitution-labile groups. Thus, (0,0 spm) refers to a non-covalent compound with a spermidine linker, *etc.*

and cisplatin stimulate the induction of polyamine catabolic enzymes such as SSAT by platinum drugs, and biochemical responses and growth inhibition can be potentiated by co-treatment with polyamine analogs.[21]

8.3 Non-Covalent Polynuclear Platinum Complexes

Many of the unique biological properties of BBR3464 and BBR3610 may have their origin in pre-association phenomena on biomolecules.[22,23] The role of pre-association in dictating the final products of reaction of PPCs with

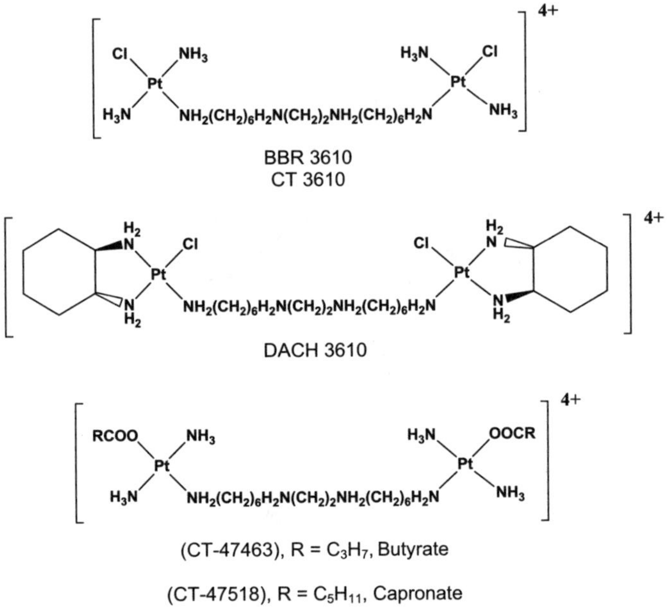

Figure 8.2 Structures of BBR3610 (CT3610) analogs of clinical interest.[14,15]

biomolecules has been assisted greatly by the synthesis and evaluation of 'noncovalent' analogs where the Pt–Cl bonds are displaced by a substitution-inert NH_3 or a 'dangling' amine $H_2N(CH_2)_6NH_3^+$ (Figure 8.1).[24,25]

8.3.1 Global DNA-Binding Profile

The stabilization of CT DNA through the hydrogen bonding and electrostatic interactions of the noncovalent PPCs has been assessed by measuring the increase in melting temperature (T_m) of DNA at various binding ratios (r). The stabilization correlates well with increasing cation charge.[24] The compounds induce B → A and B → Z transitions, which are partially irreversible in canonical DNA. The spectral changes accompanying these conformational transitions vary among the compounds, indicating that the individual linkers as well as the different charges may produce a heterogeneous population of binding sites.[24] The minimum binding ratio for the B → A transition was inversely related to the charge of the complexes. Previously, the conformational change of B → Z DNA induced by the short-chain $[\{Pt(NH_3)_3\}_2\mu\text{-}H_2N(CH_2)_nNH_2]^{4+}$ cations ($n = 2, 4$) were considered reversible by assaying competitive binding with the intercalator ethidium bromide.[26] Use of the polyamine linker affords more efficiency in converting B- to Z-form DNA, and likewise the conformational change is less easily reversed.[24]

8.3.2 Solid-State Studies: A New Mode of DNA Binding

The binding modes of the 'non-covalent' PPCs were examined with the double-stranded B-DNA dodecamer $5'-\{d(CGCGAATTCGCG)_2\}$ (the Dickerson–Drew Dodecamer, DDD).[27,28] A discrete new mode of ligand–DNA binding involving only backbone and groove-spanning interactions has been described. This 'phosphate clamp' uses the Pt(II) square planar geometry to form hydrogen-bonding networks between *cis*-oriented ammine/amine groups and the phosphate oxygen backbone of DNA (Figure 8.3). The two structures solved to date are remarkably similar. Phosphate clamps form six-membered rings, with two hydrogen bonds from *cis*-oriented Pt-a(m)-mines to a common phosphate oxygen. It is a discrete and modular DNA-binding device with high potential as a drug-design scaffold. The structural distortions caused are principally axial bending (approx. 28° compared to 13° for the 'free' DDD) and an increase in minor groove width with concomitant narrowing of the major groove.[28] The mode of binding is distinct from other well-studied ligand–DNA interactions such as intercalation and/or minor groove binding.

Because polyamines can be incorporated easily into the PPC general structure, it is of interest to note that polyamines themselves generally bind preferentially in the major groove rather than to phosphates.[29] Minor groove binding of polyamines is rarer.[30] Key features in differences in recognition

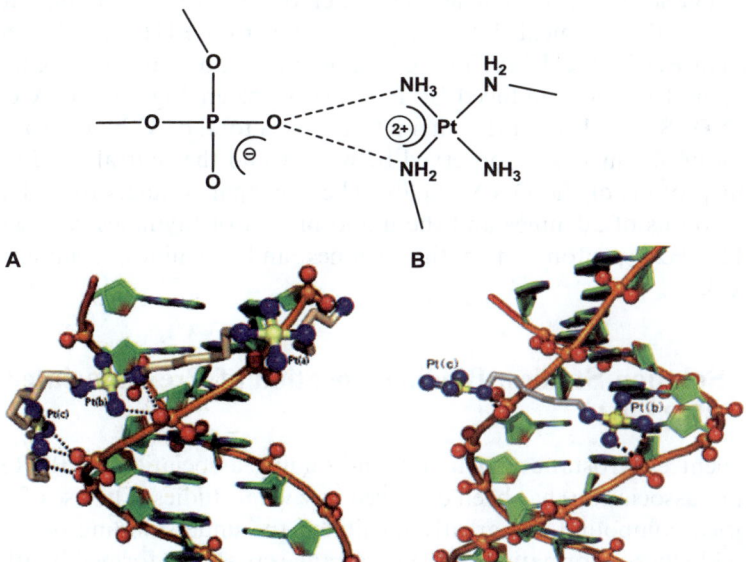

Figure 8.3 Phosphate clamp as DNA recognition motif. For minor groove spanning of compounds **1**(B) and **2**(A), see Figure 8.1.[28,29] The absence of the third Pt unit in B is due to crystallographic disorder.[29] Hydrogen bonds between Pt units and DNA are indicated by dashed lines. Views are perpendicular to the helical axis.

patterns between polyamines and PPCs (especially the non-covalent motif) are the distances between amino groups and the fact that the *cis*-{Pt(NH$_3$(NH$_2$R)} unit in fact is sterically rigid because of the strict adherence to square-planar geometry (approx. 90° bond N–Pt–N bond angle) for Pt complexes. Polyamines are conformationally flexible, and optimal distances to achieve phosphate clamp binding—provided by the platinum square planar geometry—are not as readily available.

8.3.3   Solution Studies: Comparison With Minor Groove Binders

An interesting feature of the DNA binding of non-covalent PPCs is that they induce unique cooperative binding of Hoechst Dye 33258 to DNA.[31] This feature may be related to the opening or widening of the minor groove as observed in the solid-state structures. Because non-covalent PPCs are seen to span the minor groove, it is useful to compare the similarities and differences between these compounds and a minor groove binding agent such as Hoechst 33258. We therefore have examined the interactions of 1 (see Figure 8.1) with the self-complementary duplex d(GGTAATTACC)$_2$, which is a 'high-affinity' sequence for minor groove binders such as Hoechst 33258 and has been used extensively for solution and solid-state studies of 'classical' minor groove binders.[32–34] The largest chemical-shift perturbations were observed for the resonances of the TAATTA segment of the DNA duplex and are consistent with the ligand being located at the center of the AATT binding site. The magnitude of the chemical shift changes in deoxyribose H1' and the base H6/H8 resonances of d(GGTAATTACC)$_2$ upon interaction with the Pt compound are compared to those induced by Hoechst 33258 in Figure 8.4. A comprehensive NOESY analysis of the interaction gave a total of 42 NOE cross peaks, the majority of which were observed between 1 and the central AATT/TTAA base-pair protons of the DNA duplex. The principal contacts from 1 were to the H2 protons of adenines and the imino proton of thymines, as well as the H1', H2', H2'' protons of both adenines and thymines, summarized in Figure 8.4.

8.3.4   Solution Studies: Binding Location of Pre-Associated BBR3464

Pre-covalent electrostatic and hydrogen-bonding association of BBR3464 to DNA (pre-association) has been observed in several studies. The use of the fully ^{15}N-labeled compound has greatly facilitated the understanding of the structure and kinetics of formation of the long-range crosslinks formed by BBR3464 and its congeners using {^{1}H, ^{15}N} HSQC NMR spectroscopy.[19,22,35] This technique is especially useful for following the reactions of PPCs because ^{15}N-labeling of both ^{15}NH$_2$ and ^{15}NH$_3$ groups is possible. ^{1}H and ^{15}N shifts are also sensitive to H-bonding interactions with DNA, and thus the local environment surrounding both ends of the molecule as well as the central linker can be

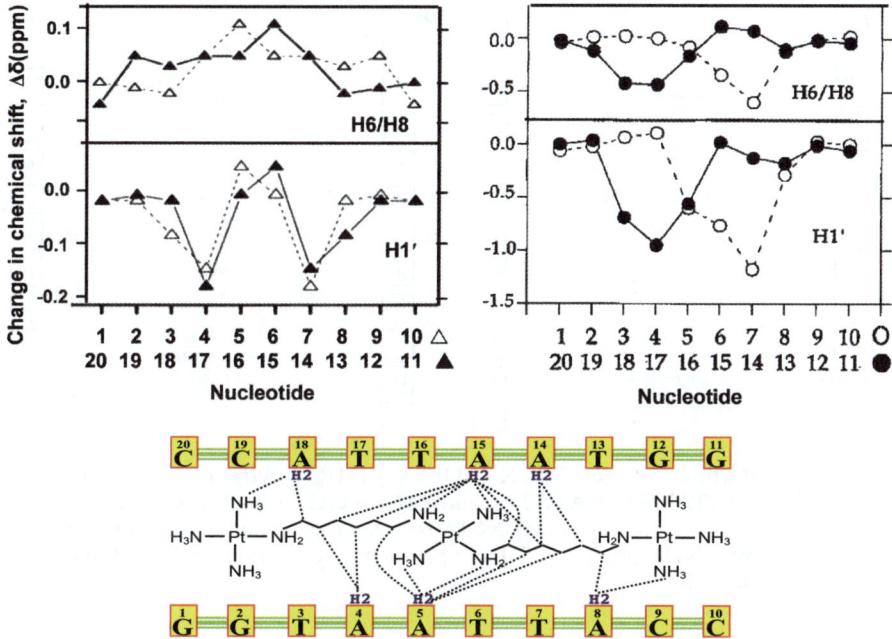

Figure 8.4 Drug-induced perturbations to ^{1}H chemical shift values of H1′ and H6/H8 of duplex d(GGTAATTACC)$_2$ by **1**(left) and Hoechst dye (right).[32] Negative shift differences correspond to resonances that move upfield on ligand binding. The schematic shows NOE connectivities observed in the NOESY spectra. Experimental conditions followed published procedures.[37]

probed throughout the process, including pre-association with the DNA.[36] A {^{1}H, ^{15}N} HSQC NMR spectroscopy examination of 1,4- and 1,6 interstrand crosslink formation in two self-complementary DNA sequences, 5′-{d(ATATGTACATAT)$_2$} (1,4-GG) and 5′-{d(TATGTATACATA)$_2$} (1,6-GG), detected initial interactions of the central platinum moiety with the DNA sequence.[22] NMR structural evidence of the site-specific 1,4-interstrand crosslink formed by BBR3464 in a 5′-{d(ATGTACAT)$_2$} sequence (the central 8 base pairs of duplex **I**) confirmed the central (non-binding) platinum moiety located in or close to the minor groove, suggestive of pre-association.[37]

Non-covalent PPCs are especially useful for comparative studies, as their interactions with DNA can be examined in the absence of any competing Pt–DNA bond formation. The chemical shift changes in 1,4-GG and 1,6-GG duplexes upon incubation with **1** showed the largest changes for the adenine H2 protons (A7 of duplex **I** and A6 and A8 of duplex **II**). Again, NOE cross-peaks are observed between the linker CH$_2$ and NH$_3$ protons of 1 and the adenine H2 protons of both duplexes. The observed connectivities are illustrated in Figure 8.5. For the 1,4-GG duplex, connectivities are seen between the majority of the protons of **1** and the A3, A7, A9 and A11 H2 protons of the DNA sequence. Based on the observed connectivities, the central {PtN$_4$} moiety can

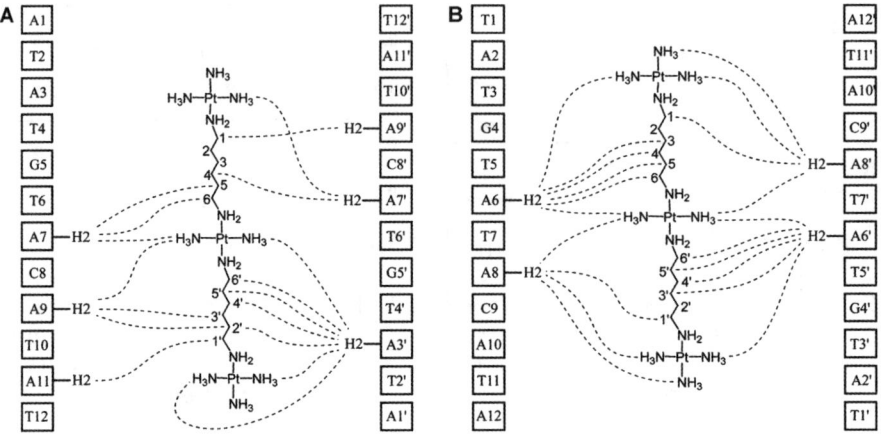

Figure 8.5   Portions of the 298 K NOESY NMR spectra of (A) 1,4-GG: 5'-{d(ATATGTACATAT)$_2$} and (B) 1,6-GG: 5'-{d(TATGTATACATA)$_2$} showing intermolecular NOEs between the adenine H$_2$ protons and CH$_2$ and NH$_3$ protons of **1**. Experimental conditions followed published procedures.[22,37]

be placed in the center of the DNA sequence, around the T7 base, as indicated also from the observed chemical shift changes in DNA protons (see above).

These studies demonstrate two main points. First, the minor-groove spanning of the phosphate clamp motif observed in the crystal structures of **1** is maintained in solution through extensive contacts in the minor groove. The phosphate clamp may give rise to a family of binding sites, as it is expected to be dynamic in nature. Second, the use of the non-covalent compound allows the details and location of the pre-association of BBR3464 to DNA to be provided with substantial evidence that the initial binding site is likely to occur in the minor groove of DNA. The interactions between PPCs and all duplexes suggest that the binding site of the complexes spans several base pairs, as shown by crystallography and molecular biology studies.[27,28,38]

8.4   Biochemical Consequences of Non-Covalent Polynuclear Platinum Association

8.4.1   Melphalan Protection Assay

Standard sequencing assays commonly used to assess binding sites of platinum complexes on DNA may not be applicable in the case of non-covalent compounds, which may simply be displaced by the probe polymerases.[39,40] The minor groove binding of I may be further probed by an adaptation of a melphalan competition assay, shown schematically in Scheme 8.1 (TPNC refers to any trinuclear platinum non-covalent compound).[41,42]

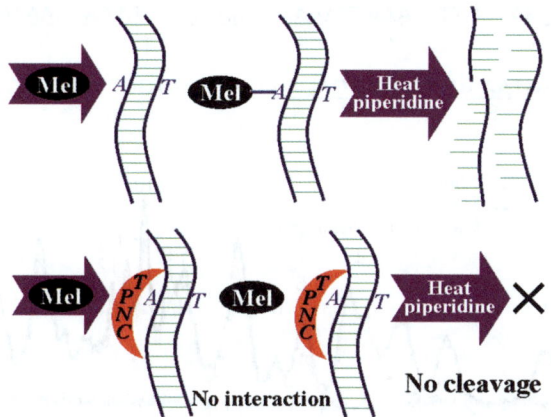

Scheme 8.1 Minor groove binding competition assay (see ref. 41 and 42).

Alkylating agents such as melphalan and chlorambucil bind to both guanine N7 and adenine N3 producing thermolabile sites which are easily visualized by gel electrophoresis as DNA cleavage sites upon treatment with piperidine.[41,42] While guanine N7 lies in the major groove of DNA, the N3 of adenine is situated in the interior of the minor groove. The presence of a ligand such as adriamycin or distamycin with a strong minor groove binding affinity may therefore protect the minor groove binding sites from the effects of alkylation.[41,42] As shown in Figure 8.6, the non-covalent PPC **1** behaves in a similar manner to distamycin, preventing melphalan binding in the minor groove of A-rich regions. The protection was concentration-dependent, and **1** was as effective as distamycin at the same 5 µM dosage (Figure 8.6).

8.4.2 Cellular Accumulation and Cellular Effects

An intriguing feature of PPC biology is that cellular accumulation actually increases with charge (Figure 8.7).[5,25] One of the original structure–activity relationships for platinum antitumor compounds stressed the need for charge neutrality with the understanding that the small molecules would enter cells by passive diffusion. This is certainly no longer the case, and the cellular accumulation of platinum compounds is affected in many ways.[43,44] Alterations in cellular accumulation (intake/efflux) are of clinical relevance in cisplatin resistance.[43,44] An early demonstration of the consequences of cellular accumulation for non-covalent PPCs was shown in an *in vitro* study of BBR3464 and its two non-covalent analogs **1** and **2**, using primary mast cells, chosen for their ability to mimic the factor-dependent and polyclonal nature of the *in vivo* environment.[45] The cellular accumulation increased with charge from 4+ to 6+ to 8+. These results were then confirmed in ovarian tumor lines.[25] Cellular accumulation was further enhanced in mastocytomas (P815 or PDMC) over primary mast cells (BMMC), suggesting a mechanism for enhancement of

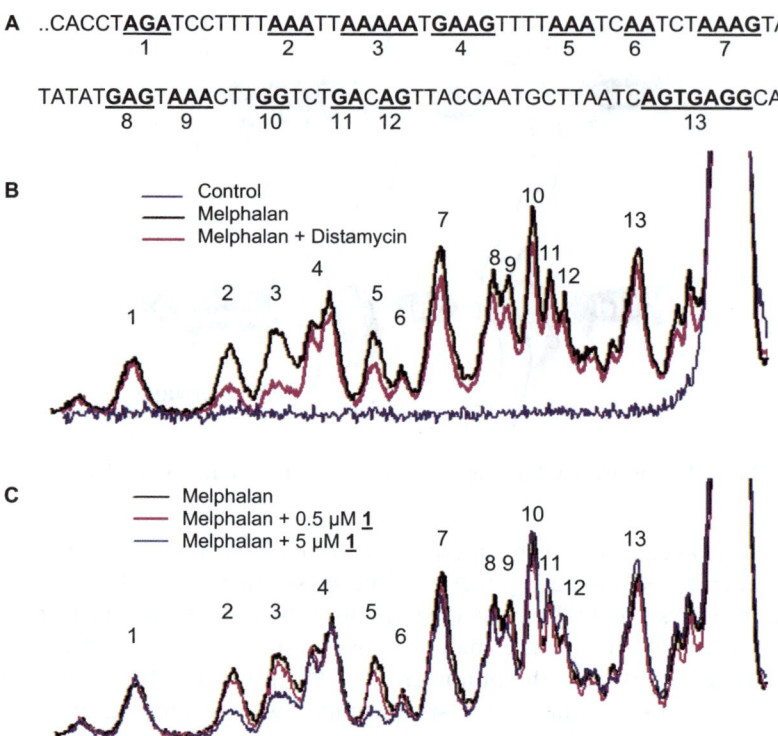

Figure 8.6   Minor groove protection by **1**. A 5′-32P-end labeled duplex (bp 3151-3351 of pBR322) was treated with melphalan in the presence of distamycin or **1** and analyzed by denaturing gel electrophoresis. (A) Sequence (bases 3218–3317) showing oligopurine tracts (underlined). (B) Phosphorimage intensity plot for untreated DNA (blue) or DNA treated with melphalan in the presence (magenta) or absence (black) or distamycin. Numbered peaks correspond to purine tracts in (A). (C) Intensity plot for DNA treated with melphalan in the presence of **1** at 0 (black), 0.5 (magenta) or 5 μM (blue). Note specific protection of oligo-A tracts 2, 3 and 5 by both distamycin and **1**. Electrophoretic migration is to the left; large peak at right is unbroken DNA.

tumor-cell selectivity.[45] In the mast cell study, cellular signaling pathways to apoptosis are affected by compound structure and are different between the covalent (BBR3464) and non-covalent compounds, suggesting that the latter may be a possible new class of antitumor agents in their own right.

8.4.3   Cytotoxicity of Non-Covalent Polynuclear Platinum Compounds

The profile of cytotoxicity of selected non-covalent compounds from Figure 8.1 confirms their interest as a new class of cytotoxic agents. The results of

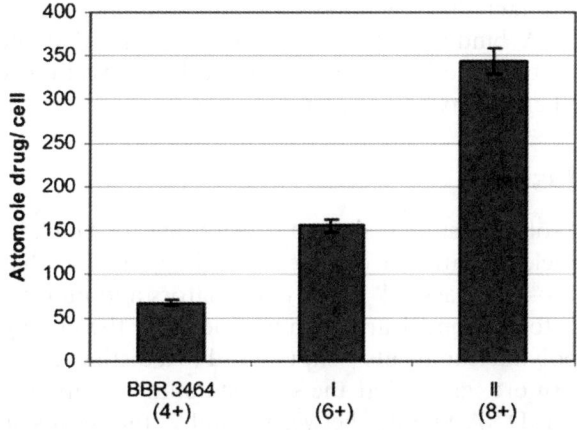

Figure 8.7 Cellular accumulation of trinuclear platinum complexes BBR3464, **1** and **2** in A2780 human ovarian cancer cells.[25] See Figure 8.1 for structures.

Table 8.1 Comparison of cytotoxicity of non-covalent PPCs with their covalently binding analogs in a human ovarian-tumor panel: effect of charge and cellular accumulation on cytotoxicity.

	A2780	A2780cisR	CH1	CH1cisR	SKOV-3
1	41.0	23.0 (0.56)	28.0	56.0 (2.0)	>100
2 (TriplatinNC)	4.3	1.6 (0.37)	3.7	7.4 (2.0)	2.6
(0,0 spm)	>100	>100	82.0	>100	>100
(0,0 spm da)	25.0	48.0 (1.9)	37.0	41.0 (1.1)	48.0
BBR 3464	0.048	0.355 (7.4)	0.016	0.019 (1.2)	0.25
BBR 3571	<0.25	<0.25	0.43	0.35 (0.8)	
c-DDP	0.76	4.2 (5.5)	0.15	0.425 (2.8)	2.4

Data were obtained using sulforhodamine B assays and 96 h incubation.[46] For structures, see Figure 8.1.

cytotoxicity tests across a panel of ovarian tumor cell lines selected for sensitivity as well as intrinsic and acquired resistance to cisplatin are shown in Table 8.1. The cytotoxicity of the non-covalent compounds in general is not as great as the covalently binding compounds, but some interesting points emerge. Incorporation of a dangling amine into the chemotype, as in the formally 8+, yields a significant increase in cytotoxicity over that of the 6+ compound for all cell lines tested. The *enhancement* of cytoxicity with the addition of the dangling amine is also noted for the dinuclear spermidine-linked compounds, but the effect is most significant for the trinuclear species. This is of interest, given that BBR3571 is very similar to BBR3464 in its activity. The vast majority of mononuclear DNA-binding platinum agents display cytotoxicity in the micromolar range, and so the low micromolar range attained for **2** is remarkable, given the charge and the non-covalent nature of the compound.

Furthermore, the fact that the solid-state and solution studies indicate little difference in DNA binding between 1 and 2 suggests that the difference in potency is a reflection of cellular accumulation, thus providing new approaches to modulating the cytotoxicity of platinum agents.

8.5   Conclusions

Solid-state and solution studies show distinct modes of DNA binding by non-covalent polynuclear platinum complexes. It is axiomatic that such distinct modes of binding—to 'classical' DNA intercalators and minor groove binding agents but also to mononuclear platinum and even the closely related covalently binding polynuclear agents—should lead to a distinct profile of biological activity. Furthermore, given that the solid-state and solution studies indicate little difference in DNA binding between 1 and 2, the results suggest that the differential potency is a reflection of cellular accumulation and suggests new approaches to modulating the cytotoxicity of platinum agents. Incorporation of a general polyamine motif into these structures can envision structural changes to further discern the biological activity of these agents.

Acknowledgements

This work was supported by the Australian Research Council and National Institutes of Health (RO1CA78754).

References

1. L. Kelland, *Nature Rev. Cancer*, 2007, **7**, 573.
2. L. R. Kelland and N. Farrell in, *Cancer Drug Discovery and Development*, B. A. Teicher, ed. Humana Press, Totowa, NJ, 2000.
3. R. C. Todd and S. J. Lippard, *Metallomics*, 2009, **1**, 280.
4. L. H. Hurley, *Nature Rev. Cancer*, 2002, **2**, 188.
5. N. Farrell, *Metal Ions Biol. Sys.*, 2004, **41**, 252.
6. J. B. Mangrum and N. P. Farrell, *Chem. Commun.*, 2010, **46**, 6640.
7. N. Farrell, N., Y. Qu, U. Bierbach, M. Valsecchi and E. Menta, in *30 Years of Cisplatin—Chemistry and Biochemistry of a Leading Anticancer Drug*, ed. B. Lippert, Wiley-VCH, Verlag GmbH, Weinheim, Germany, 1999, p. 479.
8. N. Farrell and S. Spinelli in '*Uses of Inorganic Chemistry in Medicine*', ed. N. Farrell, Royal Society of Chemistry, Cambridge, 1999, p. 124.
9. V. Vacchina, L. Torti, C. Allievi and R. Lobinski, *J. Anal. At. Spect.*, 2003, **18**, 884.
10. C. Billecke, S. Finniss, L. Tahash, C. Miller, T. Mikkelsen, N. P. Farrell and O. Bögler, *Neuro-oncology*, 2006, **8**, 215.
11. H.-S. Gwak, T. Shingu, V. C. Chumbalkar, Y.-H. Hwang, R. DeJournett, K. Latha, D. Koul, W. K. Yung, G. Powis, G., N. P. Farrell and O. Bögler, *Int. J. Cancer*, 2011, **128**, 787.

12. C. Mitchell, P. Kabolizadeh, J. J. Ryan, J. D. Roberts, A. Yacoub, D. T. Curiel, P. B. Fisher, M. P. Hagan, N. P. Farrell, N. P. S. Grant and P. Dent, *Mol. Pharmacol.*, 2007, **72**, 704.
13. J. W. Williams, Y. Qu, G. H. Bulluss, E. Alvorado and N. P. Farrell, *Inorg. Chem.*, 2007, **46**, 5820.
14. L. Zerzankova, T. Suchankova, O. Vrana, N. P. Farrell, V. Brabec and J. Kasparkova, *Biochem. Pharmacol.*, 2010, **79**, 112.
15. L. Gatti, P. Perego, R. Leone, P. Apostoli, N. Carenini, E. Corna, C. Allievi, U. Bastrup, S. De Munari, S. Di Giovine, P. Nicoli, M. Grugni, M. Natangelo, G. Pardi, G. Pezzoni, J. W. Singer and F. Zunino, *Mol. Pharm.*, 2010, **7**, 207.
16. P. M. Cullis, R. E. Green, M. E. Malone, L. Merson-Davies and R. Weaver, *Biochem. Soc. Trans.*, 1994, **22**, 402S.
17. J. L. Holley, A. Mather, R. T. Wheelhouse, P. M. Cullis, J. A. Hartley, J. P. Bingham and G. M. Cohen, *Cancer Res.*, 1992, **52**, 4190.
18. T. D. McGregor, J. Kasparkova, K. Neplechova, O. Novakova, H. Penazova, O. Vrana, V. Brabec and N. P. Farrell, *J. Biol. Inorg. Chem.*, 2002, **7**, 397.
19. R. A. Ruhayel, J. J. Moniodis, X. Yang, J. Kasparkova, V. Brabec, S. J. Berners-Price and N. P. Farrell, *Chem. Eur. J.*, 2009, **15**, 9365.
20. R. A. Casero, Jr and P. M. Woster, *J. Med. Chem.*, 2009, **52**, 4551.
21. S. Hector, R. Tummala, N. D. Kisiel, P. Diegelman, S. Vujcic, K. Clark, M. Fakih, D. L. Kramer, C. W. Porter and L. Pendyala, *Cancer. Chemother. Pharmacol.*, 2008, **62**, 517.
22. A. Hegmans, S. J. Berners-Price, M. S. Davies, D. S. Thomas, A. S. Humphreys and N. Farrell, *J. Am. Chem. Soc.*, 2004, **126**, 2166.
23. E. I. Montero, M. J. Oehlsen, B. T. Benedetti, J. B. Mangrum, Y. Qu and N. P. Farrell, *J. Chem. Soc. Dalton Trans.*, 2007, 4938.
24. Y. Qu, A. L. Harris, A. Hegmans, A. Petz, H. Penazova and N. Farrell, *J. Inorg. Biochem.*, 2004, **98**, 1591.
25. A. L. Harris, X. Yang, A. Hegmans, L. Povirk, J. J. Ryan, L. R. Kelland and N. Farrell, *Inorg. Chem.*, 2005, **44**, 9598.
26. P. Wu, M. Kharatishvili, Y. Qu and N. Farrell, *J. Inorg. Biochem.*, 1996, **63**, 9.
27. S. Komeda, T. Moulaei, K. K. Woods, M. Chikuma, N. Farrell and L. D. Williams, *J. Am. Chem. Soc.*, 2006, **128**, 16092.
28. S. Komeda, T. Moulaei, M. Chikuma, A. Odani, R. Kipping, N. P. Farrell and L. D. Williams, *Nucl. Acids Res.*, 2011, **39**, 325.
29. K. K. Woods, T. Maehigashi, S. B. Howerton, C. C. Sines, S. Tannenbaum and L. D. Williams, *J. Am. Chem. Soc.*, 2004, **126**, 15330.
30. H. Ohishi, K. Suzuki, M. Ohtsuchi, T. Hakoshima and A. Rich, *FEBS Lett.*, 2002, **523**, 29.
31. A. L. Harris, Y. Qu and N. Farrell, *Inorg. Chem.*, 2005, **44**, 1196.
32. C. E. Bostock-Smith, S. Harris, C. A. Laughton and M. S. Searle, *Nucl. Acids Res.*, 2001, **29**, 693.
33. C. E. Bostock-Smith, C. A. Laughton and M. S. Searle, *Biochem. J.*, 1999, **342**, 125.

34. K. J. Embrey, M. S. Searle, D. J. Craik and D. J. , *Eur. J. Biochem.*, 1993, **211**, 437.

35. J. W. Cox, S. J. Berners-Price, M. S. Davies, Y. Qu and N. Farrell, *J. Am. Chem. Soc.*, 2001, **123**, 1316.

36. S. J. Berners-Price, L. Ronconi and P. J. Sadler, *Prog. Nucl. Magn. Reson. Spectrosc.*, 2006, **49**, 65.

37. Y. Qu, N. J. Scarsdale, M.-C. Tran and N. P. Farrell, *J. Biol. Inorg. Chem.*, 2003, **8**, 19.

38. V. Brabec, J. Kasparkova, O. Vrana, O. Novakova, J. W. Cox, Y. Qu and N. Farrell, 1999, *Biochemistry*, **38**, 6781.

39. J. G. Collins and N. J. Wheate, *J. Inorg. Biochem.*, 2004, **98**, 1578.

40. N. J. Wheate, S. M. Cutts, D. R. Phillips, J. R. Aldrich-Wright and J. G. Collins, *J. Inorg. Biochem.*, 2001, **84**, 119.

41. P. Wang, G. B. Bauer, R. A. O. Bennett and L. F. Povirk, *Biochemistry*, 1991, **30**, 11515.

42. P. Wang, G. B. Bauer, G. E. Kellogg, D. J. Abraham and L. F. Povirk, *Mutagenesis*, 1994, **9**, 133.

43. M. D. Hall, M. Okabe, D. W. Shen, X. J. Liang and M. M. Gottesman, *Annu. Rev. Pharmacol. Toxicol.*, 2008, **48**, 495.

44. S. B. Howell R. Safaei, C. A. Larson and M. J. Sailor, *Mol. Pharmacol.*, 2010, **77**, 887.

45. A. L. Harris, J. J. Ryan and N. Farrell, *Mol. Pharmacol.*, 2006, **69**, 666.

46. A. Hegmans, J. Kasparkova, O. Vrana, V. Brabec and N. P. Farrell, *J. Med. Chem.*, 2008, **51**, 2254.

Polyamine-Based Agents for Gene and siRNA Transfer

IAN S. BLAGBROUGH*, ABDELKADER A. METWALLY
AND OSAMA A. A. AHMED[†]

Department of Pharmacy and Pharmacology, University of Bath, Bath BA2 7AY, UK

9.1 Introduction

It is still widely reported that gene therapy (delivering a missing or aberrant plasmid DNA) and siRNA therapy (silencing the over-production of an unwanted protein) will become an efficient medicine for the treatment of diseases such as cancer, cystic fibrosis and blindness, and for vaccination. For example, in cancers, cells are characterized by their hereditary properties and uncontrolled growth, often due to aberrant signaling. The current cancer-treatment strategies (surgery, chemotherapy and/or radiation) are useful for the early stages of cancer, but in the late metastasis stage, although chemotherapy is the most successful strategy, it has several side-effects, *e.g.* the suppression of bone marrow and other fast-dividing tissues, potential genesis of secondary cancers and the development of resistant phenotypes. Therefore, there is an unmet clinical need for gene or siRNA delivery.[1–3]

The essential requirements for gene therapy are the transport of DNA or siRNA through the cell membrane and ultimately to the nucleus for DNA and

† Present address: Faculty of Pharmacy, Minia University, Egypt, and King Abdulaziz University, PO Box, 80200, Jeddah 21589, KSA.

RSC Drug Discovery Series No. 17
Polyamine Drug Discovery
Edited by Patrick M. Woster and Robert A. Casero, Jr.
© Royal Society of Chemistry 2012
Published by the Royal Society of Chemistry, www.rsc.org

slightly easier to the cytosol for siRNA. Different strategies have been used for the delivery of genetic material into target cells which are classified as viral or non-viral delivery systems.[4,5] Viral delivery systems depend on the development of genetically modified viruses to utilise their capability of efficiently delivering DNA into cells through their natural infective mechanisms without their pathogenic characteristics. Although a high efficiency is achieved by viral vectors, there are concerns about their use, which include a limit to the size of the DNA to be delivered (the 'payload'), endogenous viral recombination, unexpected anti-vector immune response and oncogene activation.[6-8] Viral gene therapy suffered a major blow in 1999 with the first fatal case, the death of 18-year-old Jesse Gelsinger in a clinical trial (for ornithine transcarboxylase deficiency, OTCD) delivering the treatment using adenoviral vector, which is believed to have caused a severe immune response. Another major blow came in 2003, when a child treated in a French viral gene-therapy trial developed a leukemia-like condition that made the FDA temporarily halt all gene-therapy trials using retroviral vectors. These setbacks for viral vectors[9] have served to attract many laboratories to focus on the development of novel non-viral formulations for efficient gene delivery which must be non-toxic.

Non-viral delivery systems include the use of DNA alone, so-called 'naked DNA', or complexed with synthetic cationic lipids (lipoplexes) or cationic polymers (polyplexes). These vectors not only circumvent the drawbacks of viral vectors but also have the advantages of simplicity of use and the ease of large-scale production.[10] The major challenge for non-viral vectors is the lower transfection efficiency compared to viral vectors. In terms of concept, the naked DNA strategy is considered the safest,[10] but due to the limitations of level of expression, other procedures are being investigated to increase transfection efficiency, *e.g.* electroporation,[11] gene gun,[12] ultrasound[13] and hydrodynamic (high-pressure) injection.[14] Each of these techniques also suffers from the need for complicated medical devices for application and the limitation of only achieving drug (gene) delivery to the superficial tissues.

Synthetic non-viral gene carriers (lipoplexes and polyplexes) are being widely investigated due to their advantages: a wide variety of chemical ways of synthesizing and assembling these transfecting agents, their formulation from well-known components, their higher capability to deliver DNAs of larger sizes, their safety and ease of large-scale production. Although non-viral carriers have poor transfection efficiency in comparison with viral vectors, they have received considerable attention following the formulation of Lipofectin® by Felgner and his research group in 1987,[15] which encouraged many other laboratories to develop polyamine-based non-viral gene-delivery formulations. This research field continues to make progress with improved DNA and siRNA delivery efficiency *in vitro*, but we still need a better understanding of the molecular mechanism of polynucleotide delivery to target cells, and a more detailed understanding of the barriers to non-viral gene delivery in order to achieve efficient formulation for *in vivo* studies and eventually therapies. These barriers, which hinder the delivery of DNA and siRNA to their physical sites of action (the nucleus and the cytosol

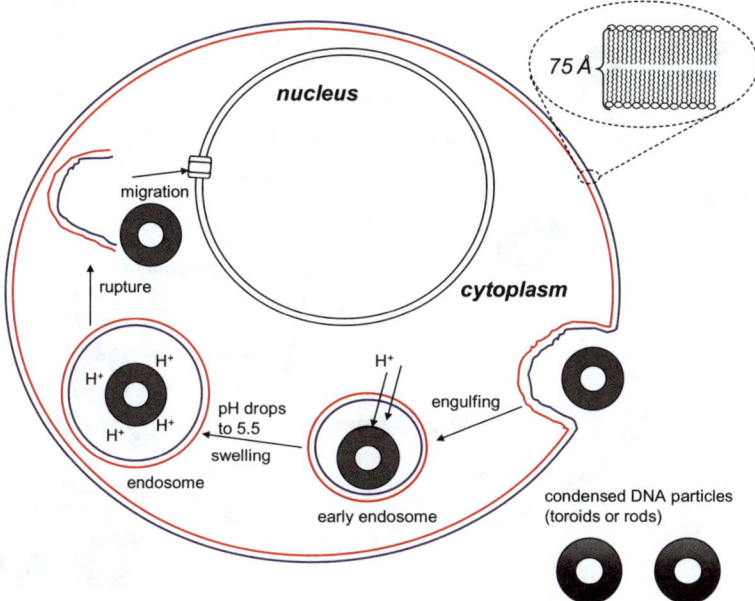

Figure 9.1 Steps in the process of non-viral gene therapy by endocytosis showing the barriers to DNA nanoparticles. This process goes from the formation of the DNA–polycation complex to protein synthesis which includes complex formation between the DNA and cationic polymer or lipid that leads to condensation of DNA into nanoparticles in the form of rods or toroids. According to the results reported by Böttcher *et al.* using cryo-electron microscopy,[28] the dimensions of toroids produced by condensation of pEG*lac*Z condensed with spermine were 15–40 nm for the central hole, with external diameter 50–110 nm and toroid thickness 11–28 nm. The calculated hollow volume was 0.8×10^4 to 1.7×10^5 nm^3. Cell-membrane entry is thought to be mediated by cationic substances which interact with DNA. This interaction causes adsorptive endocytosis and internalization of the complex. The internalized material is fused with the endosomal compartment, and then the DNA complex must escape the endosomal vesicle before the later stage of the lysosome where the DNA is degraded.

respectively) (Figures 9.1 and 9.2), and which form the backbone for the organization of this chapter, include: (1) condensation, (2) cell targeting, (3) cell membrane entry, (4) endosomal escape, (5) nuclear entry, (6) decomplexation and (7) transcription and translation.

Although DNA and siRNA are both polynucleotides, there is a large difference in the size of each. DNA used in gene therapy is in the range of 5–10 kbp plasmids, with a molecular weight of 3.3–6.6 MDa, and carrying 10 000–20 000 negative charges. siRNA is double-stranded RNA (dsRNA) typically of 21–25 nucleotides per strand. So, while there is clearly a conceptual parallel between the requirements for gene delivery and siRNA delivery, experiments from our research group and from others[16] have shown that there

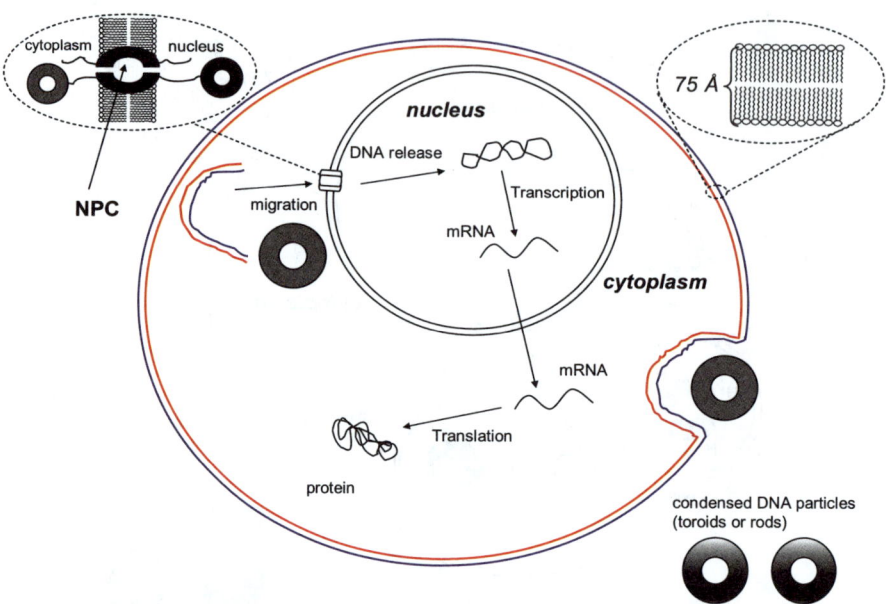

Figure 9.2   Steps in the process of non-viral gene therapy after endosomal escape, the first barrier after cellular internalization of the nanoparticles. The DNA either complexed or dissociated from the condensing agent should find its way (as DNA is subject to hydrolysis by DNase in the cytosol) to the nucleus. Also, the DNA should cross the nuclear membrane which is thought to occur through the nuclear pore complex (NPC) or by direct association with the chromatin during mitosis. After nuclear entry, the DNA should successfully be able to give the desired protein through transcription and translation processes, *e.g.* access of transcription factors may be affected by the dissociation rate of the DNA complex.

is no immediate crossover between a polyamine-based agent that can efficiently transfer a gene, or one that has been written-off as inefficient, and the structure–activity relationship profile of siRNA transfection agents. Not only does more research have to be done, but intuitive guesses and inspiration from years of work on DNA have not yet worked for siRNA delivery. Langer *et al.* have undertaken a huge (combinatorial library) study in the design and preparation of new lipid-like materials, which they termed lipidoids, in order to establish RNAi therapeutics on a certain chemical foundation.[17] They have also published a review of current advances and current understanding of barriers in siRNA delivery.[18] We have reviewed animal models for target diseases in gene therapy, using a variety of polyamine-based DNA and siRNA delivery strategies.[19] A recent review of the chemical sciences integral to RNAi is from Cosstick and co-workers.[20] This is an important research area in the basic science of polyamine–polynucleotide interactions which should find application in the therapy of difficult-to-treat diseases.

9.2 Barriers to Polynucleotide Delivery

9.2.1 DNA Condensation

For drug formulators, it is difficult to deliver a drug molecule with a molecular weight of 3.3 kDa carrying 10 negative charges, but in the case of the prodrug DNA, a 5 kbp plasmid has a molecular weight of about 3.3 MDa and carries 10 000 negative charges. Each base pair (bp) is 660 Da in an average/random DNA duplex sequence. So, the first and a key step in gene delivery is DNA condensation into a nanoparticle form by masking the negative charges of the phosphate backbone. This titration with a cationic lipid or polymer causes alleviation of charge repulsion between remote phosphates along the DNA helix leading to collapse into a more compact structure. This is a phase transition from extended conformation to a compact structure that facilitates cell entry.[21] This phase transition is favored when about 90% of the negative charges along the phosphate backbone are neutralized[22–24] based on the counterion condensation theory of Manning.[25–27]

The use of an efficient carrier for DNA is considered to be a determinant factor for the successful application of gene therapy. The carrier is responsible for the complex process of gene delivery to the nucleus.[29] The ability of multivalent cations such as polyamines to condense DNA into rods[30] or toroids[31,32] has assembled DNA in a nanoparticle form stable for gene delivery. Cationic lipids and polymers have the ability to condense DNA into particles that can be readily endocytosed by cultured cells. The formed nanoparticles and their size are essential for the development of gene-delivery vehicles.[33,34] In addition, cationic lipids and polymers increase the serum stability of DNA.[35]

DNA is condensed by naturally occurring positively charged polyamines spermine and spermidine (Figure 9.3) working together with basic proteins, called histones.[23] Histones are relatively small, highly basic (*i.e.* at physiological pH 7.4, they carry many positive charges) proteins with molecular weights in the range of 10–20 kDa.[36] The five major types of histone chains are: H1, H2A, H2B, H3 and H4 differentiated on the basis of their relative content of lysine (Lys, K) and arginine (Arg, R). The amino acid sequences of four histones (H2A, H2B, H3 and H4), from a wide variety of organisms, are similar among different species.[36] The histones appear to be involved in the supercoiling of DNA in eukaryotic chromosomes through their recurring positive charges, due to the presence of a high ratio of positively charged (basic) amino acids Lys and Arg that leads to the formation of electrostatic associations with the negatively charged phosphate groups of DNA, which renders the DNA more stable and

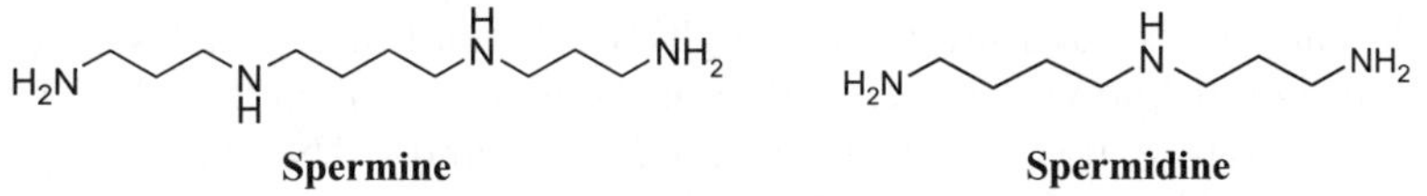

Spermine   **Spermidine**

Figure 9.3 Linear polyamines spermine and spermidine.

flexible.[36] The double helix of DNA, together with the associated histones, is supercoiled and folded back and forth many times in the chromatin fibres. Polymers of histidine (His, H) and the non-protein basic amino acid ornithine (Orn, O) have also been used for DNA condensation.[37–41]

There are few studies that utilized histones in the design of non-viral gene therapy. H1 and the cationic lipid DOSPER was formulated for the investigation of luciferase gene (pCMV Luc) delivery in ECV 304 human endothelial cells.[42] In a different investigation, Hoganson *et al.* showed that ligand-mediated endocytosis can specifically deliver DNA to the cells that have these specific receptors *in vitro*.[43] They targeted FGF receptor-bearing cells by the ligand fibroblast growth factor (FGF2) with DNA encoding saporin. Saporin is a potent ribosomal inactivating protein. In addition, FGF2 was also used in their study with DNA encoding the conditionally catatonic herpes simplex virus thymidine kinase, a protein that can kill cells by activating the prodrug ganciclovir. However, when histone H1, a ligand that binds to cell-surface heparan sulfate proteolysis ('low-affinity' FGF receptors), was used to deliver DNA encoding thymidine kinase, no ganciclovir sensitivity was observed.[43]

Chemically modified histones were also investigated in gene-delivery formulations. Galactosylated histone, a ligand for asialoglycoprotein receptors, was synthesized and constructed in an Epstein–Barr-virus-based expression vector to improve levels of the human cytokine interleukin-2 (IL-2) cDNA gene expression through targeted ligand receptor mediated endocytosis into hepatoma (HepG2) cells via the hepatocyte cell-surface receptors.[44] H1 was conjugated with nuclear localization signal (NLS) for DNA binding and formulated with dioleoyl phosphatidylethanolamine (DOPE) or phosphatidylserine (PS) as liposomes.[45]

DNA-interaction studies with polycations showed the condensation of DNA to nanoparticles with different structures and different particle sizes that affect the delivery of DNA to the cells. The investigation of DNA condensation into rods and toroids, as in most vertebrate sperm cells,[46] has attracted many research laboratories to thoroughly study the structure of DNA toroids. Studies on DNA toroids have revealed different models for the organization of DNA within toroids: the circumferential DNA wrapping (spool) model,[28,31] the constant radius-of-curvature model,[47] and the latter model assumed by Hud and Downing.[48] Hud and co-workers also showed the nucleation of DNA condensation by static loops and the formation of DNA toroids with reduced dimensions.[49] Detection and characterization of the condensed DNA nanoparticles using polycations have generated great interest in several laboratories (see Table 9.1). If we can gain control of DNA condensation with polyamines, either in a planned manner or even serendipitously, this will be a significant step forward in the development of non-viral gene therapy. This may also apply to siRNA and its condensation, although the formation of siRNA toroidal structures is not known and certainly not studied as well as their DNA counterparts. Therefore, it is important to study DNA and siRNA bending prior to condensation. The introduction of localized static curvature into an otherwise linear duplex DNA polymer can have a profound effect on the size of toroidal

Table 9.1 DNA-condensing agents and their application for nanoparticle formation and gene delivery.

Carrier	DNA	Particle size (nm)	Method	Notes	References
Spermidine	T7	45–130	EM or DLS	90% have a toroid shape when stained with uranyl acetate, while unstained show 10% toroids (outer diameter: 75–97 nm; inner diameter: 28.7–38.5 nm); collapse of DNA is a result of 89–90% neutralization of the total phosphate charges	22, 23, 51
Spermine and spermine analogues	λ-Phage, pEG*lacZ*, pSFiSVneo.pSfiSV19 or pCISfi-γIFN	50–130	SFM, DLS or EM	Distribution of positive charges along the polyamines backbone plays a major role in condensation of DNA	28, 52
Reducible lipopolyamine analogue	pCMV-Luc	100	DLS	Over +/− charge ratio of 4 gives particles about 100 nm, while at charge ratio between 1 and 3, large particles were detected	53
Cationic silanes	pBR322	40	AFM	At lower cationic silane concentrations, there are flower- and sausage-shaped structure condensates; at higher concentrations, toroids and rods are formed	54
Cobalt hexamine	λ-Phage, polynucleotides	Outer 95–185 hole, 35–85	EM	Diameter of toroids formed may be affected by poly-L-lysine during preparation; also, the volume of toroid and the amount of DNA/toroid can be calculated	48, 50, 55
Peptide and peptide conjugates	pCMV-β-gal, pCMVLuc	20–700	PCS, DLS	Short peptides with cationic lipids improve both condensation and transfection	56, 57
Chitosan and chitosan derivatives	pcRELuc, pEGFP1, pCMVCA	27–750	DLS, TEM, AFM	Particle size decreased as the charge ratio increased	58, 59
Biodegradable polymers	25-mer ODN (Phosphoro-thioates), pGL3, pBR322, pSV-β-gal	70–400	PCS, TEM, DLS, AFM	Nanoparticles <100 nm showed a higher transfection than larger nanoparticles	60–62

Table 9.1 (*Continued*)

Carrier	DNA	Particle size (nm)	Method	Notes	References
Thermo-sensitive polymers	pCMV-*LacZ*	150–200	DLS	Zeta-potential can also be used as a characteristic to predict the behaviour of this type of co-polymer/plasmid complexes in transfection	63
Cholesterol derivatives	pSV2CAT	200–400 400–1400 >2000	AFM	Zeta-potential is an important factor for controlling transfection	64
DC-Chol—non-ionic surfactants	pCMVLuc	Tween 384 Span 1365 Brij 1937	DLS	Addition of non-ionic surfactant prevents the formation of large aggregates	65
Cationic liposomes (cationic cholesterol-DOPE-protamine)	pGL3	150–200	AFM	Addition of protamine facilitates the internalization of DNA complex to the nucleus	66
Cationic liposomes (DOTMA-DOPE)	pSV2-CAT	~250	LLS	The first NVGT formulation	67
Liposome-HK-Polymer	PCI-Luc	40–100	DLS	The ratio of His to Lys in the co-polymer affects transfection efficiency	37
Gelatine	pcRELuc	200–750	Differential interference contrast microscope	Nanospheres were synthesized by salt-induced complex coacervation	68

condensates formed when the entire polymer is condensed by multivalent cations. For DNA toroid formation, with the initial step of spontaneous formation of a single loop along the DNA polymer, the loop then acted as a nucleation site for DNA condensation and was responsible for defining the size and morphology of the toroid.[47] Such loops form as a result of random polymer fluctuations in solution, with their size being a function of the persistence length of DNA. There is evidence to support this process from real-time studies of DNA condensation where toroid formation follows a nucleation-growth pathway. If a static loop or multiple loops were introduced into an otherwise linear DNA polymer, these loops would provide a site for toroid nucleation temporally favored over loops formed by random polymer fluctuations. Controlling the size of these static loops would then provide a means for altering toroid dimensions. Long-range static curvature was produced by multiple A-tracts whose incremental bends were in phase with the helical repeat of DNA (10.5 bp). The results reported showed that toroid size can be influenced by the size of the first loop (or loops) upon which DNA is condensed.[49] In addition, investigations using hexamine cobalt (III) to control the size of the condensed DNA particles revealed that the size of the initial nucleation loop in toroid formation has a major effect on the toroid size, and the ionic strength of the solution will affect the thickness of the formed toroids.[50]

9.2.2 Cell Targeting

Several experiments were undertaken to link a DNA complex with a targeting ligand in order to improve nonspecific binding of positively charged DNA nanoparticles to cells with a more specific interaction. This specific binding will lead to an increase in the efficiency of transfection and a reduction in the toxicity of the vector where DNA will be internalized by receptor-mediated endocytosis more efficiently to specific organs or cells that contain receptors for the target ligand. To utilize this improvement by targeting ligands, a number of gene-delivery systems have been formulated to include targeting ligands. These ligands were derived from several classes, including: carbohydrates, vitamins, peptides and antibodies.

From the carbohydrate class, galactose and mannose were investigated for cell-specific receptor-mediated endocytosis in gene delivery using mannosylated poly-L-lysine (PLL) and galactosylated PLL.[69–71] The asialoglycoprotein receptor on hepatocytes and the mannose receptor on macrophages and liver endothelial cells are the targets for galactosylated and mannosylated gene carriers.[70,72] Behr *et al.* designed lipopolyamine–DNA complexes with galactose,[73] and galactosylated chitosan was formulated for hepatocyte targeting.[74,75] Galactosylated histone as a ligand was synthesized and constructed in a viral vector to improve levels of gene expression into hepatoma (HepG2) cells.[44] Lactose was also utilized for hepatoma cell-targeting as a lactose-poly(ethylene glycol)-grafted PLL gene carrier, and for targeting cystic fibrosis (CF) airways with improved transfection and reduced cytotoxicity.[76,77]

B-Group vitamins have also been utilized for targeting cancer cells, and the folate receptor is a tumour marker overexpressed in most human tumours studied for targeting using folic acid (B_9).[78-80] As cancer cells have a high demand for vitamin B_{12} taken up by receptor-mediated endocytosis, this opens up the possibility for vitamin B_{12} to be used as a targeting ligand leading to preferential uptake by cancer cells, as investigated by Grissom and co-workers.[81,82]

Among the peptides class of targeting ligands, the synthetic peptide arginine–glycine–aspartate (RGD) was successfully introduced in the formulation of non-viral vectors for integrin receptor targeting.[83,84] Transferrin, a natural iron-delivery protein was also investigated for targeted gene delivery. Birnstiel *et al.* are a leading group investigating transferrin as targeting ligand to the transferrin receptor in various cell lines using a transferrin poly(L-lysine) conjugate as a carrier.[85-87] Transferrin was also incorporated with a cyclodextrin polymer-based gene-delivery system. The non-viral system is formed by condensation of a cyclodextrin polycation with nucleic acid into nanoparticles that are surface-modified to display poly(ethylene glycol) (PEG) for increasing stability in biological fluids and transferrin for targeting of cancer cells that express transferrin receptor.[88]

The ligand fibroblast growth factor (FGF2, a 155 amino acids) targeted FGF receptor-bearing cells that showed efficient ligand-mediated endocytosis.[43,89] A cell-binding ligand, the transactivator of HIV (TAT, sequence GRKKRRQRRRGYG) transcription peptide which stimulated cellular uptake, was incorporated on the particle surface of a thiocholesterol-based liposomal cationic lipid formulation and compared with DOTAP/DOPE liposomes and PEI.[90] The sequentially assembled liposomal formulation, incorporating a zwitterionic Cys disulfide, showed an 80-fold enhancement of transfection efficiency over the cationic thiocholesterol-based cationic liposomes at all concentrations tested. HIV-TAT derived peptide was efficient in targeting DNA and glycosaminoglycans for directed gene and drug delivery.[91] An antisense gene targeting formulation to brain cancer cells includes epidermal growth factor chimeric peptide (EGFR) that reduces the growth of EGFR-dependent gliomas.[92]

An antibody-based gene-delivery vector was designed for tumour-cell targeting to the G250 renal-cell carcinoma-associated antigen in tumour cells, using histone H1 to condense, protect and associate DNA with the complex.[93] Antibody-based gene targeting was designed for ovarian carcinoma cells using pegylated polyethylenimine (PEG-PEI) conjugated to the antigen-binding fragment (Fab′) of the OV-TL16 antibody directed to the OA3 surface antigen that is expressed by a majority of human ovarian carcinoma cell lines.[94] The results revealed that luciferase expression increased up to 80-fold compared to PEG-PEI and was even higher than that of PEI 25 kDa non-viral vectors. In addition, peptide NLSs were successfully applied for DNA-nuclear targeting inside the cell after endosomal escape.

9.2.3 Cell Membrane Entry

The mammalian cell membrane is a semi-permeable membrane formed from a phospholipid bilayer and containing various integral proteins. The cell membrane selectively screens the entry of foreign matter to the cell and allows the transport of macromolecules by endocytosis. Neutralization of the negative charges on the DNA by polycations will improve the delivery of DNA through the cell membrane by masking the negative charges on DNA because of the presence of negative charges on both DNA and the cell membrane that hinder the delivery of DNA alone. Also, the positively charged lipid complex will mediate transfection by fusion with the cell membrane and internalization by endocytosis and/or receptor-mediated endocytosis.[95]

9.2.4 Endosomal Escape

The major intracellular barriers that face the delivery of DNA into the nucleus are located in the cytoplasm. Although the DNA complex is efficiently inter-nalized in the cell, a small fraction will be transferred to the nucleus and expressed.[96] As the DNA complex is internalized, it is retained inside the early endosomes and then sorted to the late endosomes that are characterized by increased acidity (low pH). At this stage, the efficient escape of DNA from the endosome is an important factor for successful gene delivery before further trafficking of the trapped DNA to the lysosomal compartment where the DNA will be subject to degradation by lysosomal nucleases.

The possible mechanisms for endosomal escape are:

1. Disruption of the negatively charged endosomal membrane through mixing with the cationic lipid vector.[97]
2. Cationic polymer interaction with the negatively charged endosomal membrane.[98]
3. Proton sponge hypothesis, where a cationic polymer, such as poly-ethylenimine (PEI), has the ability of endosomal pH buffering that leads to chloride ion influx together with water molecules that leads to mem-brane swelling and disruption.[99]
4. Membrane-destabilizing peptides, including fusogenic or lytic peptides, such as the basic peptide KALA,[100] acidic peptide EGLA[101] or uncharged peptides at neutral pH such as H5WYG,[38] have the ability to disrupt the endosomal membrane and increase the release of DNA to the cytosol.
5. Compounds that act as lysosomotropic agents assist in escape from the endosome. Chloroquine inhibits lysosomal processing and sucrose causes osmotic lysis of the endosomal compartment. These compounds are used with DNA condensing agents to improve the efficiency of non-viral gene-delivery formulations.[102,103]

9.2.5   Nuclear Entry

One of the crucial steps for successful delivery of DNA into its site of action is crossing the nuclear membrane. The entry of macromolecules to the nucleus occurs through the NPC, which allows passive diffusion of molecules of up to 50 kDa molecular weight or about 10 nm in diameter, but in the case of larger molecules, NPC is capable of actively transporting particles of about 25–50 MDa or 25 nm in diameter.[104]

The utilization of NLSs through their characteristics of targeting macro-molecules to the nucleus and of increasing transfection efficiency has been applied.[105] The other way for DNA entry to the nucleus is during mitosis, when nuclear membrane is broken down so that the DNA and macromolecules can easily gain access. The first NLS to be mapped in detail was characterized by Smith and colleagues in 1984.[106] The T-antigen NLS is a seven-amino-acid sequence Pro–Lys–Lys–Lys–Arg–Lys–Val. Not only was this sequence necessary for the nuclear transport of T-antigen, but also its addition to other, normally cytoplasmic proteins was sufficient to direct their accumulation in the nucleus.[36]

The use of NLS in non-viral formulation has been investigated with encouraging results. The signal PKKKRKVEDPYC, derived from SV40, was used to increase PEI- and Transfectam-mediated transfection significantly in terms of the amount of DNA required for *in vitro* transfection.[105] In addition, the yeast transcriptional activator, GAL4, has been shown to bind DNA and act separately as an NLS.[107] NLSs are used as adjuvants in the gene-delivery formulation to improve transfection efficiency.[108] Noguchi *et al.* promoted gene transfection of the DNA–protamine–liposome complex with the derivative (I) more significantly into the nucleus of the target cells using the NLS of prota-mine.[66] Nakanishi *et al.* compared the ability of various synthetic NLS peptides to carry large molecules into nuclei.[109] They chemically modified bovine serum albumin (BSA, molecular weight 70 kDa) or IgM (molecular weight 970 kDa, diameter 33 nm) using these peptides and then estimated the propensity of these molecules to translocate into nuclei by injecting them into cytoplasm. Although many NLS can promote the nuclear transfer of BSA, only the 34-mer peptide from SV40 T-antigen could carry IgM into the nucleus.[110] A shorter peptide (12-mer) containing the NLS of T-antigen could carry BSA, but not IgM, into the nucleus, suggesting that the longer peptide contains the additional information required to transfer large molecules into nuclei.[110] However, the understanding of this difference in transport ability on a molecular basis still remains unclear.

9.2.6   Decomplexation

The dissociation of DNA from the condensing agent is an important step for the host transcription factors to initiate the transcription reaction. The decomplexation process may occur after endosomal escape in the cytoplasm[97] or after nuclear entry, as indicated by Godbey *et al.* using confocal images for tracking PEI/DNA complexes.[111] Also, the kinetics of intracellular DNA complex dissociation is an important factor for successful non-viral gene-delivery process.[112] If the DNA dissociated from the condensing agent easily and in the early stages of cellular entry, this would lead to degradation of the delivered gene by DNase in the cell. On the other hand, if the dissociation rate were slow, this would hinder the accessibility of the host transcription factors to initiate the transcription reaction that leads to low or no level of protein expression.

9.2.7 Transcription and Translation

Once inside the nucleus, the next step for gene therapy is transcription, which depends on the promoter strength and the presence of an enhancer in the vector which affects transfection efficiency.[113] Successful transcription leads to translation and the formation of the desired protein (the biological active from the 'prodrug' DNA). Kamiya *et al.* designed a simple pharmacokinetics/pharmacodynamics model for the expression of the delivered gene in the nucleus based on the general consideration that the total amount of mRNA depends on the amount of the exogenous DNA in the nucleus. They concluded from this model that the amount (efficiency) of DNA delivered to the nucleus is the only determinant for expression of the delivered gene.[112]

Unlike DNA, siRNA does not rely upon transcription/translation steps; rather, it brings about sequence-specific degradation of mRNA. The optimum length of siRNA to affect this gene silencing in mammalian cells is typically less than 30 nucleotides in each strand of the dsRNA.[114] This length does not initiate an immune response (does not induce interferon synthesis) that would lead to non-specific mRNA degradation. It has sufficient length to achieve sequence-specific mRNA degradation.[114] The core complex for mRNA degradation is the RNA-induced silencing complex (RISC), a complex of proteins and siRNA with a complementary sequence to the targeted mRNA. The argonaute family of proteins, a RISC component, contains a domain with RNAse H (endonuclease) type of activity that catalyzes cleavage of the phosphodiester bonds of the targeted mRNA.[20] The assembly of RISC and subsequently its function to mediate sequence-specific mRNA degradation occur in the cytoplasm of the cell.

9.3 Polyamines Used in Non-Viral Polynucleotide Formulation

Traditionally, the process of drug design, development and preformulation starts with a main focus on the design of the active chemical of interest (candidate drug) followed by the selection of the excipients. In the case of non-viral gene therapy, the key step is the selection of the suitable excipients (the vector) for the delivery of the DNA or RNA to their respective site of action. Due to the polyanionic nature and the size of the active ingredient, the efforts continue to design novel or improve existing non-viral gene-delivery formulations in order to increase the stability of the drug (DNA), decrease the toxicity of the excipients and/or increase transfection efficiency. Also, as a result of the improvement in DNA-delivery systems, the strategy of antisense oligonucleotide delivery to inhibit gene expression has been widely investigated.[115,116] Vectors used for non-viral gene and siRNA formulation can be classified as cationic polymers, cationic peptides and cationic lipids.

9.3.1 Cationic Polymers

One of the leading cationic polymers that is being investigated as a non-viral vector for gene delivery is PEI.[99] PEI, either linear or branched with different molecular weights (Figure 9.4), possesses a buffering capability that inhibits lysosomal enzyme activity by lowering proton concentration (the proton sponge effect) and presumably promoting the release of DNA by increasing the endosome escape efficiency that leads to improved transfection efficiency.[94,117–121] In addition, PEI has been used in conjugation with cationic lipids to utilise the characteristics of both vectors in one formula.[122]

On the other hand, among cationic polymers, the second class of non-viral vectors are the cationic polysaccharides (amino sugars) that have received increased focus as promising vectors for gene delivery, based on their well-known advantage of safety, following widespread use in drug formulations and across the pharmaceutical industry.

Chitosan (N-deacetylated chitin) is a cationic, linear biodegradable polysaccharide composed of β-(1,4)-linked glucosamine partly containing *N*-acetyl-glucosamine (Figure 9.5). Chitosan is derived by deacetylation of naturally occurring chitin in crab and shrimp shells. The pK_a value of the amino groups

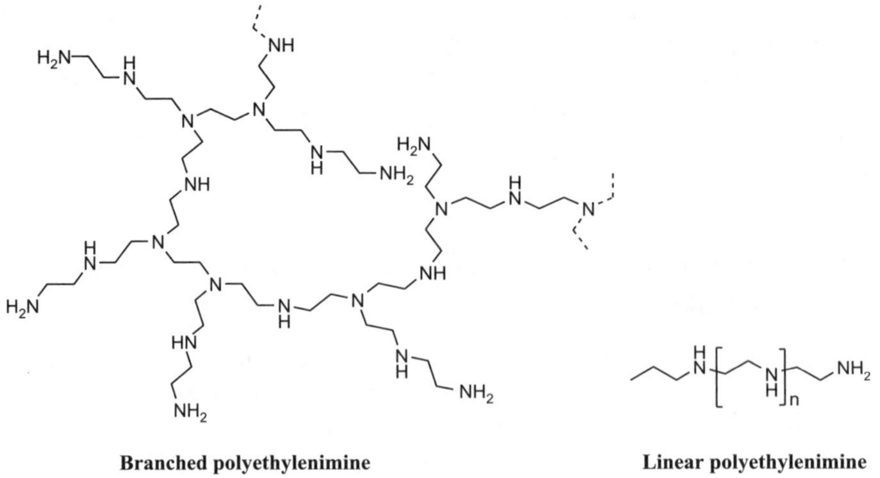

Branched polyethylenimine   **Linear polyethylenimine**

Figure 9.4   Branched and linear PEI.

of the deacetylated moieties is 6.5.[123] Rolland and co-workers were the first research group to report the use of chitosan as a condensing agent in non-viral gene therapy.[124] The biocompatibility and safety of chitosan render it potentially useful for gene delivery.[125–128] Chitosan was evaluated for its ability to transfect different cell lines in comparison with Lipofectamine™ [68,129] and Lipofectin®,[75] but the chitosan used in this paper was galactosylated for hepatocyte targeting,[75] PLL,[130] PEI[131] and *in vivo* administration with endosomolytic peptide GM225.1,[132] showing that chitosan can be used as a safe DNA carrier.

Other laboratories tried to modify chitosan chemically in order to increase the transfection efficiency, enhance cell specificity and/or increase DNA complex stability.[133–135] Chitosan was also chemically modified with hydrophobic cholesterol groups to prepare chitosan-based polymeric amphiphiles.[136] With a similar strategy, Yoo *et al.*[137] prepared a hydrophobically modified glycol chitosan with 5β-cholanic acid (Figure 9.6) based upon their earlier studies.[138,139] The *in vivo* results revealed a higher transfection efficiency over naked DNA. Chitosan microspheres were used to encapsulate two plasmids without affecting their structural and functional integrity that leads to sustained

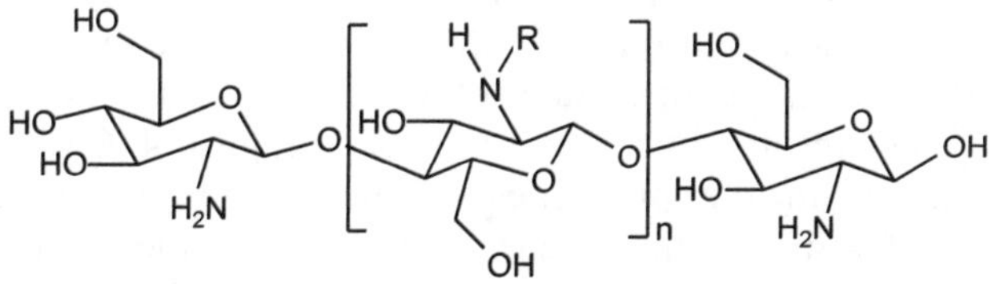

R = H glucosamine
R = COCH₃ *N*-acetyl glucosamine

Figure 9.5 Typical, idealized chitosan, where *n* may be 200–3000, giving a molecular weight of 40–600 kDa.

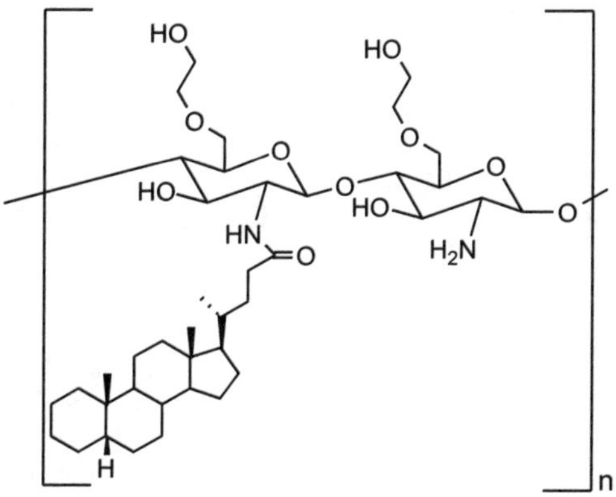

Figure 9.6 5β-Cholanic acid hydrophobically modified glycol chitosan.

and high protein production.[140] Chitosan was also combined with adenoviral vectors to increase the adenoviral infectivity to mammalian cells *in vitro* and *in vivo*.[141,142]

To demonstrate the mechanism of gene delivery by chitosan, Ishii *et al.*[143] analyzed the transfection mechanism of plasmid/chitosan complexes as well as the relationship between transfection activity and cell uptake by using fluorescein isothiocyanate-labeled plasmid and Texas Red-labeled chitosan. They found that plasmid/chitosan complexes most likely condense to form large aggregates (5–8 μm), which absorb to the cell surface. After this, plasmid/chitosan complexes are endocytosed and possibly released from endosomes due to swelling of the lysosomal compartment in addition to swelling of the plasmid/chitosan complex, causing the endosome to rupture. Ultimately, complexes can also be observed with a confocal laser scanning microscope to accumulate in the nucleus.

Chitosan was also investigated for its interaction with a lipid bilayer (didodecyl dimethylammonium bromide, DDAB).[144] The results showed that the low-molecular-weight chitosan could disrupt the lipid bilayer, and the effect seemed to be in a concentration-dependent manner. Also, the modification of chitosan using alkyl bromide leads to the formation of DNA complexes at lower charge ratios than the unmodified chitosan.[145] The long-term stability of chitosan-based polyplex solutions (25 mM sodium acetate buffer, pH 5.5) was investigated at different temperatures (4 °C, 25 °C, 45 °C) over a period of up to 1 year.[146] The results show that 1 year of storage at 4 °C did not result in any major changes in the properties of the polyplexes. At 25 °C, there were minor changes in the physicochemical characteristics of the polyplexes (particle size and zeta potential), and the *in vitro* transfection efficiency was reduced after 1 year of storage. Storage at 45 °C altered both the *in vitro* transfection efficiency and the physicochemical properties of the polyplexes after a short time.

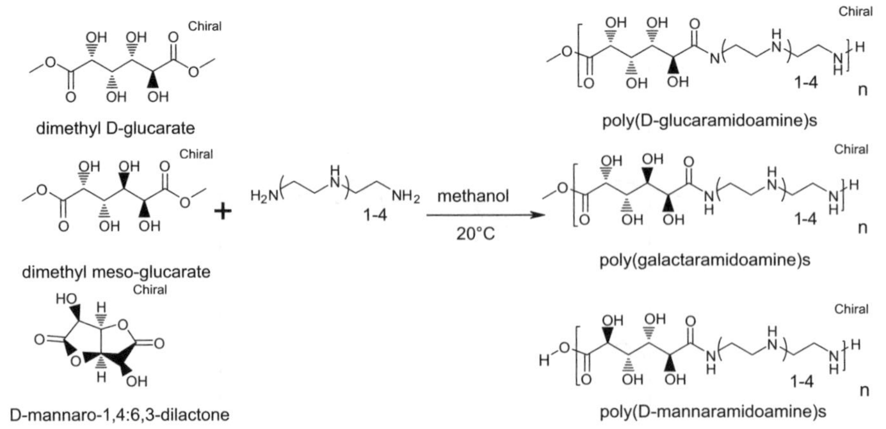

Figure 9.7 Synthesis of poly(glycoamidoamine).[148]

Novel non-viral vectors were also synthesized from a carbohydrate co-monomer (esterified D-glucaric acid) within a PEI-like backbone and also a polymer of esterified D-glucaric acid with diethylenetriamine, triethylenetetraamine, tetraethylenepentamine and pentaethylenehexamine (Figure 9.7) by Reineke and colleagues.[147,148] We have previously purified and conjugated these 2.2.2-mers.[149,150] Using the same concept, the carbohydrates D-trehalose and β-cyclodextrin (CD) were used within the polycation backbone to condense DNA into particles that can be readily endocytosed by cultured cells.[151] β-CDs continue to be promising non-toxic non-viral vectors. Their structural effects were investigated on both gene delivery and conjugation with polyamidoamine dendrimers.[151–154] β-CDs were modified to form cationic amphiphiles and polycationic (polyamino)β-CDs with different cationic groups (amino, pyridylamino or butylimidazole) that were used for efficient DNA condensation and transfection.[155–157] β-CDs were also conjugated with the targeting ligand transferrin for tumour cells and PEG to improve stability in biological fluids.[88] Azzam *et al.* have synthesized spermine conjugated with different-molecular-weight dextrans as gene-delivery vectors.[158]

9.3.2 Cationic Peptides

Cationic peptides that include the positively charged amino acids K, O and R are utilized in the design of non-viral delivery vectors. Arg has a pK_a value of 12.5, and both Lys and Orn have similar pK_a values of 10.8. Arg and Lys are the major constituents of histones (responsible for DNA condensation in the nucleus) and NLS of the delivered proteins from the cytoplasm to the nucleus where they have been used as vectors for DNA delivery.[159,160] Some research groups are interested in the use of the basic amino acid polymers such as PLL, poly-L-ornithine (PLO) and poly-L-arginine (Figure 9.8) that carry multiple positive charges at physiological pH in DNA condensation and gene delivery.[40,41] Also, Wender, Rothbard and their co-workers investigated the role of arginine oligomers as drug transporters for their ability to cross the biological barriers.[161,162] Wender, Siprashvili and others also reported that flanking a core of consecutive arginines with cysteines improves the efficiency of the delivered gene *in vitro*, but these new peptides also increased gene expression in both

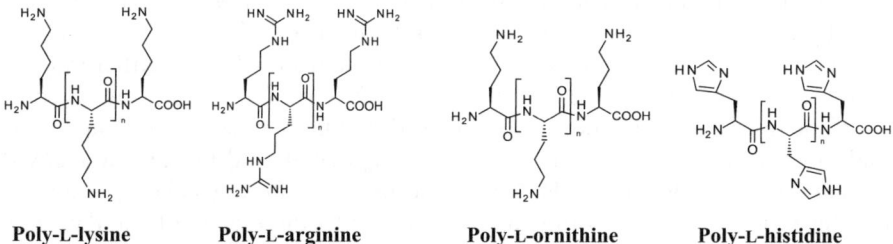

Poly-L-lysine **Poly-L-arginine** **Poly-L-ornithine** **Poly-L-histidine**

Figure 9.8 Poly-L-lysine, poly-L-arginine, poly-L-ornithine and poly-L-histidine.

murine and human tissue *in vivo*.[163] Chamarthy *et al.* examined the binding, toxicity and gene-delivery efficiency of three 16-mer peptides, K-16, K10H6 and O10H6, derived from the peptides PLO, PLL and poly-L-histidine (Figure 9.8).[39] They concluded that the *in vitro* transfection with DNA complexed with cationic peptide O10H6 resulted in the highest level of gene expression and lower toxicity in comparison with K-16 and K10H6 cationic peptides.[39] A polymer rich in imidazolium groups is likewise efficient.[164]

PLL and PLO (Figure 9.8) were converted into amphiphilic colloids by incorporation of a distearoylphosphatidylethanolamine PEG-galactosamine conjugate (with the galactosamine unit at the distal end of the PEG moiety) into the polyamino acid formulations forming molecules that reduced their toxicity.[159]

In comparison with the generally poor cellular entry of non-viral vectors, the efficient entry of viruses and toxins into cells and the action of antibacterial peptides have directed some laboratories to test membrane-active agents as components to enhance transgenic expression in non-viral gene delivery. Wagner[101,165] and Szoka[100,166] and their respective research groups have used synthetic peptides derived from the N-terminus of influenza virus haemagglutinin, for example INF6, INF7, INFA, INF10, E5CA and the artificial amphipathic peptides such as EGLA-I (sequence listed in Table 9.2) and the rhinovirus HRV2 VP-1 protein. These peptides may have specificity for endosomal pH due to acidic residues (glutamic and aspartic acids) aligning on one side of an amphipathic helix. At neutral pH, the negatively charged carboxylic groups destabilize the alpha-helical structure; acidification of the carboxylic groups shifts the equilibrium towards an amphipathic helical structure, which promotes multimerization of peptides and/or membrane interaction.

The pH specificity can be enhanced by the introduction of additional glutamic acids into the peptide sequence.[101] Additionally, the incorporation of a fusogenic peptide JTS-1 (Table 9.2), into either a DNA–liposome or a DNA liposome–polylysine–complex gave a six- to 14-fold improvement in the expression level of luciferase.[167] Also, amphiphilic basic peptides such as melittin (a bee venom peptide),[168] K5, the cationic counterpart of E5,[169] and KALA, the lysyl counterpart of GALA,[100] (Table 9.2) exhibit efficient membrane fusion and permeabilization activities at both neutral and acidic pH. KALA is cationic with its seven Lys residues. It forms a complex with pDNA that gives a 100-fold higher luciferase expression level than the PLL–DNA complex in CV-1 cells. KALA itself was used for both DNA condensation and membrane destabilization, and thus induced an increase in transfection efficiency.[170] However, the KALA peptide cannot transfect primary airway epithelial cells.[113]

Pichon *et al.* designed uncharged peptides at neutral pH that neither interact with the plasma membrane nor interact with serum proteins, but become fusogenic in endosomes, and this is achieved by making use of histidine.[38] The imidazole group in His (Figure 9.8) has a pK_a of 6.0, and in the acidic environment of the endosome (pH 5.5) it induces a buffering effect, leading to destabilization of the endosome and ultimately to the escape of the delivered gene from the endosomal compartment before the lysosomal stage that will degrade the plasmid cargo.[164]

Table 9.2 Representative membrane-active peptides that have delivery applications.

Name	Sequence	References
Anionic peptides		
INF6	GLFGAIAGFIENGWEGMIDGWYG	165
INF7	GLFEAIEGFIENGWEGMIDGWYG	165
E5CA	GLFEAIAEFIEGWEGLIEGCA	172
JTS-1	GLFEALLELLESLWELLLEA	172
GALA	WEAALAEALAEALAEHLAEALAEALEALAA	166
Cationic peptides		
HIV-1		
TAT domain	GRKKRRQRRRGYG	173
Penetratin (Antp)	RQIKIWFQNRRMKWKK	173
Melittin	CIGAVLKVLTTGLPALISWIKRKRQQ	168
KALA	WEAKLAKALAKALAKHLAKALAKALKACEA	100
K5	GLFKAIAKFIKGGWKGLIKG	169
R8	RRRRRRRR	162, 174, 175
Histidine peptides		
H5WYG	GLFHAIAHFIHGGWHGLIHGWYG	38
Histatin5	DSHAKRHHGYKRKFHEKHHSHRGY	176
Sea urchin B18	LGLLLRHLRHHSNLLANI	177
LAH$_4$	KKALLALALHHLAHLALHLALALKKA	178

H5WYG peptide (GLFHAIAHFIHGGWHGLIHGWYG) was designed as analogous to the N-terminal segment of the HA-2 subunit of the influenza virus hemagglutinin (GLFGAIAGFIEGGWTGMIDGWYG) in which G-4, G-8, E-11, T-15 and D-19 were replaced by His residues and M-17 by a Leu residue.[171] Other peptides containing His together with other peptides that have application in non-viral gene therapy are shown in Table 9.2. Lysosomal enzyme inhibitors that assist endosome escape, *e.g.* catapsin D and chloroquine, have been investigated in non-viral gene delivery formulations.[113]

9.3.3 Cationic Lipids

Cationic lipids are the most studied among the non-viral vectors, as indicated by their encouraging transfection results over cationic polymeric and naked DNA strategies. Following the design and formulation of Lipofectin by Felgner and co-workers,[15] the focus on non-viral vectors in general, particularly on cationic lipids for DNA delivery, has shown a remarkable increase worldwide in comparison with the attention to viral vectors. The focus on cationic lipids as non-viral vectors is also clear from the 10 reviews in one issue of the journal *Current Medicinal Chemistry* (Volume 10, Number 14, July 2003), as well as other RSC[20] and ACS[179] reviews, and our work dealing with cationic lipids and their roles in gene delivery.[1-3] In non-viral gene therapy, cationic lipids are potentially either liposomal or non-liposomal formulations. Liposomal formulations are composed of two types of lipid, a cationic lipid for DNA condensation and a helper lipid to improve both the stability and the

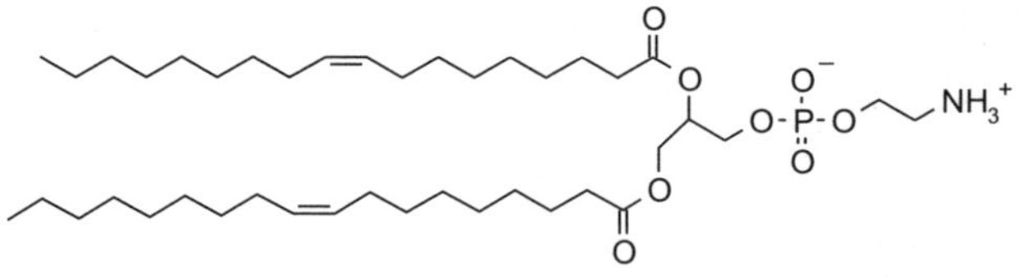

Figure 9.9   Neutral helper lipids DOPE, DOPC and cholesterol used in liposomal non-viral gene formulations, compared with DC-Chol, the first cholesterol-based cationic lipid.

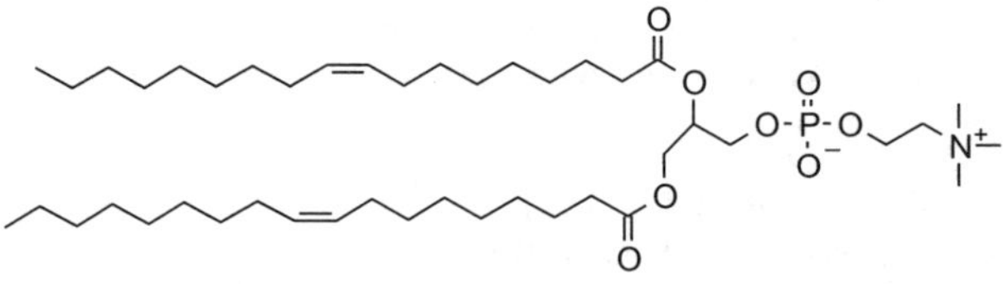

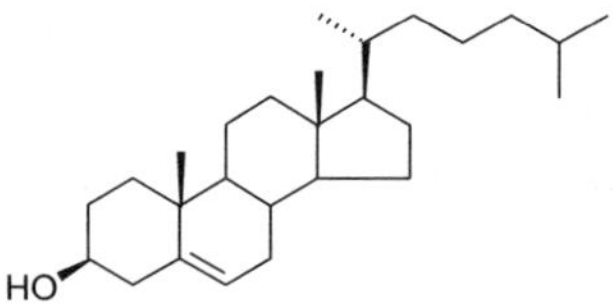

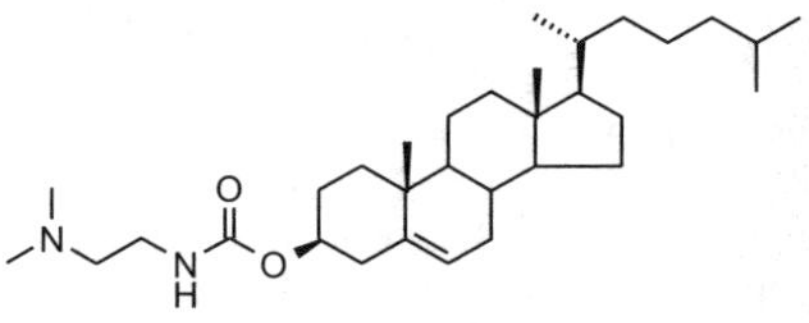

transfection efficiency of the formed liposomes. Most commonly used helper lipids are dioleoyl phosphatidylethanol-amine (DOPE), dioleoyl phosphatidylcholine (DOPC) or cholesterol (Figure 9.9). A liposomal formulation of the cationic lipid DDAB and DOPE interacted faster with a model cell membrane than the same liposomal formulation containing cholesterol as the helper lipid.[180]

Lipofectin was formulated as a liposomal formulation, a well-studied drug-delivery system that contains 1/1 w/w cationic lipid, N-[1-(2,3-dioleyloxy)-propyl]-N,N,N-trimethyl ammonium chloride (DOTMA) and the helper lipid DOPE, respectively.[15] Neutral phospholipid DOPE is used in liposomal formulation to increase transfection efficiency, as it has a membrane-destruction-promoting ability in acidic conditions.[113] Following the formulation of DOTMA, efforts continue to improve the transfection efficiency of cationic lipid formulations by modifying the structure of the cationic lipid.

This led to the synthesis of monovalent cationic lipids, DOTAP {N-[1-(2,3-dioleoyloxy)-propyl]-N,N,N-trimethylammonium}[181] and GAP-DLRIE [N-(3-aminopropyl)-N,N-dimethyl-2,3-bis(dodecyloxy)-1-propanaminium bromide] (Figure 9.10).[182] Heyes *et al.* investigated the change in degree of unsaturation of the monovalent cationic lipid, 1,2-distearyloxy-N,N-dimethyl-3-aminopropane.[183] They concluded that an increase in the degree of unsaturation improves gene silencing (siRNA delivery) efficiency.

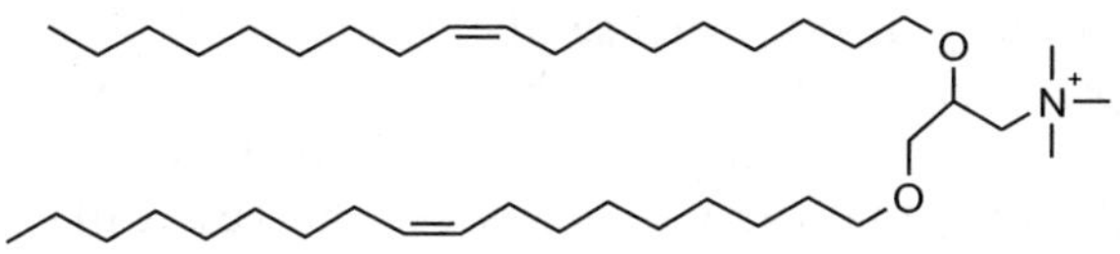

DOTMA

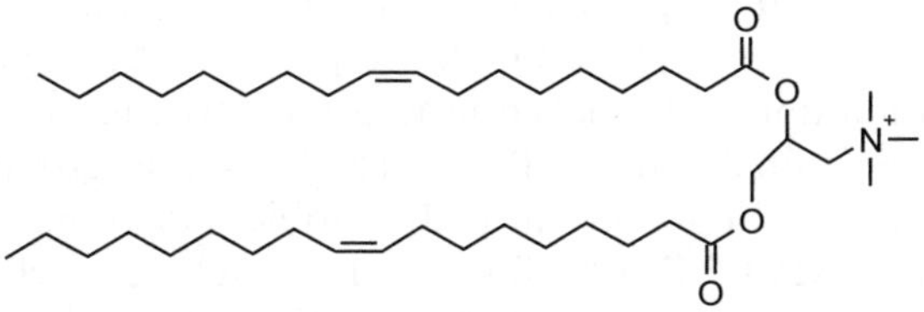

DOTAP

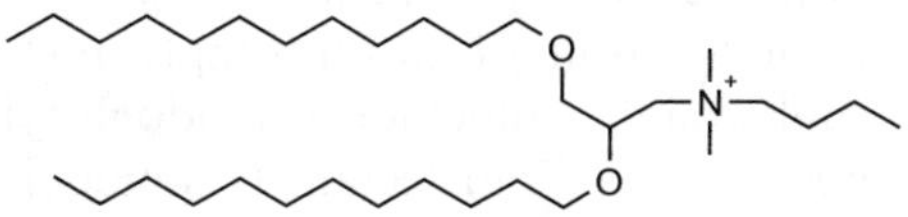

GAP-DLRIE

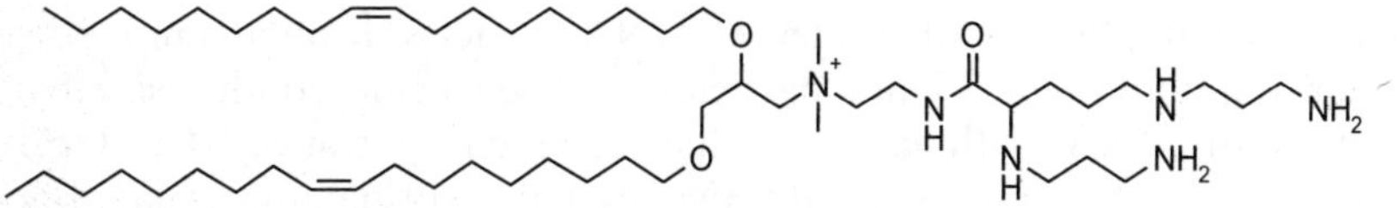

DOSPA

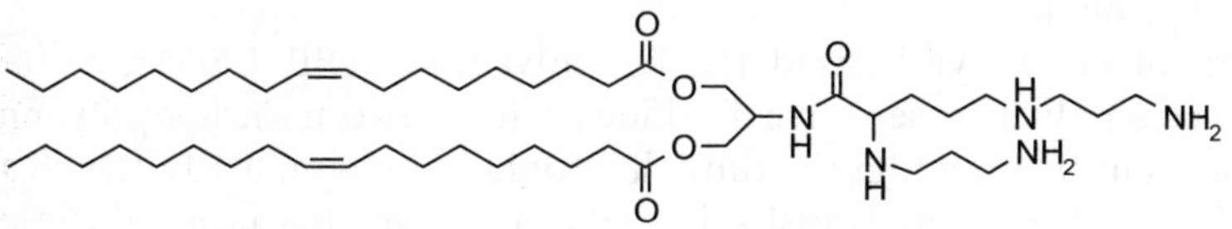

DOSPER

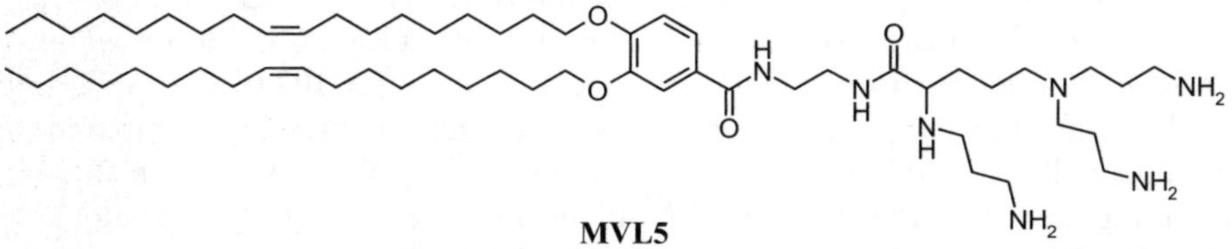

MVL5

Figure 9.10 Monovalent and multivalent cationic lipids used as non-viral gene delivery vectors in the form of liposomal formulations with the neutral helper lipids DOPE, DOPC or cholesterol.

A different strategy is to increase the number of positive charges on the cationic lipid in the liposomes. This gave rise to spermine lipid conjugates, 2,3-dioleyloxy-*N*-[2(sperminecarboxamido)ethyl]-*N*,*N*-dimethyl-1-propanaminium (DOSPA) that carry four positive charges at pH 7.2, and 1,3-dioleoyloxy-2-(6-carboxy-sperminyl)-propylamide (DOSPER) with three positive charges at pH 7.2 were synthesized (Figure 9.10).[184] Lipofectamine (Invitrogen) is a commercially available liposomal formulation containing 3/1 w/w cationic lipid, DOSPA and the neutral lipid DOPE, respectively. MVL5 is a multivalent cationic lipid with a double-branched head-group structure (Figure 9.10) formulated as a cationic lipid with DOPC.[185] The lipid head group is attached to an unsaturated *cis*-C18 double-chain hydrophobic moiety based on 3,4-dihydroxybenzoic acid. MVL5/DOPC liposomal formulation achieved a higher transfection efficiency when compared with the monovalent cationic lipid DOTAP/DOPC system.[185]

On the other hand, some research groups including ours are focusing on the design, synthesis and formulation of novel simple non-liposomal cationic lipids.[1–3] The lipopolyamine dioctadecylamidoglycylspermine DOGS (Figure 9.11), commercialized as Transfectam (Promega), was the first non-liposomal formulation to be synthesized by Behr and co-workers.[186] In the same work, they synthesized the lipospermine dipalmitoyl phosphatidylethanolamidospermine (DPPES) (Figure 9.11).[186] Also, Dauty and Behr investigated controlling the size of condensed DNA particles through a monodisperse lipoplex population of 35 nm particles.[187] They achieved this size by condensation of DNA with cationic thiol-detergent, tetradecyl-Cys-Orn. The stability of DNA complexes with two cysteine surfactants (guanidino-Cys *N*-decylamide and Orn-Cys-tetradecyl amide) was also investigated. These complexes were able to convert themselves, via oxidative dimerization, into cationic cystine lipids.[188]

Scherman *et al.* synthesized the lipopolyamine RPR-120535;[189] they also introduced a reduction sensitive disulfide bridge between the lipopolyamine and side-chain entity by adding a third hydrophobic chain, RPR132688 (Figure 9.11), that improves the transfection efficiency over the non-reducible RPR-120535.[190] Vierling and co-workers have synthesized fluorinated lipospermines that have hydrophobic and lipophobic chains.[191,192] They synthesized close analogs to DOGS, *e.g.* [F4C11][C14]-GS and [F4C11][F8C2]-GS (Figure 9.11), to improve the transfection efficiency of non-viral vectors. Camilleri *et al.* have designed a series of cationic Gemini surfactants, *e.g.* GSC103 (Figure 9.11) and GS11, that show an increase in the level of gene expression compared to commercially available non-viral formulations.[193,194] We have shown the diamine Lipogen (Figure 9.11), an N^4,N^9-diacylated spermine, to be a particularly effective delivery agent for genes[2] and for siRNA.[195]

Gao and Huang synthesized the first cholesterol-based cationic lipid, {3β[*N*-(*N'*,*N'*-dimethylaminoethane)carbamoyl] cholesterol} DC-Chol (Figure 9.9).[196] Cholesterol derivatives can be used as liposomal formulation with DOPE or as a non-liposomal formulation without the use of DOPE. In an effort to modify the structure of cationic cholesterol lipids to improve the transfection efficiency,

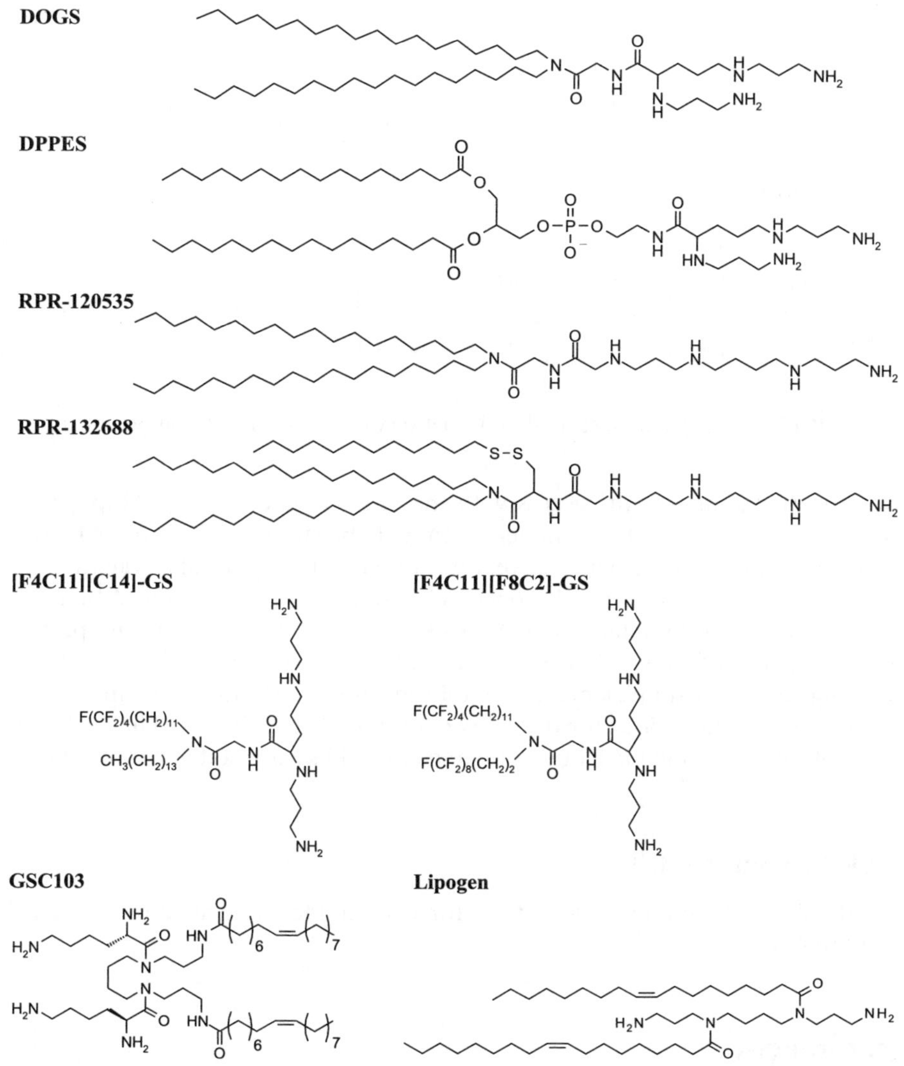

DOGS

DPPES

RPR-120535

RPR-132688

[F4C11][C14]-GS

[F4C11][F8C2]-GS

GSC103

Lipogen

Figure 9.11 Spermine-based lipopolyamines that can be used as non-liposomal formulations for non-viral gene delivery.

Lehn and co-workers have synthesized guanidinium cholesterol cationic lipids, bis-guanidinium spermidine cholesterol (BSGC) and bis-guanidinium tren cholesterol (BGTC) (Figure 9.12) by replacing spermidine in BSGC with tris(2-aminoethyl)amine (tren) (Figure 9.12).[197] BGTC shows a high efficiency when used as a non-liposomal cationic lipid. In addition, both BGTC and BGSC are efficient when used as liposomal formulations in combination with DOPE in a variety of cell lines.[197] Atufect (Figure 9.12) is a recent siRNA non-viral vector based upon a polyamine incorporating a guanidine functional group.[198]

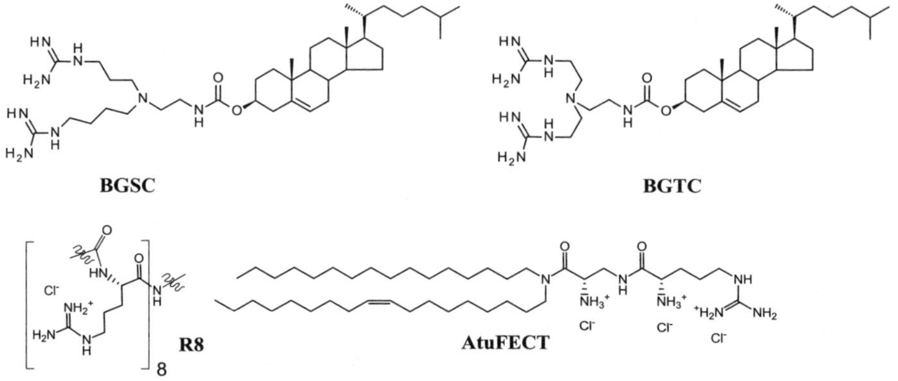

Figure 9.12 Some guanidines used in the delivery of genes, siRNA and other cargoes.

We have had some recent success with the di-guanidine analog of Lipogen, a diamidine derivative of spermine.[199] In collaboration with Prof. M. Hof (Prague Academy of Science), we are studying the molecular interactions between lipopolyamines, *e.g.* Lipogen N^4,N^9-dioleoylspermine and N^1-cholesteryl spermine carbamate, and DNA using fluorescence correlation spectroscopic studies of single lipopolyamine–DNA nanoparticles.[200–202] We are working to design simpler medicines that are non-viral, non-liposomal, polyamine-based formulations for the delivery of pDNA and siRNA. They will self-assemble with the polynucleotide of choice to yield serum-stable nanoparticles.

Acknowledgments

We thank the Egyptian Government for fully funded studentships (to AAM and OAAA).

References

1. I. S. Blagbrough, A. J. Geall and A. P. Neal, *Biochem. Soc. Trans.*, 2003, **31**, 397–406.
2. I. S. Blagbrough, N. Adjimatera, O. A. A. Ahmed, A. P. Neal and C. Pourzand, in *Neurotox '03: Neurotoxicological Targets from Functional Genomics and Proteomics*, ed. D. J. Beadle, I. R. Mellor and P. N. R. Usherwood, SCI, London, 2004, pp. 147–159.
3. I. S. Blagbrough and H. M. Ghonaim, in *Biological Aspects of Biogenic Amines, Polyamines and Conjugates*, ed. G. Dandrifosse, Research Signpost, India, 2009, pp. 81–112.
4. T. Friedmann and R. Roblin, *Science*, 1972, **175**, 949–955.
5. T. Merdan, J. Kopecek and T. Kissel, *Adv. Drug Deliv. Rev.*, 2002, **54**, 715–758.

6. R. E. Donahue, S. W. Kessler, D. Bodine, K. Mcdonagh, C. Dunbar, S. Goodman, B. Agricola, E. Byrne, M. Raffeld, R. Moen, J. Bacher, K. M. Zsebo and A. W. Nienhuis, *J. Exp. Med.*, 1992, **176**, 1125–1135.
7. N. Somia and I. M. Verma, *Nat. Rev. Genet.*, 2000, **1**, 91–99.
8. H. Nakai, E. Montini, S. Fuess, T. A. Storm, M. Grompe and M. A. Kay, *Nat. Genet.*, 2003, **34**, 297–302.
9. Y. Lu and C. O. Madu, *Exp. Opin. Drug Del.*, 2010, **7**, 19–35.
10. T. Niidome and L. Huang, *Gene Ther.*, 2002, **9**, 1647–1652.
11. R. Heller, M. Jaroszeski, A. Atkin, D. Moradpour, R. Gilbert, J. Wands and C. Nicolau, *FEBS Lett.*, 1996, **389**, 225–228.
12. N. S. Yang, J. Burkholder, B. Roberts, B. Martinell and D. McCabe, *Proc. Natl. Acad. Sci. U. S. A.*, 1990, **87**, 9568–9572.
13. C. M. Newman, A. Lawrie, A. F. Brisken and D. C. Cumberland, *Echocardiography*. 2001, **18**, 339–347.
14. S. Li and Z. Ma, *Curr. Gene Ther.*, 2001, **1**, 201–226.
15. P. L. Felgner, T. R. Gadek, M. Holm, R. Roman, H. W. Chan, M. Wenz, J. P. Northrop, G. M. Ringold and M. Danielsen, *Proc. Natl. Acad. Sci. U. S. A.*, 1987, **84**, 7413–7417.
16. D. J. Gary, N. Puri and Y.-Y. Won, *J. Control. Release*, 2007, **121**, 64–73.
17. A. Akinc, A. Zumbuehl, M. Goldberg, E. S. Leshchiner, V. Busini, N. Hossain, S. A. Bacallado, D. N. Nguyen, J. Fuller, R. Alvarez, A. Borodovsky, T. Borland, R. Constien, A. de Fougerolles, J. R. Dorkin, K. N. Jayaprakash, M. Jayaraman, M. John, V. Koteliansky, M. Manoharan, L. Nechev, J. Qin, T. Racie, D. Raitcheva, K. G. Rajeev, D. W. Y. Sah, J. Soutschek, I. Toudjarska, H.-P. Vornlocher, T. S. Zimmermann, R. Langer and D. G. Anderson, *Nature Biotechnol.*, 2008, **26**, 561–569.
18. K. A. Whitehead, R. Langer and D. G. Anderson, *Nature Rev. Drug Discovery*, 2009, **8**, 129–138.
19. I. S. Blagbrough and C. Zara, *Pharm. Res.*, 2009, **26**, 1–18.
20. J. W. Gaynor, B. J. Campbell and R. Cosstick, *Chem. Soc. Rev.*, 2010, **39**, 4169–4184.
21. V. A. Bloomfield, *Biopolymers*, 1997. **44**, 269–282.
22. L. C. Gosule and J. A. Schellman, *Nature*, 1976, **259**, 333–335.
23. R. W. Wilson and V. A. Bloomfield, *Biochemistry*, 1979, **18**, 2192–2196.
24. V. A. Bloomfield, *Biopolymers*, 1991, **31**, 1471–1481.
25. G. S. Manning, *Biopolymers*, 1972, **11**, 937–949.
26. G. S. Manning, *Biophys. Chem.*, 1977, **7**, 141–145.
27. G. S. Manning, *Q. Rev. Biophys.*, 1978, **11**, 179–246.
28. C. Böttcher, C. Endisch, J. H. Fuhrhop, C. Catterall and M. Eaton, *J. Am. Chem. Soc.*, 1998, **120**, 12–17.
29. E. Dauty, J. S. Remy, T. Blessing and J. P. Behr, *J. Am. Chem. Soc.*, 2001, **123**, 9227–9234.
30. P. G. Arscott, C. L. Ma, J. R. Wenner and V. A. Bloomfield, *Biopolymers*, 1995, **36**, 345–364.
31. K. A. Marx and G. C. Ruben, *Nucleic Acids Res.*, 1983, **11**, 1839–1854.

32. R. Golan, L. I. Pietrasanta, W. Hsieh and H. G. Hansma, *Biochemistry*, 1999, **38**, 14069–14076.

33. G. Zuber, E. Dauty, M. Nothisen, P. Belguise and J. P. Behr, *Adv. Drug Deliv. Rev.*, 2001, **52**, 245–253.

34. T. Blessing, J. S. Remy and J. P. Behr, *Proc. Natl. Acad. Sci. U. S. A.*, 1998, **95**, 1427–1431.

35. M. Ruponen, S. Yla-Herttuala and A. Urtti, *Biochim. Biophys. Acta*, 1999, **1415**, 331–341.

36. G. M. Cooper, *The Cell: A Molecular Approach*, ASM Press, Washington, DC, 2000.

37. Q. R. Chen, L. Zhang, S. A. Stass and A. J. Mixson, *Nucleic Acids Res.*, 2001, **29**, 1334–1340.

38. C. Pichon, C. Goncalves and P. Midoux, *Adv. Drug Deliv. Rev.*, 2001, **53**, 75–94.

39. S. P. Chamarthy, J. R. Kovacs, E. McClelland, D. Gattens and W. S. Meng, *Mol. Immunol.*, 2003, **40**, 483–490.

40. Y. Dong, A. I. Skoultchi and J. W. Pollard, *Nucleic Acids Res.*, 1993, **21**, 771–772.

41. E. Ramsay and M. Gumbleton, *J. Drug Target.*, 2002, **10**, 1–9.

42. A. Haberland, T. Knaus, S. V. Zaitsev, B. Buchberger, A. Lun, H. Haller and M. Bottger, *Pharm. Res.*, 2000, **17**, 229–235.

43. D. K. Hoganson, L. A. Chandler, G. A. Fleurbaaij, W. Ying, M. E. Black, J. Doukas, G. F. Pierce, A. Baird and B. A. Sosnowski, *Hum. Gene Ther.*, 1998, **9**, 2565–2575.

44. H. Junbo, Q. Li, W. Zaide and H. Yunde, *Int. J. Mol. Med.*, 1999, **3**, 601–608.

45. J. E. Hagstrom, M. G. Sebestyen, V. Budker, J. J. Ludtke, J. D. Fritz and J. A. Wolff, *Biochim. Biophys. Acta*, 1996, **1284**, 47–55.

46. N. V. Hud, M. J. Allen, K. H. Downing, J. Lee and R. Balhorn, *Biochem. Biophys. Res. Commun.*, 1993, **193**, 1347–1354.

47. N. V. Hud, K. H. Downing and R. Balhorn, A constant radius of curvature model for the organization of DNA in toroidal condensates. *Proc. Natl. Acad. Sci. U. S. A.*, 1995, **92**, 3581–3585.

48. N. V. Hud and K. H. Downing, *Proc. Natl. Acad. Sci. U. S. A.*, 2001, **98**, 14925–14930.

49. M. R. Shen, K. H. Downing, R. Balhorn and N. V. Hud, *J. Am. Chem. Soc.*, 2000, **122**, 4833–4834.

50. C. C. Conwell, I. D. Vilfan and N. V. Hud, *Proc. Natl. Acad. Sci. U. S. A.*, 2003, **100**, 9296–9301.

51. D. K. Chattoraj, L. C. Gosule and J. A. Schellman, *J. Mol. Biol.*, 1978, **121**, 327–337.

52. V. Vijayanathan, T. Thomas, A. Shirahata and T. J. Thomas, *Biochemistry*, 2001, **40**, 13644–13651.

53. G. Byk, B. Wetzer, M. Frederic, C. Dubertret, B. Pitard, G. Jaslin and D. Scherman, *J. Med. Chem.*, 2000, **43**, 4377–4387.

54. Y. Fang and J. H. Hoh, *FEBS Lett.*, 1999, **459**, 173–176.

55. J. Widom and R. L. Baldwin, *J. Mol. Biol.*, 1980, **144**, 431–453.
56. Y. Park, K. Y. Kwok, C. Boukarim and K. G. Rice, *Bioconjug. Chem.*, 2002, **13**, 232–239.
57. B. Schwartz, M. A. Ivanov, B. Pitard, V. Escriou, R. Rangara, G. Byk, P. Wils, J. Crouzet and D. Scherman, *Gene Ther.*, 1999, **6**, 282–292.
58. I. K. Park, T. H. Kim, Y. H. Park, B. A. Shin, E. S. Choi, E. H. Chowdhury, T. Akaike and C. S. Cho, *J. Control. Release*, 2001, **76**, 349–362.
59. K. Y. Lee, I. C. Kwon, Y. H. Kim, W. H. Jo and S. Y. Jeong, *J. Control. Release*, 1998, **51**, 213–220.
60. S. Prabha, W. Z. Zhou, J. Panyam and V. Labhasetwar, *Int. J. Pharm.*, 2002, **244**, 105–115.
61. C. H. Ahn, S. Y. Chae, Y. H. Bae and S. W. Kim, *J. Control. Release*, 2002, **80**, 273–282.
62. A. L. Martin, M. C. Davies, B. J. Rackstraw, C. J. Roberts, S. Stolnik, S. J. B. Tendler and P. M. Williams, *FEBS Lett.* **480**, 106–112 (2000).
63. W. L. J. Hinrichs, N. M. E. Schuurmans-Nieuwenbroek, P. van de Wetering and W. E. Hennink, *J. Control. Release*, 1999, **60**, 249–259.
64. C. Kawaura, A. Noguchi, T. Furuno and M. Nakanishi, *FEBS Lett.*, 1998, **421**, 69–72.
65. T. Hara, F. Liu, D. X. Liu and L. Huang, *Adv. Drug Deliv. Rev.*, 1997, **24**, 265–271.
66. A. Noguchi, N. Hirashima and M. Nakanishi, *Pharm. Res.*, 2002, **19**, 933–938.
67. P. L. Felgner and G. M. Ringold, *Nature*, 1989, **337**, 387–388.
68. K. W. Leong, H. Q. Mao, V. L. Truong-Le, K. Roy, S. M. Walsh and J. T. August, *J. Control. Release*, 1998, **53**, 183–193.
69. M. Nishikawa, S. Takemura, F. Yamashita, Y. Takakura, D. K. F. Meijer, M. Hashida and P. J. Swart, *J. Drug Target.*, 2000, **8**, 29–38.
70. M. Hashida, M. Nishikawa, F. Yamashita and Y. Takakura, *Adv. Drug Deliv. Rev.*, 2001, **52**, 187–196.
71. J. C. Perales, T. Ferkol, H. Beegen, O. D. Ratnoff and R. W. Hanson, *Proc. Natl. Acad. Sci. U. S. A.*, 1994, **91**, 4086–4090.
72. K. Fabio, J. Gaucheron, C. Di Giorgio and P. Vierling, *Bioconjug. Chem.*, 2003, **14**, 358–367.
73. J. S. Remy, A. Kichler, V. Mordvinov, F. Schuber and J. P. Behr, *Proc. Natl. Acad. Sci. U. S. A.*, 1995, **92**, 1744–1748.
74. Y. K. Park, Y. H. Park, B. A. Shin, E. S. Choi, Y. R. Park, T. Akaike and C. S. Cho, *J. Control. Release*, 2000, **69**, 97–108.
75. S. Y. Gao, J. N. Chen, X. R. Xu, Z. Ding, Y. H. Yang, Z. C. Hua and J. F. Zhang, *Int. J. Pharm.*, 2003, **255**, 57–68.
76. Y. H. Choi, F. Liu, J. S. Park and S. W. Kim, *Bioconjug. Chem.*, 1998, **9**, 708–718.
77. R. Kircheis, S. Schuller, S. Brunner, M. Ogris, K. H. Heider, W. Zauner and E. Wagner, *J. Gene Med.*, 1999, **1**, 111–120.

78. J. M. Benns, R. I. Mahato and S. W. Kim, *J. Control. Release*, 2002, **79**, 255–269.
79. E. Dauty, J. S. Remy, G. Zuber and J. P. Behr, *Bioconjug. Chem.*, 2002, **13**, 831–839.
80. S. H. Kim, J. H. Jeong, K. W. Chun and T. G. Park, *Langmuir*, 2005, **21**, 8852–8857.
81. J. D. Bagnato, A. L. Eilers, R. A. Horton and C. B. Grissom, *J. Org. Chem.*, 2004, **69**, 8987–8996.
82. M. Lee and C. B. Grissom, *Org. Lett.*, 2009, **11**, 2499–2502.
83. R. P. Harbottle, R. G. Cooper, S. L. Hart, A. Ladhoff, T. McKay, A. M. Knight, E. Wagner, A. D. Miller and C. Coutelle, *Hum. Gene Ther.*, 1998, **9**, 1037–1047.
84. H. Hosseinkhani and Y. Tabata, *J. Control. Release*, 2004, **97**, 157–171.
85. M. Cotten, F. Langle-Rouault, H. Kirlappos, E. Wagner, K. Mechtler, M. Zenke, H. Beug and M. L. Birnstiel, *Proc. Natl. Acad. Sci U. S. A.*, 1990, **87**, 4033–4037.
86. E. Wagner, M. Zenke, M. Cotten, H. Beug and M. L. Birnstiel, *Proc. Natl. Acad. Sci. U. S. A.*, 1990, **87**, 3410–3414.
87. M. Zenke, P. Steinlein, E. Wagner, M. Cotten, H. Beug and M. L. Birnstiel, *Proc. Natl. Acad. Sci. U. S. A.*, 1990, **87**, 3655–3659.
88. N. C. Bellocq, S. H. Pun, G. S. Jensen and M. E. Davis, *Bioconjug. Chem.*, 2003, **14**, 1122–1132.
89. B. A. Sosnowski, A. M. Gonzalez, L. A. Chandler, Y. J. Buechler, G. F. Pierce and A. Baird, *J. Biol. Chem.*, 1996, **271**, 33647–33653.
90. Z. Huang, W. Li, J. A. Mackay and F. C. Szoka, Jr, *Mol. Ther.*, 2005, **11**, 409–417.
91. S. Sandgren, F. Cheng and M. Belting, *J. Biol. Chem.*, 2002, **277**, 38877–38883.
92. W. M. Pardridge, *Jpn. J. Pharmacol.*, 2001, **87**, 97–103.
93. O. Deas, E. Angevin, C. Cherbonnier, A. Senik, B. Charpentier, J. P. Levillain, E. Oosterwijk, F. Hirsch and A. Durrbach, *Hum. Gene Ther.*, 2002, **13**, 1101–1114.
94. T. Merdan, J. Callahan, H. Peterson, U. Bakowsky, P. Kopeckova, T. Kissel and J. Kopecek, *Bioconjug. Chem.*, 2003, **14**, 989–996.
95. X. H. Zhou and L. Huang, *Biochim. Biophys. Acta Biomembr.*, 1994, **1189**, 195–203.
96. M. R. Capecchi, *Cell*, 1980, **22**, 479–488.
97. Y. Xu and F. C. Szoka, Jr, *Biochemistry*, 1996, **35**, 5616–5623.
98. Z. Y. Zhang and B. D. Smith, *Bioconjug. Chem.*, 2000, **11**, 805–814.
99. O. Boussif, F. Lezoualc'h, M. A. Zanta, M. D. Mergny, D. Scherman, B. Demeneix and J. P. Behr, *Proc. Natl. Acad. Sci. U. S. A.*, 1995, **92**, 7297–7301.
100. T. B. Wyman, F. Nicol, O. Zelphati, P. V. Scaria, C. Plank and F. C. Szoka, Jr, *Biochemistry*, 1997, **36**, 3008–3017.
101. E. Wagner, *Adv. Drug Deliv. Rev.*, 1999, **38**, 279–289.

102. J. H. Felgner, R. Kumar, C. N. Sridhar, C. J. Wheeler, Y. J. Tsai, R. Border, P. Ramsey, M. Martin and P. L. Felgner, *J. Biol. Chem.*, 1994, **269**, 2550–2561.

103. K. Ciftci and R. J. Levy, *Int. J. Pharm.*, 2001, **218**, 81–92.

104. K. J. Ryan and S. R. Wente, *Curr. Opin. Cell Biol.*, 2000, **12**, 361–371.

105. M. A. Zanta, P. Belguise-Valladier and J. P. Behr, *Proc. Natl. Acad. Sci. U. S. A.*, 1999, **96**, 91–96.

106. D. Kalderon, B. L. Roberts, W. D. Richardson and A. E. Smith, *Cell*, 1984, **39**, 499–509.

107. C. K. Chan, S. Hubner, W. Hu and D. A. Jans, *Gene Ther.*, 1998, **5**, 1204–1212.

108. E. Hebert, *Biol. Cell*, 2003, **95**, 59–68.

109. M. Nakanishi, H. Mizuguchi, K. Ashihara, T. Senda, A. Eguchi, A. Watabe, T. Nakanishi, M. Kondo, T. Nakagawa, A. Masago, J. Okabe, S. Ueda, T. Mayumi and T. Hayakawa, *Mol. Membr. Biol.*, 1999, **16**, 123–127.

110. Y. Yoneda, T. Semba, Y. Kaneda, R. L. Noble, Y. Matsuoka, T. Kurihara, Y. Okada and N. Imamoto, *Exp. Cell Res.*, 1992, **201**, 313–320.

111. W. T. Godbey, K. K. Wu and A. G. Mikos, *Proc. Natl. Acad. Sci. U. S. A.*, 1999, **96**, 5177–5181.

112. H. Kamiya, H. Akita and H. Harashima, *Drug Discov. Today*, 2003, **8**, 990–996.

113. W. C. Tseng and L. Huang, *Pharm. Sci. Technol. Today*, 1998, **1**, 206–213.

114. S. M. Elbashir, J. Harborth, W. Lendeckel, A. Yalcin, K. Weber and T. Tuschl, *Nature*, 2001, **411**, 494–498.

115. J. Weyermann, D. Lochmann and A. Zimmer, *J. Control. Release*, 2004, **100**, 411–423.

116. F. Shi and D. Hoekstra, *J. Control. Release*, 2004, **97**, 189–209.

117. S. Choosakoonkriang, B. A. Lobo, G. S. Koe, J. G. Koe and C. R. Middaugh, *J. Pharm. Sci.*, 2003, **92**, 1710–1722.

118. M. L. Forrest, J. T. Koerber and D. W. Pack, *Bioconjug. Chem.*, 2003, **14**, 934–940.

119. K. Kunath, A. von Harpe, D. Fischer, H. Peterson, U. Bickel, K. Voigt and T. Kissel, *J. Control. Release*, 2003, **89**, 113–125.

120. X. T. Shuai, T. Merdan, F. Unger, M. Wittmar and T. Kissel, *Macromolecules*, 2003, **36**, 5751–5759.

121. M. Thomas and A. M. Klibanov, *Proc. Natl. Acad. Sci. U. S. A.*, 2003, **100**, 9138–9143.

122. C. H. Lee, Y. H. Ni, C. C. Chen, C. K. Chou and F. H. Chang, *Biochim. Biophys. Acta Biomembr.*, 2003, **1611**, 55–62.

123. N. G. Schipper, K. M. Varum and P. Artursson, *Pharm. Res.*, 1996, **13**, 1686–1692.

124. R. J. Mumper, J. J. Wang, J. M. Clapsell and A. P. Rolland, *Proc. Int. Symb. Control. Rel. Bioactive Mater.*, 1995, **22**, 178–179.

125. S. C. W. Richardson, H. J. V. Kolbe and R. Duncan, *Int. J. Pharm.*, 1999, **178**, 231–243.
126. G. Borchard, *Adv. Drug Deliv. Rev.*, 2001, **52**, 145–150.
127. W. G. Liu and K. De Yao, *J. Control. Release*, 2002, **83**, 1–11.
128. S. Mansouri, P. Lavigne, K. Corsi, M. Benderdour, E. Beaumont and J. C. Fernandes, *Eur. J. Pharm. Biopharm.*, 2004, **57**, 1–8.
129. K. Corsi, F. Chellat, L. Yahia and J. C. Fernandes, *Biomaterials*, 2003, **24**, 1255–1264.
130. M. Lee, J. W. Nah, Y. Kwon, J. J. Koh, K. S. Ko and S. W. Kim, *Pharm. Res.*, 2001, **18**, 427–431.
131. P. Erbacher, S. M. Zou, T. Bettinger, A. M. Steffan and J. S. Remy, *Pharm. Res.*, 1998, **15**, 1332–1339.
132. F. C. MacLaughlin, R. J. Mumper, J. J. Wang, J. M. Tagliaferri, I. Gill, M. Hinchcliffe and A. P. Rolland, *J. Control. Release*, 1998, **56**, 259–272.
133. I. K. Park, T. H. Kim, S. I. Kim, T. Akaike and C. S. Cho, *J. Disper. Sci. Technol.*, 2003, **24**, 489–498.
134. B. I. Florea, P. G. M. Ravenstijn, H. E. Junginger and G. Borchard, *STP Pharma Sci.*, 2002, **12**, 243–249.
135. M. Thanou, B. I. Florea, M. Geldof, H. E. Junginger and G. Borchard, *Biomaterials*, 2002, **23**, 153–159.
136. S. H. Son, S. Y. Chae, C. Y. Choi, M. Y. Kim, V. G. Ngugen, M. K. Jang and J. W. Nah, *Macromol. Res.*, 2004, **12**, 573–580.
137. H. S. Yoo, J. E. Lee, H. Chung, I. C. Kwon and S. Y. Jeong, *J. Control. Release*, 2005, **103**, 235–243.
138. S. Kwon, J. H. Park, H. Chung, I. C. Kwon, S. Y. Jeong and I. S. Kim, *Langmuir*, 2003, **19**, 10188–10193.
139. J. H. Park, S. G. Kwon, J. O. Nam, R. W. Park, H. Chung, S. B. Seo, I. S. Kim, I. C. Kwon and S. Y. Jeong, *J. Control. Release*, 2004, **95**, 579–588.
140. S. Ozbas-Turan, C. Aral, L. Kabasakal, M. Keyer-Uysal and J. Akbuga, *J. Pharm. Pharm. Sci.*, 2003, **6**, 27–32.
141. Y. Kawamata, Y. Nagayama, K. Nakao, H. Mizuguchi, T. Hayakawa, T. Sato and N. Ishii, *Biomaterials*, 2002, **23**, 4573–4579.
142. M. A. Croyle, X. Cheng, A. Sandhu and J. M. Wilson, *Mol. Ther.*, 2001, **4**, 22–28.
143. T. Ishii, Y. Okahata and T. Sato, *Biochim. Biophys. Acta Biomembr.*, 2001, **1514**, 51–64.
144. F. Yang, X. Q. Cui and X. R. Yang, *Biophys. Chem.*, 2002, **99**, 99–106.
145. W. G. Liu, X. Zhang, S. J. Sun, G. J. Sun, K. De Yao, D. C. Liang, G. Guo and J. Y. Zhang, *Bioconjug. Chem.*, 2003, **14**, 782–789.
146. K. Romoren, A. Aaberge, G. Smistad, B. J. Thu and O. Evensen, *Pharm. Res.*, 2004, **21**, 2340–2346.
147. Y. M. Liu, L. Wenning, M. Lynch and T. M. Reineke, *J. Am. Chem. Soc.*, 2004, **126**, 7422–7423.
148. Y. M. Liu and T. M. Reineke, *J. Am. Chem. Soc.*, 2005, **127**, 3004–3015.

149. A. J. Geall, R. J. Taylor, M. E. Earll, M. A. W. Eaton and I. S. Blagbrough, *Chem. Commun.*, 1998, 1403–1404.

150. A. J. Geall, R. J. Taylor, M. E. Earll, M. A. W. Eaton and I. S. Blagbrough, *Bioconjug. Chem.*, 2000, **11**, 314–326.

151. T. M. Reineke and M. E. Davis, *Bioconjug. Chem.*, 2003, **14**, 247–254.

152. J. J. Cheng, K. T. Khin, G. S. Jensen, A. J. Liu and M. E. Davis, *Bioconjug. Chem.*, 2003, **14**, 1007–1017.

153. S. R. Popielarski, S. Mishra and M. E. Davis, *Bioconjug. Chem.*, 2003, **14**, 672–678.

154. F. Kihara, H. Arima, T. Tsutsumi, F. Hirayama and K. Uekama, *Bioconjug. Chem.*, 2003, **14**, 342–350.

155. S. A. Cryan, R. Donohue, B. J. Ravo, R. Darcy and C. M. O'Driscoll, *STP Pharma Sci.*, 2004, **14**, 57–62.

156. S. A. Cryan, A. Holohan, R. Donohue, R. Darcy and C. M. O'Driscoll, *Eur. J. Pharm. Sci.*, 2004, **21**, 625–633.

157. J. F. Guo, K. A. Fisher, R. Darcy, J. F. Cryan and C. O'Driscoll, *Mol. Biosyst*, 2010, **6**, 1143–1161.

158. T. Azzam, A. Raskin, A. Makovitzki, H. Brem, P. Vierling, M. Lineal and A. J. Domb, *Macromolecules*, 2002, **35**, 9947–9953.

159. H. Hosseinkhani and Y. Tabata, *J. Control. Release*, 2003, **86**, 169–182.

160. M. Nakanishi, A. Eguchi, T. Akuta, E. Nagoshi, S. Fujita, J. Okabe, T. Senda and M. Hasegawa, *Curr. Protein Pept. Sci.*, 2003, **4**, 141–150.

161. J. B. Rothbard, E. Kreider, C. L. VanDeusen, L. Wright, B. L. Wylie and P. A. Wender, *J. Med. Chem.*, 2002, **45**, 3612–3618.

162. J. B. Rothbard, T. C. Jessop, R. S. Lewis, B. A. Murray and P. A. Wender, *J. Am. Chem. Soc.*, 2004, **126**, 9506–9507.

163. Z. Siprashvili, F. A. Scholl, S. F. Oliver, A. Adams, C. H. Contag, P. A. Wender and P. A. Khavari, *Hum. Gene Ther.*, 2003, **14**, 1225–1233.

164. M. H. Allen, M. D. Green, H. K. Getaneh, K. M. Miller and T. E. Long, *Biomacromolecules*, 2011, **12**, 2243–2250.

165. C. Plank, B. Oberhauser, K. Mechtler, C. Koch and E. Wagner, *J. Biol. Chem.*, 1994, **269**, 12918–12924.

166. R. A. Parente, S. Nir and F. C. Szoka, Jr, *Biochemistry*, 1990, **29**, 8720–8728.

167. S. Gottschalk, J. T. Sparrow, J. Hauer, M. P. Mims, F. E. Leland, S. L. Woo and L. C. Smith, *Gene Ther.*, 1996, **3**, 48–57.

168. C. E. Dempsey, *Biochim. Biophys. Acta*, 1990, **1031**, 143–161.

169. M. Murata, S. Takahashi, S. Kagiwada, A. Suzuki and S. Ohnishi, *Biochemistry*, 1992, **31**, 1986–1992.

170. J. G. Duguid, C. Li, M. Shi, M. J. Logan, H. Alila, A. Rolland, E. Tomlinson, J. T. Sparrow and L. C. Smith, *Biophys. J.*, 1998, **74**, 2802–2814.

171. P. Midoux, A. Kichler, V. Boutin, J. C. Maurizot and M. Monsigny, *Bioconjug. Chem.*, 1998, **9**, 260–267.

172. L. Vaysse, I. Burgelin, J. P. Merlio and B. Arveiler, *Biochim. Biophys. Acta*, 2000, **1475**, 369–376.

173. B. R. Meade and S. F. Dowdy, *Adv. Drug Deliv. Rev.*, 2007, **59**, 134–140.
174. J. B. Rothbard, T. C. Jessop and P. A. Wender, *Adv. Drug Deliv. Rev.*, 2005, **57**, 495–504.
175. P. A. Wender, W. C. Galliher, E. A. Goun, L. R. Jones and T. H. Pillow, *Adv. Drug Deliv. Rev.*, 2008, **60**, 452–472.
176. S. Melino, S. Rufini, M. Sette, R. Morero, A. Grottesi, M. Paci and R. Petruzzelli, *Biochemistry*, 1999, **38**, 9626–9633.
177. R. W. Glaser, M. Grune, C. Wandelt and A. S. Ulrich, *Biochemistry*, 1999, **38**, 2560–2569.
178. T. C. Vogt and B. Bechinger, *J. Biol. Chem.*, 1999, **274**, 29115–29121.
179. M. A. Mintzer and E. E. Simanek, *Chem. Rev.*, 2009, **109**, 259–302.
180. P. Callow, G. Fragneto, R. Cubitt, D. J. Barlow, M. J. Lawrence and P. Timmins, *Langmuir*, 2005, **21**, 7912–7920.
181. R. Leventis and J. R. Silvius, *Biochim. Biophys. Acta*, 1990, **1023**, 124–132.
182. C. J. Wheeler, P. L. Felgner, Y. J. Tsai, J. Marshall, L. Sukhu, S. G. Doh, J. Hartikka, J. Nietupski, M. Manthorpe, M. Nichols, M. Plewe, X. W. Liang, J. Norman, A. Smith and S. H. Cheng, *Proc. Natl. Acad. Sci. U. S. A.*, 1996, **93**, 11454–11459.
183. J. Heyes, L. Palmer, K. Bremner and I. MacLachlan, *J. Control. Release*, 2005, **107**, 276–287.
184. B. Buchberger, E. Fernholz, H. v.d. Eltz and M. Hinzepeter, *Biochem. Inform.*, 1996, **98**, 27–29.
185. K. Ewert, A. Ahmad, H. M. Evans, H. W. Schmidt and C. R. Safinya, *J. Med. Chem.*, 2002, **45**, 5023–5029.
186. J. P. Behr, B. Demeneix, J. P. Loeffler and J. P. Mutul, *Proc. Natl. Acad. Sci. U. S. A.*, 1989, **86**, 6982–6986.
187. E. Dauty and J. P. Behr, *Polym. Int.*, 2003, **52**, 459–464.
188. D. Lleres, J. P. Clamme, E. Dauty, T. Blessing, G. Krishnamoorthy, G. Duportail and Y. Mely, *Langmuir*, 2002, **18**, 10340–10347.
189. G. Byk, C. Dubertret, V. Escriou, M. Frederic, G. Jaslin, R. Rangara, B. Pitard, J. Crouzet, P. Wils, B. Schwartz and D. Scherman, *J. Med. Chem.*, 1998, **41**, 224–235.
190. G. Byk, B. Wetzer, M. Frederic, C. Dubertret, B. Pitard, G. Jaslin and D. Scherman, *J. Med. Chem.*, 2000, **43**, 4377–4387.
191. J. S. Remy, C. Sirlin, P. Vierling and J. P. Behr, *Bioconjug. Chem.*, 1994, **5**, 647–654.
192. J. Gaucheron, C. Santaella and P. Vierling, *Bioconjug. Chem.*, 2001, **12**, 114–128.
193. C. McGregor, C. Perrin, M. Monck, P. Camilleri and A. J. Kirby, *J. Am. Chem. Soc.*, 2001, **123**, 6215–6220.
194. G. Ronsin, C. Perrin, P. Guedat, A. Kremer, P. Camilleri and A. J. Kirby, *Chem. Commun.*, 2001, 2234–2235.
195. A. A. Metwally, C. Pourzand and I. S. Blagbrough, *Pharmaceutics*, 2011, **3**, 125–140.
196. X. Gao and L. Huang, *Biochem. Biophys. Res. Commun.*, 1991, **179**, 280–285.

197. J. P. Vigneron, N. Oudrhiri, M. Fauquet, L. Vergely, J. C. Bradley, M. Basseville, P. Lehn and J. M. Lehn, *Proc. Natl. Acad. Sci. U. S. A.*, 1996, **93**, 9682–9686.
198. A. Santel, M. Aleku, O. Keil, J. Endruschat, V. Esche, G. Fisch, S. Dames, K. Loffler, M. Fechtner, W. Arnold, K. Giese, A. Klippel and J. Kaufmann, *Gene Therapy*, 2006, **13**, 1222–1234.
199. A. A. Metwally and I. S. Blagbrough, *Pharmaceutics*, 2011, **3**, 406–424.
200. N. Adjimatera, A. P. Neal and I. S. Blagbrough, in *Fluorescence Spectroscopy in Biology. Advanced Methods and Their Applications to Membranes, Proteins, DNA and Cells*, ed. M. Hof, R. Hutterer and V. Fidler, *Fluorescence Methods and Applications* (series ed. O. S. Wolfbeis), Springer, Berlin, 2005, **3**, pp. 201–228.
201. N. Adjimatera, T. Kral, M. Hof and I. S. Blagbrough, *Pharm. Res.*, 2006, **23**, 1564–1573.
202. N. Adjimatera, A. Benda, I. S. Blagbrough, M. Langner, M. Hof and T. Kral, in *Fluorescence of Supermolecules, Polymers, and Nanosystems*, ed. M. Berberan-Santos, Fluorescence (series ed. O. S. Wolfbeis), Springer, Berlin, 2008, **4**, pp. 381–413.

The Design and Development of Polyamine-Based Analogs with Epigenetic Targets

YI HUANG,*[a] LAURENCE J. MARTON[b] AND
PATRICK M. WOSTER[c]

[a] Department of Pharmacology and Chemical Biology, University of
Pittsburgh Cancer Institute, 5117 Centre Avenue, University of Pittsburgh,
Pittsburgh, PA 15213, USA; [b] Department of Laboratory Medicine,
University of California at San Francisco, San Francisco, CA 94143, USA;
[c] Department of Pharmaceutical Sciences, Wayne State University, Detroit,
MI 48202, USA

10.1 Polyamine–Nucleic Acid Interaction as a Potential Epigenetic Target for Cancer Therapy

Extensive work has suggested that DNA is one of the major targets for natural polyamines.[1,2] The polyamine–nucleic acid interaction plays an important role in DNA conformational transition and stabilization.[3–5] Functional interaction between nucleic acids and polyamines extends to DNA–protein binding, particularly those interactions involving gene-regulatory proteins such as histone and transcription factors.[4,6] Due to their structural similarity with natural polyamines, polyamine analogs might follow the same course and therefore alter chromatin structure, histone-modifying enzymes, specific gene expression, cell-cycle regulation and apoptosis. This chapter focuses on the potential for

RSC Drug Discovery Series No. 17
Polyamine Drug Discovery
Edited by Patrick M. Woster and Robert A. Casero, Jr.
© Royal Society of Chemistry 2012
Published by the Royal Society of Chemistry, www.rsc.org

polyamine analogs to inhibit the activity of histone-modifying enzymes and thus affect epigenetic gene regulation.

Epigenetic modifications refer to heritable and reversible changes in chromatin structure that are not due to alterations in primary DNA sequence.[7] The modifications that constitute epigenetic regulatory marks include methylation of cytosine residues in CpG dinucleotides and post-translational modifications of the histone tails such as acetylation, methylation, phosphorylation, ubiquitination, sumoylation and ADP ribosylation.[8,9] Among these alterations, histone acetylation is one of the major regulators of chromatin remodeling and hence gene expression.

The unique structure of natural polyamines has been used as a backbone to conjugate with specific chemical groups such as alkylating agents, spermine-intercalator, DNA intercalators and other antiproliferative agents. These modified molecules, which maintain the cationic nature of the polyamine backbone, typically possess a high affinity for the energy-dependent polyamine transport system and thus are accumulated by cells and target negatively charged chromatin.[10–14] The findings from these aforementioned studies provide clear evidence that the polyamine structure can be used effectively to transport specific active moieties into cells and target the chromatin. By taking advantage of this unique feature, we have recently attempted to design and develop a new generation of agents targeting chromatin-modifying enzymes to alter its regulation of gene expression in cancer cells.

10.2 Polyamine Analogs as HDAC Inhibitors

10.2.1 HDAC Inhibitors

The dynamic nature of histone acetylation is determined by the counter-balancing activity of histone acetyltransferases (HATs) and histone deacetylases (HDACs). The HDAC family is divided into zinc-dependent enzymes, of which there are 11 subtype enzymes (class I, II and IV), and zinc-independent enzymes (class III, also called Sirtuins) that require NAD^+ for their catalytic activities.[15] Acetylation of histone residues by HATs is generally associated with transcriptionally active chromatin, and deacetylation via HDACs usually leads to chromatin condensation and inhibition of transcription. Abnormal activity of HDACs has been implicated in tumorigenesis, and so considerable effort has been put into the development of inhibitors of HDACs as a means of cancer therapy. A number of HDAC inhibitors have been rationally developed in the past decade. Many of the synthetic HDAC inhibitors have been examined for their ability to alter chromatin structures and induce re-expression of aberrantly silenced genes, which in turn leads to growth inhibition and/or apoptosis in cancer cells.[16,17] Based on their chemical structures, HDAC inhibitors are divided into four groups: hydroxamic acids, cyclic tetrapeptides, short-chain fatty acids and benzamides (Table 10.1). Among the most interesting inhibitors are those that have been designed to target primarily the zinc

Table 10.1   Characteristics of leading histone deacetylase inhibitors.

Class	Compound	Clinical trial stage
Hydroxamate	Suberoylanilide hydroxamic acid (Vorinostat)	US Food and Drug Administration approved for cutaneous T-cell lymphoma
	LBH-589 (Panobinostat)	
	PXD-101 (Belinostat)	Phase III
	ITF2357	Phase II
		Phase I
Cyclic peptide	Romidepsin (FK/228)	US Food and Drug Administration approved for cutaneous T-cell lymphoma
Short-chain	Valproic acid	Phase II
Fatty acids	Phenylbutyrate	Phase I, II
Benzamide	MS-275 (Entinostat)	Phase II
	MGC0103	Phase II

cofactor at the active site of the class I/II HDACs.[18,19] The anti-tumor activity of well-studied HDAC inhibitors, such as trichostatin A (TSA), vorinostat (SAHA), romidepsin (FK228), LBH589 and MS-275, has been demonstrated in a wide range of cancer cell lines as well as in animal models.[20–23] Among these HDAC inhibitors, vorinostat and romidepsin have already been approved by the US FDA for the clinical treatment of cutaneous T-cell lymphoma (CTCL). A number of other promising HDAC inhibitors are currently in advanced clinical trials (Table 10.1).

10.2.2   Polyaminohydroxamic Acid (PAHA) and Polyaminobenzamide (PABA) Polyamine Derivatives as HDAC Inhibitors

Based on the observation that polyamine analogs have high affinities for DNA and enter cells using the cellular polyamine transport system, we recently developed a series of polyamine derivatives by combining a polyamine backbone with the active site-directed inhibitor moieties found in class I/II HDACs, such as hydroxamic acids and benzamides.[24,25]

A series of PAHA polyamine analogs were first synthesized to incorporate structural features of the polyamines spermidine and spermine, and the hydroxamic acid moiety commonly found in active HDAC inhibitor molecules such as TSA and SAHA.[24] We hypothesized that these compounds would utilize the polyamine transporter to cross the cell membrane and inhibit HDAC activity via the hydroxamic acid moiety incorporated into the analogs. Sixteen synthetic PAHA analogs were initially screened for their inhibitory activity against a mixture of HDACs, and three lead compounds were successfully identified from the library (Figure 10.1A).[24] Compounds **1** and **2** potently inhibit HDAC activity (75%) at a concentration of 1 μM. In ML-1 mouse leukemia cells, compound **2** produced significant induction of pan acetylated histone H3 (AcH3) and acetylated histone H4 (AcH4) and induced the re-expression of the CDK inhibitor p21[Wafl]. Interestingly, compound **3** displayed

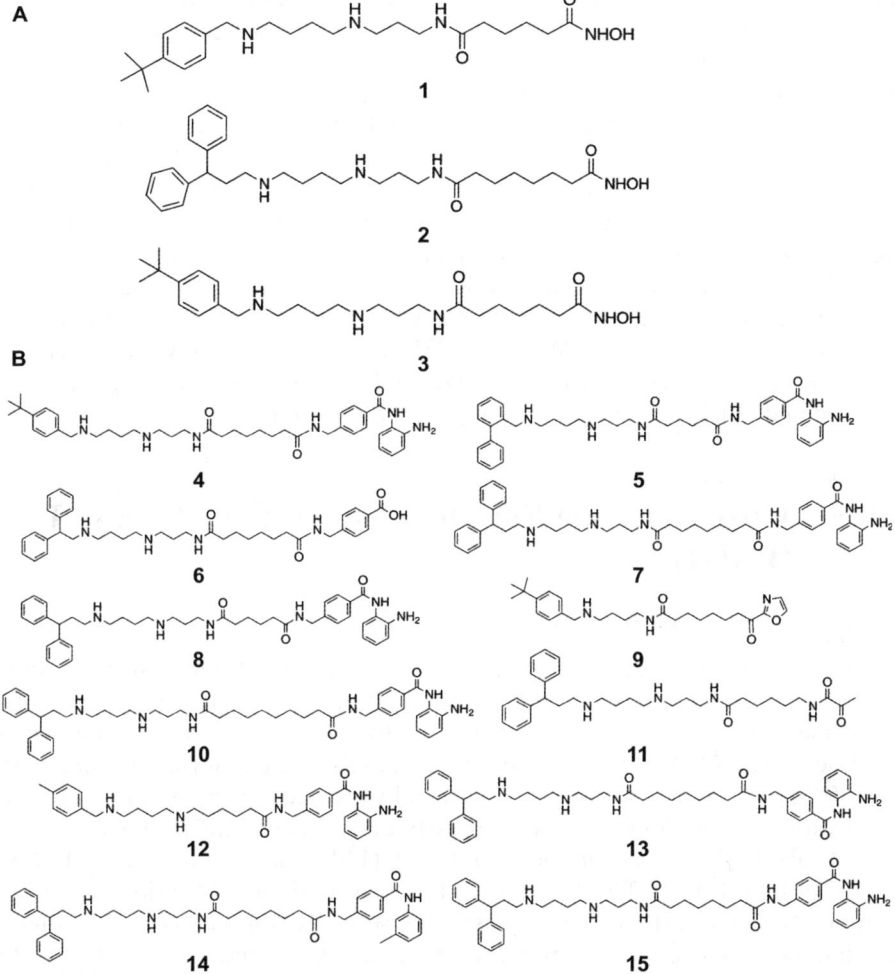

Figure 10.1 Structures of PAHAs **1–3** (A) and PABAs **4–15** (B).

minimal effects on global AcH3 or AcH4 but demonstrated potent *in situ* inhibition of HDAC6 leading to a substantial increase in α-acetyltubulin in treated cells.[25]

By using similar design strategies, another class of 40 PABA analogs were synthesized and evaluated as potential inhibitors of HDACs (Figure 10.1B).[25] Structural modifications, such as the linker chain length, polyamine substituent and metal binding moiety, were introduced into these analogs.[25] Interestingly, these compounds displayed selectivity among specific HDACs, demonstrating that global HDAC inhibition is the sum of individual effects on the specific HDACs inhibited.[26] For example, compound **4** effectively inhibits global HDAC activity but has minor effects on HDAC 3 or 6, suggesting that the inhibitory activity of compound **4** on global HDACs results from the inhibition of one or more of the other HDACs.[25] In human breast-cancer MCF7 cells, most of the PABA analogs did not produce significant cytotoxicity, although these compounds show interesting selectivity among different HDACs. In nontumorigenic human breast epithelial MCF10A cells, compound **12** exhibited a much higher IC_{50} for growth inhibition than that found for MCF7 cells; demonstrating significant selectivity for cancer cells.[25] Further study showed that compound **12** promoted the induction of annexin 1, an early regulatory gene of apoptosis, in human breast-cancer cells (data not shown).

Although the exact molecular mechanisms underlying HDAC inhibition by polyamine analogs are still not fully understood, the initial effort in development of the PAHAs and PABAs as HDAC inhibitors shows considerable promise. This interesting work has laid a foundation for developing more specific and efficient polyamine analogs as HDAC inhibitors.

10.3   Histone Lysine-Specific Histone Demethylase 1 (LSD1)

10.3.1   Discovery of LSD1

Although the dynamic nature of histone acetylation has been known for some time, it was only recently found that histone lysine methylation marks are also dynamically regulated by both histone methyltransferases and histone demethylases. Ground-breaking work from Yang Shi's group reported the discovery of the first enzyme capable of specifically demethylating mono- and dimethylated lysine 4 of histone H3 (H3K4me1 and H3K4me2). This enzyme is called histone lysine-specific demethylase 1 (LSD1, also known as BHC110, AOF2 or KDM1).[27,28] The discovery of LSD1 led to the identification of another family of human histone demethylases, the so-called Jumonji (JmjC) domain-containing proteins. These enzymes use α-ketoglutarate and iron as cofactors to demethylate histone lysine residues through a hydroxylation reaction.[29–32] The discovery of LSD1 and JmjC containing histone demethylases has revolutionized the concept of histone methylation as a dynamically regulated process under enzymatic control, rather than chromatin marks that could only be changed by histone replacement.

Structural analysis demonstrates that LSD1 is an evolutionarily conserved enzyme across species and consists of three major domains: a C-terminal amine oxidase-like domain, an N-terminal SWIRM domain and a central protruding tower domain. The demethylation reaction catalyzed by LSD1 occurs through the reduction of FAD to $FADH_2$ that is then reoxidized by oxygen to produce hydrogen peroxide (H_2O_2), leading to the generation of an imine intermediate. The resultant imine intermediate is subsequently hydrolyzed to generate demethylated H3K4me and formaldehyde (Figure 10.2A). Since there is a requirement for the presence of a lone pair of electrons in the unprotonated nitrogen of the methylated lysine, LSD1 is not able to demethylate

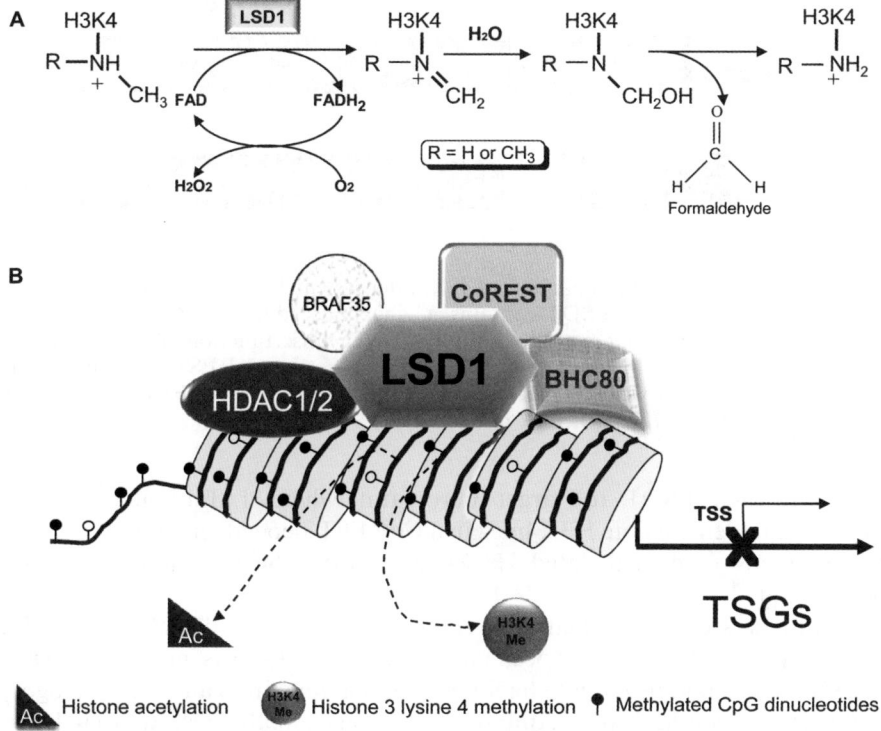

Figure 10.2 Mechanism of histone demethylation by LSD1. (A) The demethylation reaction of mono- (R = H) or di-methyl (R = CH3) Lys4 of histone 3 is catalyzed by LSD1 through the reduction of FAD to FADH$_2$, which is then reoxidized by oxygen to produce hydrogen peroxide (H$_2$O$_2$), leading to the generation of an imine intermediate. The resultant imine intermediate is subsequently hydrolyzed to generate demethylated H3k4me and release of formaldehyde. (B) LSD1 forms a transcriptional repressor complex with CoREST, HDAC1/2, BRAF35 and BHC80. LSD1-mediated demethylation of H3K4me2 involves the deacetylation of the target site by HDAC1/2, BRAF35 association, CoREST binding, favin-dependent demethylation catalyzed by LSD1 and BHC80 activity. The enhanced LSD1 activity and low level H3K4me2 at promoters may lead to the suppressed transcription of specific tumor-suppressor genes (TSGs) in cancer cells. The CpG dinucleotide islands at the promoter regions of some of these TSGs are found to be heavily methylated.

trimethylated histone lysine residues. LSD1 has been typically found in association with a transcriptional repressor complex comprising transcriptional corepressor for element 1-silencing transcription factor (CoREST), HDAC1/2, BHC80 and BRAF35 (Figure 10.2B).[27] The LSD1 complex demethylates its histone substrate in a stepwise manner. First, HDAC1/2 deacetylates histone 3 to create a hypoacetylated histone tail for CoREST binding. Subsequently,

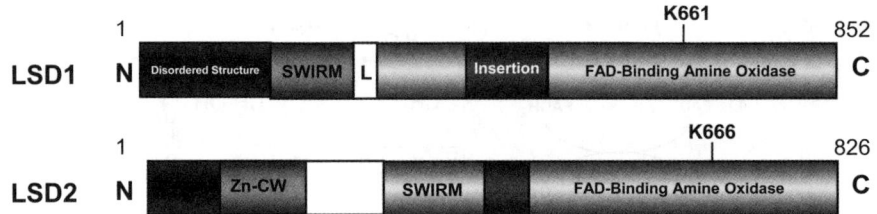

Figure 10.3   Structural comparison of flavin-dependent histone demethylases LSD1
and LSD2. LSD2 contains a zinc finger domain (Zn-CW) at the N
terminal that is not found in LSD1. The insertion domain that is present
in LSD1 for interaction with the corepressor CoREST is not present in
LSD2. Lysine 661 (K661) and lysine 666 (K666) are FAD binding sites
for LSD1 and LSD2, respectively.

BRAF35 leads LSD1 to target chromatin where LSD1 interacts with
HDAC1/2 to demethylate H3K4me2 in a CoREST-dependent manner. Finally,
BHC80 binds to demethylated H3K4, preventing the restoration of H3K4
methylation.[28,33,34]

Recently, a second mammalian FAD-dependent histone demethylase, LSD2
(also known as KDM1B), has been identified.[35–37] LSD2 was initially known as
the AOF1 gene (amine oxidase flavin-containing domain 1) and is homologous
to the gene encoding LSD1. Amino acid sequence analysis shows that LSD1
and LSD2 share 33% overall identity in the amine oxidase domain that contains the enzymatic active site for FAD association and in the SWIRM domain
for protein–protein interaction. Despite the structural and catalytic similarities,
LSD2 is associated with chromatin-remodeling complexes that are different
from those involving LSD1. LSD2 contains a CW-type zinc finger domain with
zinc-binding sites that are not found in LSD1, and it lacks the insertion
structure that is required for interaction with CoREST and HDAC1/2
(Figure 10.3). Therefore, LSD2 represents a new member of the flavin-
dependent histone lysine demethylase family similar to LSD1 in structure and
chemical reactions, but likely to be part of chromatin-remodeling complexes
that are different from those involving the LSD1-CoRest-HDAC1/2 complex.

10.3.2   LSD1 is an Amine Oxidase Homolog

Structural analysis revealed that LSD1 shares considerable homology with
monoamine oxidases (MAOs) and the polyamine oxidases, N^1-acetylpolyamine
oxidase (APAO) and spermine oxidase (SMO).[27,38] In the FAD-binding
domain, 60% of the amino acids are identical or similar in LSD1 and the
polyamine oxidases (Figure 10.4). Despite the similarity of structure and
chemistry between these amine oxidases, LSD1 exhibits completely different
biological activity and/or cellular localization as compared to the monoamine
and polyamine oxidases (Table 10.2). Similar to LSD1, both APAO and SMO

FAD Binding Site
*

LSD1	635	PAVQFVPPLPEWKTSAVQRMGFGNLNKVVLCFDRVFWDPSV	675
SMO	342	TSF–FRPGLPTEKVAAIHRLGIGTTDKIFLEEEEPFWGPEC	381
APAO	296	LDTFFDPPLPAEKAEAIRKIGFGTNNKIFLEFEEPFWEPDC	336

Figure 10.4 Structural comparison of LSD1 and polyamine oxidases. The catalytic domains of LSD1, SMO and APAO possess over 60% similarity in amino acid sequences.

are FAD-dependent enzymes and produce H_2O_2 in their oxidative reactions. However, none of the natural monoamines or polyamines serve as substrates for LSD1, and neither monoamine oxidase nor polyamine oxidase functions as histone demethylases. Monoamine oxidases are bound to the outer mitochondrial membrane in association with a FAD molecule covalently bound to a cysteinyl residue.[39,40] APAO is a peroxisomal enzyme, whereas SMO is found in both the cytoplasm and nucleus.[41]

10.3.3 LSD1 Complex is Implicated in Tumorigenesis

The activity of the LSD1-CoREST-HDAC complex has been implicated in tumorigenesis. In human breast-cancer MCF7 cells, 42% of all Pol II bound promoters were also found to be bound by LSD1. About 58% of ERα-enriched promoters exhibited LSD1 recruitment. LSD1 recruitment on most of the promoters of LSD1 + /ERα + target genes is stimulated by E2. ERα, ER coativator, LSD1 and AcH3K9 exhibit similar profiles in ERα target genes.[42] Lim reported that LSD1 is highly expressed in ER-negative breast cancers and may provide a predictive marker for aggressive biology and a novel attractive therapeutic target for treatment of ER-negative breast cancers.[43] Metzger *et al.* demonstrated that in prostate cancer cells, LSD1 directly associates with the androgen receptor (AR) and acts as a coactivator for transcriptional activation by AR.[44] In addition, LSD1 expression has been associated with poorly differentiated neuroblastomas, suggesting that targeting LSD1 may provide a novel option for treatment of this disease.[29] Recent studies revealed that LSD1 is also able to demethylate non-histone substrates such as p53 and DNA methyltransferase 1 (DNMT1) and to regulate their cellular activities, indicating broader biological functions for LSD1 by directly acting on both histone and non-histone proteins.[45,46]

10.4 Identification of Polyamine Analogs as LSD1 Inhibitors

10.4.1 Bisguanidine and Biguanide Polyamine Analogs as LSD1 Inhibitors

Since the discovery of LSD1, there is increasing interest and effort to identify known compounds or to design new chemical entities that can inhibit LSD1

Table 10.2 Comparison of biological functions of flavin-dependent amine oxidases.

Enzyme	Substrate	Subcellular localization	Flavin-dependent	Complex	H_2O_2 product	Cellular function
Monoamine oxidase	Arylalkyl amines	Mitochondrial membrane	Yes	No	Yes	Neurological activities
Acetylpolyamine oxidase	Ac-spermine, Ac-spermidine	Peroxisome	Yes	No	Yes	Polyamine catabolism
Spermine oxidase	Spermine, spermidine	Cytoplasm, Nucleus	Yes	No	Yes	Polyamine homeostasis
Listone lysine-specific demethylase 1	H3K4me2, H3K4me1, p53, DNA methyltransferase 1	Nucleus	Yes	CoREST, histone deacetylase, BCH80	Yes	Transcription regulation, chromatin remodeling
Listone lysine-specific demethylase 2	H3K4me2, H3K4me1	Nucleus	Yes	Transcription factors	Yes	Transcription regulation

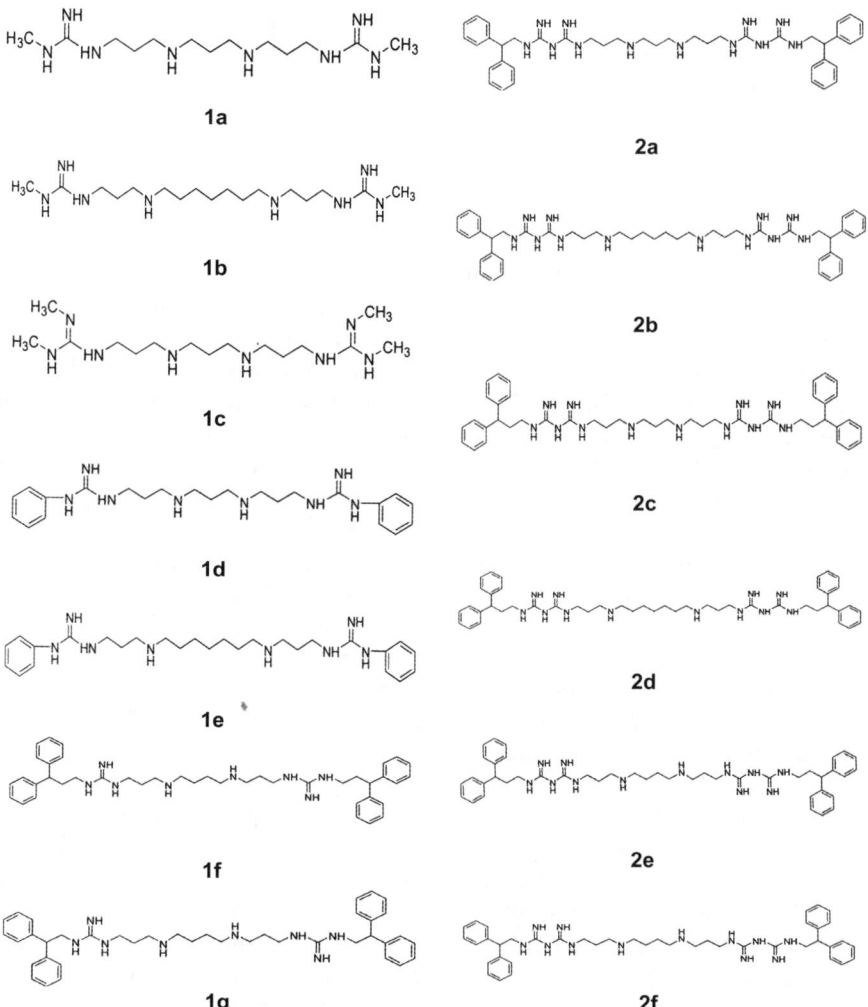

Figure 10.5 Biguanide and bisguanidine polyamine analogs. **1c** and **2d** are potent inhibitors of LSD1 activity and re-activate aberrantly silenced genes in cancer cells.

activity and function as potential novel therapeutic agents for cancer therapy. The structural and catalytic similarities of polyamine oxidases and LSD1 provided the rationale to investigate whether existing polyamine oxidase inhibitors might also act as inhibitors of LSD1. We first tested a small library of bisguanidine and biguanide polyamine analogs (Figure 10.5) for their ability to inhibit purified recombinant LSD1 *in vitro*.[47] Nine out of 13 tested compounds inhibit demethylase activity by $> 50\%$ at concentrations less than 1 μM.[48] In inhibition kinetics studies, the two most potent analogs, **1c** (1,11-bis{N^2,N^3-dimethyl-N^1-guanidino}-4,8-diazaundecane) and **2d** (1,15-bis{N^5-[3,3–(diphenyl)propyl]-N^1-biguanido}-4,12-diazapentadecane), exhibited noncompetitive inhibition kinetics at concentrations

<2.5 μM. These results suggest that polyamine compounds do not appear to directly compete with H3K4me2 at the active site of LSD1. It is likely that these analogs may block the FAD binding site, or that they induce a fundamental change in enzyme structure such that LSD1 is no longer able to bind to its substrate correctly.

10.4.2 Oligoamine Polyamine Analogs as LSD1 Inhibitors

Recently, a novel class of polyamine analogs has been developed, which includes conformationally restricted polyamine analogs and long-chain oligoamine analogs.[49,50] (Figure 10.6) Some of these compounds have demonstrated effectiveness in inhibiting growth and inducing apoptosis in cultured cancer cells and animal studies.[51,52] A number of these analogs, particularly the long-chain oligoamines, initiate DNA aggregation at much lower concentrations than do the natural polyamines, and therefore are more cytotoxic against several different types of cultured cancer cells.[1,49,51–53] We previously demonstrated that specific oligoamine analogs were not substrates of either N^1-acetylpolyamine oxidase or spermine oxidase but that they are potent inhibitors of these purified polyamine oxidases.[54] Because of the structural similarity of these FAD-dependent enzymes and the multivalent cationic structure of oligoamine analogs, it is likely that these compounds would target chromatin, the site of action for LSD1, and act as inhibitors of this histone demethylase. To test this hypothesis, we selected a series of conformationally restricted and saturated polyamine and oligoamine analogs, including pentamines, hexamines, octamines and decamines (Figure 10.6), to test their effects on recombinant LSD1 activity. Most of the selected compounds have four CH_2 residues between each imine nitrogen containing incorporated molecular alterations that restrict the free rotation of specific bonds, and are longer than the natural polyamines. Whereas the pentamine and hexamine analogs marginally affect LSD1 activity, octamine and decamine analogs inhibited this activity by >50% at 10 μM. Two decamine analogs, PG-11144 and PG-11150, which are *cis* and *trans* isomers, were found to be the most potent oligoamines tested and chosen for further functional studies in cancer cells.[55] Both analogs inhibit recombinant LSD1 activity in a dose-dependent manner, and PG-11144 exhibited competitive inhibition kinetics at concentrations of <10 μM, suggesting that oligoamine analogs may compete with substrate at the LSD1 active site.[55]

10.5 Inhibition of LSD1 by Polyamine Analogs Reactivates Aberrantly Silenced Gene Expressions in Cancer Cells

Cancer cells simultaneously show genome-wide hypomethylation and local gene-promoter-specific hypermethylation.[56,57] Abnormal promoter CpG

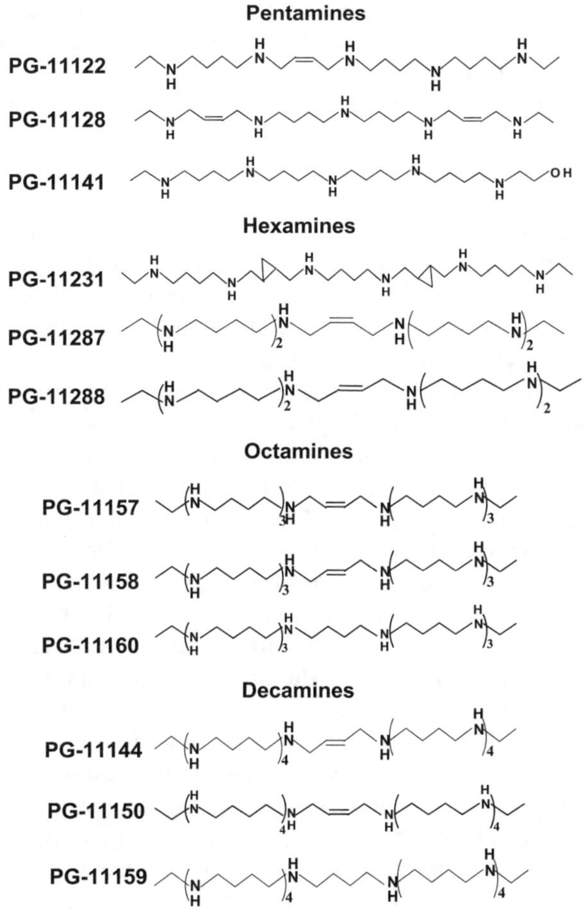

Figure 10.6 Structures of conformational restricted and long-chain polyamine analogs. The decamines PG-11144 and PG-11150, a *cis/trans* pair with double bonds in the center of their structure, competitively inhibit LSD1 activity *in vitro* and re-activate aberrantly silenced genes in cancer cells.

hypermethylation, and concomitant loss of gene expression, has been the recent focus of epigenetic research.[58] Epigenetic silencing of important genes, such as tumor suppressor genes, has been shown to play an important role in tumorigenesis. DNA promoter hypermethylation frequently acts in collaboration with histone modifications, leading to decreased chromatin activating marks such as H3K4me and AcH3K9, and increased repressive marks such as H3k9me and H3K27me that result collectively in the aberrant silencing of specific genes.[7,58] A number of hypermethylated and aberrantly silenced genes have been identified in different types of cancer cells.[58] These genes include members of the WNT signaling pathway antagonists secreted frizzle-related protein family (*SFRPs*), *GATA* family transcription factors, the mismatch repair gene *MLH1*, the cell-cycle regulator *CDKN2A*, the p53 responsive gene *HIC1* and the cell invasive regulator *TIMP3*.[58–60] Increasing effort is ongoing to target this aberrant epigenetic regulation for therapeutic benefit, and clinical trials are presently focusing on combining agents that inhibit DNA CpG methylation and histone deacetylases.

To investigate whether the *in vitro* inhibition of LSD1 activity by these specific polyamine analogs translates into cellular response, the effects on global H3K4me2 status were examined in multiple human cancer cell lines including colorectal cancer, breast cancer, lung cancer and leukemia cells. Exposure of colorectal cancer HCT 116 or RKO cells to these polyamine analogs leads to a remarkable increase in global H3k4me2 and H3K4me1.[48,55] Treatment of HCT116 cells with **1c** or **2d** for 48 h re-expressed three members of the secreted frizzle-related protein family, *SFRP1, SFRP4* and *SFRP5*, and the *GATA5* transcription factor.[48,55] Similarly, treatment with the oligoamine analogs PG-11144 or PG-11150 for 24 h led to the obvious re-expression of *SFRP1, SFRP2, SFRP4* and *GATA5* in HCT116 cells.[55] In acute myeloid leukemia cells, **2d** treatment increased global H3K4me2 concurrent with re-expression of silenced *E*-cadherin.[61] The silencing of *E*-cadherin is thought to contribute to progression in cancer by increasing proliferation, invasion and/or metastasis. Treatment with **2d** or PG-11144 significantly enhanced global H3K4me2 and altered gene expression in human breast-cancer MDA-MB-231 cells. Microarray studies in MDA-MB-231 cells showed there are 274 and 625 genes that changed at least twofold after treatment with **2d** or PG-11144, respectively.[62] These results show that specific polyamine analogs are effective LSD1 inhibitors in cancer cells and that they induce significant re-expression of multiple epigenetically silenced genes important in tumorigenesis.

The inhibition of LSD1 by polyamine analogs in cancer cells has raised an important question about the specificity of these analogs against LSD1 and about whether the effects noted are secondary to effects on chromatin caused by alterations of polyamine metabolism and polyamine pools. To address these questions, the effect of RNAi-mediated knockdown of LSD1 expression was compared with the pharmacological inhibition of LSD1 within HCT116 cells. Knocking down 85% of LSD1 protein expression by RNAi resulted in remarkably increased H3K4me2 at the promoters of reactivated genes, and modest re-expression of the examined genes.[48] Interestingly, pharmacologic inhibition of LSD1 with the analogs was found to be more effective than LSD1 RNAi with respect to re-expression of silenced genes, and this phenomenon may reflect inherent difference in chromatin structure resulting from inhibitor/LSD1 complexes versus RNAi-induced decreases in LSD1 protein. In addition, no detectable changes in polyamine pools were observed after **1c** or **2d** treatment, suggesting that the changes in chromatin marks by these polyamine analogs are not a result of changes in polyamine pools, but are due to inhibition of LSD1.[48]

10.6 Polyamine Analogs Increase Activating Chromatin Marks and Decrease Repressive Marks at the Promoters of Re-Expressed Genes with Retention of DNA Hypermethylation

Site-specific chromatin immunoprecipitation (ChIP) demonstrated that re-expression of aberrantly silenced genes in HCT116 cells treated with polyamine analogs was associated with increased H3K4me1 and H3K4me2 marks at the promoters of all re-expressed genes, along with concurrent decreases in the repressive marks H3K9me1 and H3K9me2.[48,55] Consistent with the results observed for global methylation, no changes in promoter H3K4me3 levels were detected, suggesting that the polyamine analogs specifically target LSD1 without affecting the activity of the Jumonji (JmjC) domain-containing histone demethylase that act on H3K4me3 at the same lysine residue. The increased H3K4 methylation caused by treatment with **1c** or **2d** was accompanied by increased acetyl-H3K9, a chromatin mark associated with active transcription. H3K9me3 levels and the H3K27 methylation status remained unchanged after treatment with **1c** or **2d**, similar to the findings observed in the re-expression of silenced genes in cells treated with the DNMT1 inhibitor 5-aza-deoxycytidine (DAC).[63] In HCT116 cells treated with oligoamine analogs, ChIP analyses of the *SFRP2* gene were extended to cover the proximal promoter region from approximately –1000 bp to +300 bp relative to the transcriptional start site (TSS). Through this analysis, the effects of oligoamines on the LSD1 activity were precisely described in the critical proximal promoter regions that are likely responsible for the oligoamine-induced re-expression of *SFRP2*.[55]

In colorectal cancer cells, promoter CpG dinucleotide hypermethylation collaborates with specific histone marks in the epigenetic silencing of tumor suppressor genes.[64] Bisulfite sequencing analysis after treatment with **2d** or **PG-11144** showed marginal changes of DNA methylation at the promoter regions of *SFRP* genes.[48,55] These results indicate that analog-induced increases in H3K4 methylation alone might be potent enough to drive some extent of re-expression of even heavily methylated genes without affecting the DNA methylation status.

10.7 Combination of LSD1 Inhibitors with Other Agents Targeting Epigenetic Regulation of Gene Expression

Combination of agents targeting multiple epigenetic modifiers may provide for the possibility of using lower and better tolerated doses of the individual agents while achieving improved clinical efficacy. The combination of DNA methyltransferase and HDAC inhibitors has demonstrated synergistic effects in re-expressing epigenetically silenced genes in cancer cells and has produced clinical responses in patients with leukemia.[65,66] To determine if the LSD1-inhibiting polyamine analogs would be more effective when combined with DNMT inhibitors in the re-expression of silenced genes and in tumor-growth inhibition, the expression status of several epigenetically silenced genes in HCT116 cells following a 24 h treatment with low doses of PG-11144 or PG-11150 and the DNMT inhibitor 5-Aza-2-deoxycitidine (DAC), alone and in combination, was investigated. These studies demonstrate that the combination of low-dose oligo-amines with DAC results in synergistic expression of the *SFRP2* gene.[55] These results further suggest that the combination of LSD1 inhibitor and a DNA methylation inhibitor can collaborate in the re-expression of specific silenced genes, thus providing a useful strategy for the clinical utility of these two agents that target different epigenetic entities.

10.8   *In Vivo* Effects of Polyamine Analogs on LSD1 and Tumor Growth

The *in vitro* synergy in inducing gene re-expression caused by the LSD1-inhibiting oligoamine analogs in combination with the DNMT inhibitor DAC, raised the important question as to whether such results might translate into therapeutic efficacy. In a recent experiment using human colorectal cancer cell HCT116 xenografts in athymic nude mice, the *in vivo* therapeutic effect of polyamine analogs as LSD1 inhibitors alone or in combination with the DNMT inhibitor 5-Aza was assessed. Both PG-11144 and **2d** significantly increase H3K4me2 levels in treated HCT116 tumors. Treatment with either PG-11144 or 5-Aza alone both generally displayed significant antitumor effects against growth of HCT116 xenografts, whereas treatment with **2d** alone did not. However, for both polyamine analogs, a marked increase in the inhibition of tumor growth was noted when combined with 5-Aza.[55] Importantly, there was no overall toxicity observed in the treatment groups. These results indicate that LSD1 inhibitors may be most effective when used in combination with agents that target other epigenetic regulatory mechanisms. These *in vivo* results provide the basis for clinical trials combining agents targeting these two important epigenetic regulatory pathways.

10.9   Conclusion

Over the last three decades, increasing understanding about the molecular mechanisms of polyamine metabolism in cancer has provided unique oppor-tunities for the design, synthesis and development of promising polyamine-based antitumor drugs. The interaction of polyamine or polyamine-based analogs with nucleic acids has received increasing attention recently. Specific polyamine analogs, PAHAs and PABAs, were rationally designed for the purpose of enhancing transport or DNA affinity and have been shown to be potent inhibitors of the HDACs. These polyamine-based HDAC inhibitors exhibit potent inhibitory activities against specific HDACs. Treatment with those compounds in cancer cells results in the re-expression of tumor-suppressor genes, and in some cases significant inhibition of tumor-cell growth. The identification of LSD1 and Jmjc domain-containing histone demethylases has made a major impact on the field of epigenetics. These findings demonstrate that histone methylation, which was previously considered to be a stable mark, is dynamic. The implication of LSD1 in tumorigenesis has led to tremendous interest in the search for LSD1 inhibitors as a new generation of epigenetic drugs in the treatment of cancer and other diseases. Our recent studies demonstrate that a series of novel polyamine analogs are powerful inhibitors of LSD1, both alone and in combination with other epigenetic agents, and that they are capable of inducing re-expression of several aberrantly silenced genes important in tumorigenesis. These advances show the promise of using polyamine-based HDAC inhibitors and/or LSD1 inhibitors as a novel approach to cancer treatment and are anticipated to lead to the development of a new generation of therapeutically effective agents with considerable clinical potential.

Acknowledgment

We thank Dr Robert A. Casero, Jr, professor of oncology at Johns Hopkins University Sidney Kimmel Comprehensive Cancer Center for helpful comments on the manuscript.

References

1. R. A. Casero, Jr and L. J. Marton, *Nat. Rev. Drug Discov.*, 2007, **6**, 373–390.
2. T. J. Thomas and R. P. Messner, *J. Mol. Biol.*, 1988, **201**, 463–467.
3. B. G. Feuerstein, L. D. Williams, H. S. Basu and L. J. Marton, *J. Cell. Biochem.*, 1991, **46**, 37–47.
4. C. A. Panagiotidis, S. Artandi, K. Calame and S. J. Silverstein, *Nucleic Acids Res.*, 1995, **23**, 1800–1809.
5. B. G. Feuerstein, H. S. Basu and L. J. Marton, *Adv. Exp. Med. Biol.*, 1988, **250**, 517–523.
6. Y. Maeda, C. Rachez, L. Hawel, 3rd, C. V. Byus, L. P. Freedman and F. M. Sladek, *Mol. Endocrinol.*, 2002, **16**, 1502–1510.
7. P. A. Jones and S. B. Baylin, *Cell*, 2007, **128**, 683–692.
8. T. Jenuwein and C. D. Allis, *Science*, 2001, **293**, 1074–1080.
9. L. Ho and G. R. Crabtree, *Nature*, 2010, **463**, 474–484.
10. J. Holley, A. Mather, P. Cullis, M. R. Symons, P. Wardman, R. A. Watt and G. M. Cohen, *Biochem. Pharmacol.*, 1992, **43**, 763–769.
11. O. I. Phanstiel, H. L. Price, L. Wang, J. Juusola, M. Kline and S. M. Shah, *J. Org. Chem.*, 2000, **65**, 7710.
12. C. Wang, J. G. Delcros, J. Biggerstaff and O. t. Phanstiel, *J. Med. Chem.*, 2003, **46**, 2663–2671.
13. C. Wang, J. G. Delcros, J. Biggerstaff and O. t. Phanstiel, *J. Med. Chem.*, 2003, **46**, 2672–2682.

14. R. D. Verschoyle, P. Carthew, J. L. Holley, P. Cullis and G. M. Cohen, *Cancer Lett.*, 1994, **85**, 217–222.
15. S. Minucci and P. G. Pelicci, *Nat. Rev. Cancer*, 2006, **6**, 38–51.
16. C. M. Marson, *Anticancer Agents Med. Chem.*, 2009, **9**, 661–692.
17. O. Khan and N. B. La Thangue, *Nat. Clin. Pract. Oncol.*, 2008, **5**, 714–726.
18. P. A. Marks, V. M. Richon, T. Miller and W. K. Kelly, *Adv. Cancer Res.*, 2004, **91**, 137–168.
19. R. Ficner, *Curr. Top. Med. Chem.*, 2009, **9**, 235–240.
20. P. A. Marks and W. S. Xu, *J. Cell. Biochem.*, 2009, **107**, 600–608.
21. C. Campas-Moya, *Drugs Today (Barc)*, 2009, **45**, 787–795.
22. L. Stimson, V. Wood, O. Khan, S. Fotheringham and N. B. La Thangue, *Ann. Oncol.*, 2009, **20**, 1293–1302.
23. P. A. Marks and R. Breslow, *Nat. Biotechnol.*, 2007, **25**, 84–90.
24. S. Varghese, D. Gupta, T. Baran, A. Jiemjit, S. D. Gore, R. A. Casero, Jr and P. M. Woster, *J. Med. Chem.*, 2005, **48**, 6350–6365.
25. S. Varghese, T. Senanayake, T. Murray-Stewart, K. Doering, A. Fraser, R. A. Casero, Jr and P. M. Woster, *J. Med. Chem.*, 2008, **51**, 2447–2456.
26. R. A. Casero, Jr and P. M. Woster, *J. Med. Chem.*, 2009, **52**, 4551–4573.
27. Y. Shi, F. Lan, C. Matson, P. Mulligan, J. R. Whetstine, P. A. Cole and R. A. Casero, *Cell*, 2004, **119**, 941–953.
28. M. G. Lee, C. Wynder, N. Cooch and R. Shiekhattar, *Nature*, 2005, **437**, 432–435.
29. J. H. Schulte, S. Lim, A. Schramm, N. Friedrichs, J. Koster, R. Versteeg, I. Ora, K. Pajtler, L. Klein-Hitpass, S. Kuhfittig-Kulle, E. Metzger, R. Schule, A. Eggert, R. Buettner and J. Kirfel, *Cancer Res.*, 2009, **69**, 2065–2071.
30. Y. Huang, J. C. Keen, A. Pledgie, L. J. Marton, T. Zhu, S. Sukumar, B. H. Park, B. Blair, K. Brenner, R. A. Casero, Jr and N. E. Davidson, *J. Biol. Chem.*, 2006, **281**, 19055–19063.
31. K. Yamane, C. Toumazou, Y. Tsukada, H. Erdjument-Bromage, P. Tempst, J. Wong and Y. Zhang, *Cell*, 2006, **125**, 483–495.
32. Z. Chen, J. Zang, J. Whetstine, X. Hong, F. Davrazou, T. G. Kutateladze, M. Simpson, Q. Mao, C. H. Pan, S. Dai, J. Hagman, K. Hansen, Y. Shi and G. Zhang, *Cell*, 2006, **125**, 691–702.
33. Y. J. Shi, C. Matson, F. Lan, S. Iwase, T. Baba and Y. Shi, *Mol. Cell*, 2005, **19**, 857–864.
34. F. Lan, R. E. Collins, R. De Cegli, R. Alpatov, J. R. Horton, X. Shi, O. Gozani, X. Cheng and Y. Shi, *Nature*, 2007, **448**, 718–722.
35. A. Karytinos, F. Forneris, A. Profumo, G. Ciossani, E. Battaglioli, C. Binda and A. Mattevi, *J. Biol. Chem.*, 2009, **284**, 17775–17782.
36. D. N. Ciccone, H. Su, S. Hevi, F. Gay, H. Lei, J. Bajko, G. Xu, E. Li and T. Chen, *Nature*, 2009, **461**, 415–418.
37. Z. Yang, J. Jiang, D. M. Stewart, S. Qi, K. Yamane, J. Li, Y. Zhang and J. Wong, *Cell Res.*, **20**, 276–287.
38. M. G. Lee, C. Wynder, D. M. Schmidt, D. G. McCafferty and R. Shiekhattar, *Chem. Biol.*, 2006, **13**, 563–567.

39. C. Binda, A. Mattevi and D. E. Edmondson, *J. Biol. Chem.*, 2002, **277**, 23973–23976.
40. E. B. Kearney, J. I. Salach, W. H. Walker, R. L. Seng, W. Kenney, E. Zeszotek and T. P. Singer, *Eur. J. Biochem.*, 1971, **24**, 321–327.
41. T. Murray-Stewart, Y. Wang, A. Goodwin, A. Hacker, A. Meeker and R. A. Casero, Jr, *Febs J*, 2008, **275**, 2795–2806.
42. I. Garcia-Bassets, Y. S. Kwon, F. Telese, G. G. Prefontaine, K. R. Hutt, C. S. Cheng, B. G. Ju, K. A. Ohgi, J. Wang, L. Escoubet-Lozach, D. W. Rose, C. K. Glass, X. D. Fu and M. G. Rosenfeld, *Cell*, 2007, **128**, 505–518.
43. S. Lim, A. Janzer, A. Becker, A. Zimmer, R. Schule, R. Buettner and J. Kirfel, *Carcinogenesis*, **31**, 512–520.
44. E. Metzger, M. Wissmann, N. Yin, J. M. Muller, R. Schneider, A. H. Peters, T. Gunther, R. Buettner and R. Schule, *Nature*, 2005, **437**, 436–439.
45. J. Huang, R. Sengupta, A. B. Espejo, M. G. Lee, J. A. Dorsey, M. Richter, S. Opravil, R. Shiekhattar, M. T. Bedford, T. Jenuwein and S. L. Berger, *Nature*, 2007, **449**, 105–108.
46. J. Wang, S. Hevi, J. K. Kurash, H. Lei, F. Gay, J. Bajko, H. Su, W. Sun, H. Chang, G. Xu, F. Gaudet, E. Li and T. Chen, *Nat. Genet.*, 2009, **41**, 125–129.
47. X. Bi, C. Lopez, C. J. Bacchi, D. Rattendi and P. M. Woster, *Bioorg. Med. Chem. Lett*, 2006, **16**, 3229–3232.
48. Y. Huang, E. Greene, T. Murray Stewart, A. C. Goodwin, S. B. Baylin, P. M. Woster and R. A. Casero, Jr, *Proc. Natl. Acad. Sci. U. S. A.*, 2007, **104**, 8023–8028.
49. A. Valasinas, V. K. Reddy, A. V. Blokhin, H. S. Basu, S. Bhattacharya, A. Sarkar, L. J. Marton and B. Frydman, *Bioorg. Med. Chem.*, 2003, **11**, 4121–4131.
50. V. K. Reddy, A. Valasinas, A. Sarkar, H. S. Basu, L. J. Marton and B. Frydman, *J. Med. Chem.*, 1998, **41**, 4723–4732.
51. Y. Huang, A. Pledgie, R. Casero, Jr and N. Davidson, *Anticancer Drugs*, 2005, **16**, 229–241.
52. Y. Huang, E. R. Hager, D. L. Phillips, V. R. Dunn, A. Hacker, B. Frydman, J. A. Kink, A. L. Valasinas, V. K. Reddy, L. J. Marton, R. A. Casero, Jr and N. E. Davidson, *Clin Cancer Res.*, 2003, **9**, 2769–2777.
53. Y. Huang, A. Pledgie, E. Rubin, L. J. Marton, P. M. Woster, S. Sukumar, R. A. Casero, Jr and N. E. Davidson, *Cancer Biol. Ther.*, 2005, **4**, 1006–1013.
54. Y. Wang, T. Murray-Stewart, W. Devereux, A. Hacker, B. Frydman, P. M. Woster and R. A. Casero, Jr, *Biochem. Biophys. Res. Commun.*, 2003, **304**, 605–611.
55. Y. Huang, T. M. Stewart, Y. Wu, S. B. Baylin, L. J. Marton, B. Perkins, R. J. Jones, P. M. Woster and R. A. Casero, Jr, *Clin Cancer Res.*, 2009, **15**, 7217–7228.
56. A. P. Feinberg and B. Tycko, *Nat. Rev. Cancer*, 2004, **4**, 143–153.
57. Y. Akiyama, N. Watkins, H. Suzuki, K. W. Jair, M. van Engeland, M. Esteller, H. Sakai, C. Y. Ren, Y. Yuasa, J. G. Herman and S. B. Baylin, *Mol. Cell. Biol.*, 2003, **23**, 8429–8439.

58. S. B. Baylin and J. E. Ohm, *Nat. Rev. Cancer*, 2006, **6**, 107–116.
59. H. Suzuki, E. Gabrielson, W. Chen, R. Anbazhagan, M. van Engeland, M. Weijenberg, J. Herman and S. Baylin, *Nat. Genet.*, 2002, **31**, 141–149.
60. Y. Akiyama, N. Watkins, H. Suzuki, K. Jair, M. van Engeland, M. Esteller, H. Sakai, C. Ren, Y. Yuasa, J. Herman and S. Baylin, *Mol. Cell. Biol.*, 2003, **23**, 8429–8439.
61. T. Murray-Stewart, Y. Huang, P. Woster and R. Casero, *AACR Annual Meeting*, 2008.
62. Q. Zhu, Huang, Y., Marton L. J., Woster, P. M., Davidson, N. E. and Casero, R. A., AACR Annual Meeting, 2010.
63. K. M. McGarvey, J. A. Fahrner, E. Greene, J. Martens, T. Jenuwein and S. B. Baylin, *Cancer Res.*, 2006, **66**, 3541–3549.
64. H. Suzuki, D. N. Watkins, K. W. Jair, K. E. Schuebel, S. D. Markowitz, W. D. Chen, T. P. Pretlow, B. Yang, Y. Akiyama, M. Van Engeland, M. Toyota, T. Tokino, Y. Hinoda, K. Imai, J. G. Herman and S. B. Baylin, *Nat. Genet.*, 2004, **36**, 417–422.
65. E. E. Cameron, K. E. Bachman, S. Myohanen, J. G. Herman and S. B. Baylin, *Nat. Genet.*, 1999, **21**, 103–107.
66. W. J. Whish, M. I. Davies and S. Shall, *Biochem. Biophys. Res. Commun.*, 1975, **65**, 722–730.

CHAPTER 11

Clinical Applications of Polyamine-Based Therapeutics

ANDRÉ S. BACHMANN*[a,b] AND VICTOR A. LEVIN[c,d]

[a] Department of Pharmaceutical Sciences, College of Pharmacy, University of Hawaii at Hilo, Hilo, HI 96720, USA; [b] Cancer Research Center of Hawaii, University of Hawaii at Manoa, Honolulu, HI 96813, USA; [c] Department of Neuro-Oncology, Division of Cancer Medicine, The University of Texas MD Anderson Cancer Center, Houston, TX 77030, USA; [d] Department of Neurosurgery and Neurosciences, Kaiser Permanente, Redwood City, CA 94063, USA

11.1 Introduction

Research over the past 30 years has shown that the polyamines putrescine, spermidine and spermine are essential for normal cell growth and differentiation, and depletion of polyamines inhibits the growth of neoplastic cells both *in vitro*[1-3] and in animal models.[1-5] It has been well established that polyamine levels in mammalian cells correlate with the rate of cell growth and that high concentrations are found in rapidly proliferating cells, while low concentrations are detected in slow-growing or quiescent cells.[6-8] In addition, polyamine-metabolizing enzymes are activated by a variety of growth factors, carcinogens, viruses and oncogenes contributing to their importance in the regulation of cell growth.

Given the absolute requirement of polyamines for cell growth and the potentially oncogenic consequences of their overproduction, it is not surprising that the polyamine-metabolizing enzymes and polyamine levels are subjected to extensive regulation.

For example, ODC levels are tightly controlled in normal cells and regulated at the levels of transcription, translation and degradation,[8-10] and it is perhaps

RSC Drug Discovery Series No. 17
Polyamine Drug Discovery
Edited by Patrick M. Woster and Robert A. Casero, Jr.
© Royal Society of Chemistry 2012
Published by the Royal Society of Chemistry, www.rsc.org

not surprising that dysregulation of ODC homeostasis leads to unbalanced cell growth and cancer. Indeed, ODC has oncogenic and transforming abilities, and ODC over-expression has been observed in many tumor types.[9,11,12] ODC is also a favorite drug target for the treatment of Human African Trypanosomiasis, a disease caused by the parasite *Trypansoma brucei*, which is transmitted by the tsetse fly (reviewed by Bacchi[13]).

For these reasons, the polyamine biosynthetic pathway has provided interesting targets for the development of agents that prevent carcinogenesis, tumor growth and replicating parasites. In this chapter, we describe the importance of the polyamine biosynthetic enzymes ODC and AdoMetDC and the therapeutic potential of inhibitors that block these enzymatic activities. Ornithine decarboxylase (ODC) is a rate-limiting enzyme of polyamine biosynthesis, catalyzing the conversion of ornithine to the diamine putrescine, which is the basic building block of the higher-order polyamines, such as spermidine and spermine. Spermidine and spermine are anabolized by the spermidine and spermine synthases, respectively.[14] These catalytic events occur by the sequential transfer of aminopropyl groups to putrescine and spermidine, respectively.[8] The aminopropyl group is donated by decarboxylated *S*-adenosylmethionine (dcAdoMet), which is converted from *S*-adenosylmethionine (AdoMet) by *S*-adenosylmethionine decarboxylase (AdoMetDC, also known as SAMDC or AMD1).[15] In this fashion, ODC and AdoMetDC are two critical enzymes necessary for the biosynthesis of the cellular polyamine pool.[8,14,15]

While there are currently a number of promising new types of polyamine inhibitors and polyamine analogs (including uptake/transport inhibitors) assessed in pre-clinical studies, our discussion focuses primarily on those compounds that have been or are currently in clinical trials. While cancer is a main focus, the chapter also discusses a variety of other diseases and conditions that have benefited from this class of drugs, such as trypanosomiasis. The chapter concludes with a brief discussion of other potential applications for this class of inhibitors. We have made every effort to acknowledge the many important contributions of our colleagues, and we sincerely apologize for any oversight in the referenced material.

11.2 Polyamine Inhibitors in Therapeutic Clinical Trials

11.2.1 Cancer

11.2.1.1 ODC As a Primary Drug Target in Human Malignancies

The drug most associated with ODC is α-difluromethylornithine (DFMO, also referred to as eflornithine, ornidyl). DFMO is an irreversible suicide inhibitor of ODC and one of the first rationally designed anti-tumor agents.[16] Aside from DFMO, several other drugs have been synthesized and tested in

Table 11.1 Collated responses to DFMO in phase I and II studies.

Cancer type	Dose $(g\ m^{-2}\ day^{-1})$	PR+CR (%)	SD (%)	R+SD (%)
Leukemia (with MGBG)[34,94]		40–100	0	56
Cervical cancer[95]	0.5–1	50	33	88
Melanoma[36,37,96,97]	2–10	0–29	0–53	
Pooled melanoma >7 days[36,37,97]	4–10	14	35	49
Colon cancer ×21 days[39]	9	0	14	14
Breast cancer ×14 days[41]	9	0	17	17
SCLC ×21 days[39]	9	4	11	15
Anaplastic glioma ×14 days[98]	10.8	29	16	45
Glioblastoma ×14 days[98]	10.8	12	6	18

preclinical models, and in phase I and phase II clinical trials in cancer patients.[17,18] For example, methylacetylenic putrescine (MAP) is another enzyme-activated irreversible inhibitor of ODC, and a phase I study was carried out in patients with advanced malignancies.[19] In the main, these drugs were synthesized to be analogs of DFMO, spermidine or spermine.[20]

A variety of preclinical studies, over the years, added to a substantial literature showing that ODC inhibition could slow tumor growth, reduce tumor incidence and interact in an additive manner with some cytotoxic chemotherapeutic agents. As expected, some of the effect of DFMO on cultured cancer cells could be reversed by the addition of putrescine or spermidine to the media and, at times, to rodent chow when DFMO was studied in rodent tumor models. Nonetheless, even though there is a theoretical risk that polyamines from foods in our diet would interfere with the anticancer potential of DFMO, clinical trials have shown efficacy against recurrent tumors in phase II studies and with cytotoxic drugs in upfront phase III studies (Table 11.1).

11.2.1.2 AdoMetDC As an Alternative Drug Target in Human Malignancies

Although ODC has generally been regarded as the enzyme catalyzing the rate-limiting step in polyamine biosynthesis, it has been shown that the supply of decarboxylated *S*-adenosylmethionine represents a second rate-limiting factor in polyamine biosynthesis.[21] Thus, the enzyme *S*-adenosylmethionine decarboxylase (AdoMetDC) is a possible second rational target for anti-cancer agent development.

Methylglyoxal bis(guanylhydrazone) or mitoguazone (MGBG, also known as methyl-GAG) was synthesized in 1898,[22] but was not introduced into clinical use in the USA until the work of Freedlander and French in 1958.[23–25] When first introduced on a daily schedule, MGBG had serious problems with mucositis, myelosuppression, mitochondrial and other toxicities, which, in some instances, were fatal. Subsequent research showed that MGBG inhibited

AdoMetDC.[26] After a weekly schedule of IV MGBG was introduced, renewed interest led to a large number of clinical trials during the 1980s and 1990s (38 trials referenced in Nishioka & Ajani).[27] Some of these trials are summarized in Table 11.2. While only weak clinical activities were observed in solid tumors, responses against recurrent and refractory lymphomas and Hodgkin's disease were excellent.

In an attempt to improve MGBG efficacy and to reduce the unfavorable safety profile associated with toxicities, a series of sterically hindered molecules related to MGBG were synthesized which led to the more potent and more selective second-generation inhibitor SAM486A (also known as CGP48664). SAM486A is the free base of 4-(aminoimino-methyl)-2,3-dihydro-1*H*-inden-1-one-diaminomethylenehydrazone and entered several phase I and II clinical studies in breast cancer and advanced solid malignancies.[28,29] A phase II trial with relapsed or refractory non-Hodgkin's lymphoma reported promising clinical activity in some patients,[30] while others found that the tested dose and schedule of administration revealed no significant therapeutic potential of SAM486A in a phase II study with metastatic melanoma.[31]

Since ODC and AdoMetDC are co-regulated by intracellular polyamine pools so that inhibition of one results in a compensatory increase in the other, it follows that targeted interference with a drug cocktail composed of ODC and AdoMetDC inhibitors could be especially useful and may sharpen the antiproliferative effects by causing complete depletion of the polyamine pools. Over the years, a number of DFMO combinations have been evaluated in clinical studies, including the combination of DFMO with MGBG[32–35] or interferon,[36–38] but to our knowledge, there are no reports of human clinical studies that assessed the efficacy of DFMO combined with second-generation inhibitor SAM486A.

11.2.1.3   *Examples of Human Cancer Chemotherapy Studies*

When used as single therapeutic agents, polyamine inhibitors such as DFMO have shown unpredictable clinical activity against human tumors. From the literature, it appears that, outside of gliomas, most trials have been small. For instance, in trials with DFMO in adencarcinoma, 89 patients were studied in phase I/II; in melanoma, 59 patients were studied in phase I/II; in leukemia, only 25 children and adults were studied in phase II. This compares to glioma trials where over 100 patients were studied in phase II as a single agent, while over 250 patients were studied in phase II and III in combination with nitrosoureas. The schedule and doses used in single agent phase I and II studies are summarized in Table 11.1.

Phase II (efficacy) studies have been conducted in non-small cell lung cancer and metastatic colon cancer.[39] Against metastatic melanoma, DFMO alone produced disease stabilization in 33% of patients36 and also in 33% when combined with IFN-alpha.[37] However, no survival advantage was observed.[40] Against breast cancer, no consistent benefit was seen, although an occasional

Table 11.2 Single agent MGBG in phase II clinical trials.

Investigator (first author)/study phase	Patient population	Study end-points	Regimen	Number
Bukowski et al.[99]	Pancreatic carcinoma	CR + PR 6%; MS 7.6 months	60 mg m^{-2} weekly	32
Winter et al.[100]	Multiple myeloma	No responses	500–600 mg m^{-2} on days 1 and 8; then every 2 weeks	14
Simon et al.[101]	Small-cell carcinoma	PR 4%, SD 12%	500 mg m^{-2} per wk for 4 cycles; then every 2 weeks thereafter	24
Barrett et al.[102]	Transitional-cell carcinoma of urinary tract	PR 11%; MS 14 weeks	600 mg m^{-2} weekly	18
Douglass et al.[103]	Colorectal carcinoma	PR 6%; median TTP 8 weeks	500–700 mg m^{-2} weekly; escalated if tolerated	77
Knight et al.[104]	Colorectal carcinoma	PR 6%, SD 9%	500–600 mg m^{-2} weekly	32
Wiernik et al.[105]	Rai stages 2–4 CLL	No responses	500 mg m^{-2} weekly	13
Knight et al.[106]	Lymphoma	PR 30%; median TTP 12 weeks	500–600 mg m^{-2} weekly	46
Knight et al.[106]	Hodgkins lymphoma	CR 2%, PR 22%; median TTP 6 weeks	500–600 mg m^{-2} weekly	10
Levine et al.[107]	High-grade AIDS-lymphoma	CR 12%, PR 12%, SD 23%; MS 2.6 months, MS 22 months for CR and MS 5,6 months for PR	600 mg m^{-2} on days 1 and 8; then every 2 weeks	26
Scher et al.[108]	Hormone-resistant prostate adenocarcinoma	PR 24%, SD 8%; median TTP 3 months	500–600 mg m^{-2} weekly	25
Knight et al.[109]	Renal cell carcinoma	CR 1%, PR 3%, SD 5%	500–600 mg m^{-2} weekly	87

patient appeared to benefit by slowed progression.[41] Studies against intrae-pithelial cervical intraepithelial neoplasia found that at lower DFMO doses, there was no activity.[42]

The situation with malignant gliomas differed from prior trials in that efficacy was seen in both phase II and phase III studies (Table 11.3). DFMO was evaluated alone,[43] with MGBG,[43] with BCNU[44] and in a randomized study with PCV (procarbazine, CCNU, vincristine).[45] DFMO produced an improvement in time to tumor progression, progression-free survival and patient survival with only modest toxicity, thus showing promising clinical results in this patient population. The studies indicated that patients with anaplastic glioma (AG, WHO grade 3) tumors benefited more than those with glioblastoma (GBM, WHO grade 4) tumors.[45] The implications of these studies are important to consider to better understand the importance of ODC and polyamine synthesis as targets for cancer chemotherapy. From the literature and our own studies, it appears that for most cancers, as malignancy increases (tumor-grade increase) so do polyamine levels.[46–48] Furthermore, we (VL) have demonstrated that for gliomas, the response to DFMO is coupled to the grade of malignancy and to tissue levels of ODC[43–45] and that progression-free survival of these patients was inversely related to the tumor ODC level.[49] In a Cox proportional hazards model, progression-free survival was found to be inversely related to median tumor ODC activity, with an unadjusted hazard ratio for the median ODC group (>3.3 nmol 30 min^{-1} μg protein^{-1} *vs.* ≤ 3.3) of 5.8 ($p < 0.0001$); a median progression-free survival of 120 months for patients with AGs with a median ODC activity of ≤ 3.3 nmol 30 min^{-1} μg protein^{-1} and 9 months for the eight AG and 10 glioblastoma patients with an ODC activity of >3.3 nmol 30 min^{-1} μg protein^{-1}. Of AG tumors in which ODC activity was evaluated, 26% had ODC levels of >3.3 nmol 30 min^{-1} μg protein^{-1}, indicating that it is not unique to glioblastoma tumors. This led to the expectation that patients with glioma lineage tumors with low ODC activities may be expected to have a more durable response to DFMO than those with higher ODC activity levels.

The phase II trials of DFMO in recurrent anaplastic glioma patients were encouraging and led investigators to continue exploring the role of DFMO in the management of patients with anaplastic glioma. In particular, results from preclinical and clinical phase I/II studies supported a potential synergism between DFMO and nitrosoureas such as BCNU and the CCNU combination, PCV. A randomized, controlled phase III study by one of our groups (VL) comparing the benefits of DFMO as sole therapy with DFMO in combination with PCV demonstrated a sustained benefit in survival probability for AG patients.[45] The survival advantage was observed throughout the 10-year trial and was reflected in actual survival gains of greater than 1 year for patients taking DFMO-PCV compared with those taking PCV alone. Survival analysis showed a DFMO-PCV median survival of 6.3 years (49 of 114 events) from study entry compared to 5.1 years for PCV. The hazard function for progression-free survival demonstrated a difference over the first 20 months of study (hazard ratio 0.53, p = 0.02), supporting the conclusion that DFMO adds to the

Table 11.3 DFMO trials in patients with gliomas.

Investigator/study phase	Patient Population	Study end-points	Regimen	Number
DFMO/PCV (CCNU, Procarbazine, and Vincristine) combination studies				
Levin *et al.* (unpublished study T89-0213) Phase I/II	Malignant gliomas	Determination of myelotoxicity of DFMO doses of 2 to 3 g/m² in combination with PCV	DFMO in combination with PCV (CCNU, Procarbazine, and Vincristine)	$N = 16$
	11 anaplastic glioma; 5 glioblastoma multiforme		DFMO: 2, 2.5 and 3 g/m² PCV (CCNU 110 mg m⁻² *p.o.* on day 15, vincristine 1.4 mg m⁻² iv, days 22 and 43; and precarbazine 60 mg m⁻² *p.o.* on days 22–35)	
Levin *et al.*,[45] 2003 Phase III	Arm A: 89 anaplastic astrocytoma; 4 anaplastic oligoastrocytoma; 16 anaplastic oligodendroglioma; 5 other malignant glioma	Primary: patient survival Secondary: progression-free survival	Arm A: PCV and DFMO DFMO: 3 g/m² *p.o.* every 8 h days 1–14, 29–42 PCV: CCNU 110 mg m⁻², *p.o.* day 15; Procarbazine 60 mg m⁻² day⁻¹ *p.o.* days 22–35; Vincristine 1.4 mg m⁻², iv (max 2 mg) days 8, 29 The cycle was repeated at 8-week intervals for a total of 7 cycles	$N = 249$ total Arm A: 125 with 114 evaluable for response
	Arm B: 79 anaplastic astrocytoma; 8 anaplastic oligoastrocytoma; 24 anaplastic oligodendroglioma; 3 other malignant glioma		Arm B: PCV alone PCV: CCNU 110 mg m⁻², *p.o.* day 1; Procarbazine 60 mg m⁻² day⁻¹ *p.o.* days 8–21; Vincristine 1.4 mg m⁻², iv (max 2 mg) days 8, 29 The cycle was repeated at 6-week intervals for a total of 7 cycles	Arm B: 124 patients with 114 patients evaluable for response.

Table 11.3 (*Continued*)

Investigator/study phase	Patient Population	Study end-points	Regimen	Number
DFMO/mitoguazone (MGBG) combination studies				
Levin *et al.*,[110] 1987 Phase I/II	Recurrent primary malignant glioma	Primary: Time to tumor progression from initiation of DFMO therapy	Group I: DFMO: 1 g m^{-2}, *p.o.* every 6 h, days 1–42;	$N = 45$ total (33 evaluable) 19 AG, 9 GM, 8 NOS
	Anaplastic glioma (AG)		MGBG 350 mg m^{-2} iv over 45 min, days 14, 28 and 42	
	Glioblastoma multiforme[111]		Group II: DFMO 1.33 g m^{-2} *p.o.* every 8 h, days 1–14, 21–35, and 42–56;	
	Other unspecified gliomas (NOS)		MGBG 200 mg m^{-2} iv over 45 min, days 14, 35 and 56. Courses repeated until tumor progression or for up to 1 year	
Levin *et al.*,[112] 1992 Phase II	Recurrent malignant glioma	Primary: Time to tumor progression from initiation of DFMO therapy	Group I: DFMO as sole therapy DFMO 3.6 g m^{-2} *p.o.* every 8 h on study days 1–14, 22–35 and 43–56. (first year); or	$N = 121$ total (101 evaluable)
	Anaplastic glioma (AG)		DFMO 3.6 g/m^2 *p.o. every 8 h* on study days 1–14, 29–42 and 57–70. (second year)	$N = 98$ patients DFMO only (80 evaluable)
	Glioblastoma multiforme[111]		Group II: DFMO/MGBG DFMO 1.8 g m^{-2} *p.o.* every 6 h on study days 1–14, 22–35 and 43–56; MGBG 200 mg m^{-2} i.v. on study days 14, 35 and 56. The first year of treatment utilized a 63-day cycle. The second year extended the cycle to 84 days.	$N = 23$ DFMO/MGBG combo (21 evaluable);

DFMO/BCNU (1,3-bis(2-chloroethyl)-1-nitrosourea) combination studies

Prados et al.,[113] 1989 Phase I/II	Recurrent malignant glioma	Primary: Time to tumor progression from initiation of DFMO	DFMO: 2.0 g/m^2 *p.o.* every 8 hours on d 1–14 and d 29–42, BCNU: 210 mg m^{-2} i.v. on d 15.	$N=38$
	5 brainstem 1 cerebellar 32 supratentorial gliomas		Treatment was repeated every 8 weeks for six courses or until the disease progressed.	

Other studies

Buckner et al.,[114] 1998 Phase II	Recurrent malignant glioma	Primary: Time to tumor progression from initiation of DFMO	Recombinant α-Interferon with DFMO	29 evaluable patients

survival advantage of PCV chemotherapy for AG patients by direct temporal interaction with PCV. Clinical safety of DFMO, when used over the dose range and regimen intended for clinical study, has been shown to be quite acceptable in this patient population and actually appears mild in comparison with most cytotoxic anticancer drugs.

Neuroblastoma (NB) is a highly malignant tumor of childhood accounting for 15% of all pediatric cancer-related deaths. The *MYCN* amplification in NB is associated with advanced stage disease, rapid tumor progression, poor prognosis and resistance to therapy. ODC is a transcriptional target of c-Myc and MYCN. While there is extensive literature on both ODC and NB, the role of ODC in NB and its potential relevance as a novel therapeutic target have only recently become of interest. This is critical given the importance of *MYCN* amplification in NB and its direct connection to ODC.

Based on this MYCN-ODC connection in NB, our group (AB) at the NCI-designated Cancer Research Center in Hawaii began in 2002 to investigate the effect of DFMO and SAM486A on ODC and AdoMetDC, respectively, and the role of polyamines in *MYCN*-amplified and non-amplified NB to verify whether the polyamine biosynthetic pathway is a potential alternative target for NB treatment. The primary goal in this endeavor was to re-purpose clinically well-characterized polyamine inhibitors, such as DFMO and SAM486A, in order to move them forward to the clinic more rapidly, in the event that *in vitro* (cell culture) and *in vivo* (mice) results are encouraging for this newly tested cancer type. We have discovered significant anti-proliferative activities with DFMO and SAM486A in tumor-derived NB cell lines.[50–54] In addition, this group found that an elevated ODC expression level correlates with several unfavorable genetic and clinical features in tumor samples of NB patients,[55] thus further demonstrating the importance of this target in NB. Interestingly, high ODC expression also showed a significant correlation with poor survival prognosis in Kaplan–Meier analyses stratified for patients without *MYCN* amplification, suggesting an additional role for ODC independent of *MYCN*. Most recently, Giselle Sholler and colleagues at the Vermont Children's Hospital in Burlington, Vermont (now at Van Andel Institute and Helen DeVos Children's Hospital, Grand Rapids, Michigan) in collaboration with our group demonstrated that DFMO/etoposide exhibits synergistic activities in cell culture and in mouse NB tumor xenografts.[56] Similar findings were confirmed by John Cleveland and colleagues at Scripps Florida and Michael Hogarty and colleagues at the Children's Hospital of Philadelphia (CHOP) with DFMO using an elegant NB transgenic (TH-MYCN) mouse model.[57,58]

Since the launch of our initial NB therapy studies with DFMO in 2002, the investigation of ODC and polyamines in the pediatric cancer NB has gained significant momentum, and several laboratories are now actively pursuing this line of research at the basic and pre-clinical research levels.[50–55,57,58] Further testament to the growing interest in polyamine depletion strategies for NB is the recent publication of an excellent review by Evageliou and Hogarty.[59]

Based on the current lack of effective therapies for relapsed/refractory NB patients, the preclinical effectiveness of DFMO and its high safety profile, we decided to advance DFMO into an FDA-approved phase I clinical trial for relapsed/refractory NB patients. This trial is headed by pediatric oncologist Dr Giselle Sholler, and comparative biological analyses of tumors and bone marrow aspirates are performed by Dr Bachmann's lab. The trial first opened in February 2010 at the Vermont Children's Hospital in Burlington, Vermont (recently transferred to the Helen DeVos Children's Hospital, Grand Rapids, Michigan), and later at the Arnold Palmer Hospital for Children in Orlando, Florida and Children's Hospital of Orange County, California. The primary objective of the trial is to monitor for safety and to establish the maximum tolerated dose in these children. The secondary objectives are to study the efficacy and pharmacokinetics of DFMO alone and in combination with etoposide. The dosing regimen is a dose escalation beginning at 500 mg m^{-2} DFMO orally, twice a day for each day during the study, and the safety of the dosing regimen in this trial is tested by a continuous risk/benefit assessment. The study design is based on a total of 5 cycles, and each cycle is 21 days in length. In the first cycle, DFMO is given alone by oral administration (dissolved in liquid). In the second cycle, 50 mg m^{-2} day^{-1} etoposide is added to the regimen, also by oral administration, once a day for the first 14 days of each 21-day cycle (cycles 2–5). Dose escalation of DFMO (750–1500 mg m^{-2}) is pursued after each cohort (consisting of three patients) has been evaluated for the safety and risk/benefit. With the oral formulation of both drugs, the patients do not require expensive hospitalization and are able to take the treatment cocktail in the comfort of their home. This approach is a significant advantage for the pediatric patient population.

In a phase II multicenter study, Miklos Pless at the University of Basel and Renaud Capdeville of Novartis Pharma in Switzerland and coworkers examined the clinical efficacy, tolerability and safety of SAM486A monotherapy in patients with relapsed or refractory non-Hodgkin's lymphoma (NHL).[30] Forty-one previously treated patients with diffuse large cell, follicular or peripheral T-cell NHL were treated *i.v.* with 100 mg m^{-2} SAM486A. The treatment was daily (1 h infusion) for 5 consecutive days, repeated every 3 weeks, and was continued for a total of eight cycles or until disease progression. Two patients showed a complete response at cycle 3 and were stable for ≥13 and ≥28 months. Five patients had a partial response, and three patients had stable disease at their last follow-up. The overall response rate was 18.9% for all patients that were evaluated. The most frequent side effects were nausea (39%), vomiting (22%), diarrhea (19.5%), asthenia (17.1%), abdominal pain (14.6%) and flushing (9.8%). Based on the overall response, SAM486A exhibited promising clinical activity with manageable side effects in patients with poor-prognosis NHL.

11.2.2 Other Diseases

11.2.2.1 Human African Trypanosomiasis

Human African Trypanosomiasis (HAT), also known as African sleeping sickness, has been endemic in sub-Saharan Africa for thousands of years, and currently around 50 million people are at risk for this disease.[13] DFMO was approved by the US Food and Drug Administration (FDA) in 1990 (Ornidyl®), and by the European Community regulatory authority in 1991 for the treatment of HAT. As a consequence, DFMO has been vigorously investigated for the treatment of this disease. Other HAT drugs in use are pentamidine, berenil (diminazene aceturate), suramin (sulfonated naphtylamine) and melarsoprol, all drugs synthesized in the 1920s to 1940s, making DFMO the most recently developed agent for this disease. DFMO cured children, adults, patients with melarsoprol-refractory strains and patients with late-stage disease, which makes this drug superior in efficacy to melarsoprol.[60,61] A number of clinical drug-combination studies assessed the combinations DFMO–melarsoprol, DFMO–nifurtimox and melarsoprol–nifurtimox revealing that the DFMO–nifurtimox regimen was far more effective than the other two combinations, with significantly reduced adverse effects.[60–63] Remarkably, one clinical study found a 100% cure rate with this regimen, and associated side-effects were significantly reduced compared to melarsoprol-based therapy.[60] Of note, nifurtimox was also evaluated in patients with neuroblastoma,[64] and a phase II clinical trial is currently under way. The encouraging results with DFMO–nifurtimox in HAT suggest that DFMO–nifurtimox may be a potential combination worth considering for the NB patient population.

11.3   Polyamine Inhibitors in Chemoprevention Trials

11.3.1   Cancer

In addition to direct therapeutic trials with DFMO, studies have been conducted using chronic daily low-dose DFMO in an effort to prevent malignancy in patients predisposed by virtue of associated morbid history or an existing precancerous lesion. In these 'prevention' trials, preliminary studies were encouraging in bladder, colon, prostate and skin cancers,[65–78] but to date, the best evidence appears to be from trials conducted by Frank Meyskens, Jr (UC Irvine, California) and Eugene Gerner (University of Arizona) that showed that DFMO with sulindac was able to reduce malignant colon polyp formation in people predisposed to develop cancer from colon polyps.[71,72,79] Of note is that clinical trials of DFMO as a chemoprevention agent for Barrett's esophagus were negative. Barrett's esophagus is a premalignant lesion in the lower esophagus, which can lead to adenocarcinoma of the esophagus. The ODC level of the Barrett's lesions were significantly higher than the normal mucosa, and the trial of DFMO was not effective.[80] Extrapolating from the anticancer studies in gliomas the fact that DFMO was not effective in this precancerous condition might reflect the inability of drugs such as DFMO to effectively inhibit ODC when levels are elevated. The use of DFMO to inhibit elevated ODC levels continues to be evaluated as a prevention strategy for many cancers, including superficial bladder cancer, cervical cancer, colorectal cancer, breast cancer, prostate cancer and nonmelanoma skin cancer. The value of DFMO as a chemopreventive agent in these various malignancies has been the subject of several recent reviews.[10] At this juncture, we await the completion of all DFMO chemoprevention trials before concluding whether partial ODC inhibition can be viewed as a successful cancer-prevention strategy in humans.

11.4 Polyamine Analogs in Therapeutic Clinical Trials

11.4.1 Cancer

In addition to inhibitors of polyamine biosynthesis, there are a number of polyamine analogs that act as antagonists of polyamine function, modulate polyamine metabolism and present an important alternative category that may develop into effective anticancer drugs. The theory behind polyamine analogs is that they reduce the production of downstream polyamines by interfering at multiple sites in the pathway, thereby preventing the upregulation of compensatory reactions seen with the single enzyme inhibitors.[20] For example, early studies with bis(ethyl)polyamine analogs Casero and colleagues (John Hopkins University) demonstrated that specific polyamine analogs can superinduce spermidine/spermine N^1-acetyltransferase (SSAT),[81] a rate-limiting factor in polyamine catabolism and a determining factoring in tumor-cell sensitivity to these analogs.

A first generation of polyamine analogs included deoxyspergualin (spergualin analog), BES (spermidine analog) or DESpm and DENSpm (spermine analogs), followed by second-generation polyamine analogs such as CHENSpm, IPENSpm and CPENSpm, reviewed by Wallace and Niiranen.[20] Most recent variations include PG-11093 and PG-11047analog from Progen Pharmaceuticals, Inc. To this end, at least four analogs have entered clinical trials; namely DENSpm (also referred to as BENSpm or BE-3-3-3), DEHSpm (also referred to as BEHSpm or BE-4-4-4), PG-11047 and PG-11093.

A phase I study with DEHSpm was performed in 15 patients with advanced solid tumors in order to determine the maximum tolerated dose (MTD) and dose-limiting toxicities (DLT). DEHSpm was administered by subcutaneous injection daily for 5 consecutive days, repeated every 4 weeks, and three dose levels were tested, starting from 12.5 mg m^{-2} day^{-1} and escalating to 37.5 mg m^{-2} day^{-1}. While DEHSpm was well tolerated at lower doses, significant toxicities were observed at the higher dose level with an MTD at 25 mg m^{-2} day^{-1}, and further investigation with this drug was not recommended due to potential neurotoxicities and hepatic damage.[82]

While BENSpm was better tolerated than BEHSpm, and phase I studies revealed that the drug appeared safe in patients with advanced malignancies,[83,84] a phase II study with patients with previously treated metastatic breast cancer revealed very little clinical activity, and all patients had disease progression by 4 months.[85] Clinical trials were also performed with the polyamine analogs PG-11093 and PG-11047, but the results have not yet been published (Laurence Marton, Progen Pharmaceuticals, Inc., personal communication).

Although many pre-clinical reports over the years have provided strong evidence for the effectiveness of polyamine analogs as anticancer drugs *in vitro* or in animal systems, it remains to be seen whether this interesting drug category will live up to its promise and will produce the anticipated antitumor response concomitant with a manageable safety profile in patients with cancer. One area that needs greater study is the appropriate dosing and dose scheduling for polyamine analogs. Early trials with these agents may not have used optimal scheduling, and thus the lack of significant positive outcomes with the analogs may be a result of poor scheduling rather than inherent inactivity of the compounds.

11.5 Future Directions

The recent chemopreventive phase III studies with DFMO/sulindac in colorectal cancer by Meyskens and Gerner as well as the ongoing phase I study with DFMO/etoposide in pediatric neuroblastoma by our group have revitalized the potential use and excitement for the old drug DFMO and repositioned its existence to the forefront in cancer treatment. The repurposing of already-established, old or forgotten drugs for new applications has recently become of interest and has perhaps been stimulated in response to an economically challenging environment, in an effort to reduce the high costs associated with clinical drug development. It is possible that additional cancer forms for which the polyamine depletion strategy has not been explored in the clinical setting may benefit from this approach, for example, most Myc-driven malignancies, medulloblastoma, pheochromocytoma, non-melanoma skin cancers and also mesothelioma. Clearly, we know from past experiences that DFMO monotherapy is not a feasible therapeutic approach for established cancers; rather, drug combinations are required in order to achieve a successful therapeutic outcome. Such regimens may include the combination of polyamine inhibitors with standard, front-line chemotherapeutic drugs, non-steroidal anti-inflammatory drugs (NSAIDs) and also biologically targeted inhibitors, such as Akt/PKB, TORC and proteasome inhibitors. Animal studies by Cleveland and colleagues demonstrated that the polyamine analog PG-11093 enhances the anti-myeloma activity of bortezomib (Velcade®), the first FDA-approved proteasome inhibitor on the market.[86] Therapies may also include the combination of polyamine inhibitors and polyamine analogs, including polyamine transport inhibitors as recently suggested by Burns and O'Brien.[87] In all those instances, it will be important in the clinical setting to carefully consider various drug-administration schedules for each drug in order to optimize for drug synergisms that are most effective in reducing the tumor burden of patients.

Besides the treatment of cancer and HAT, polyamine inhibitors and analogs may become beneficial to the prevention or treatment of other types of diseases, including Chagas' disease and leishmaniasis, malaria or diseases which involve disrupted or dysfunctional cell growth (hyperplasia).[20,88] Remarkably, ODC activity and elevated polyamine levels are also associated with a number of mental disorders including schizophrenia, anxiety and mood disorders.[89] Neuropathological and clinical studies further show that the ODC/polyamine system is also heavily involved in other human brain diseases, including human stroke and epilepsy as well as disorders with a clear-cut neurodegenerative and/or neurodevelopmental background, such as Alzheimer's disease (AD) and schizophrenia.[90,91] In AD, amyloid-β (Aβ) aggregation and insoluble filament formation of tau are well-established characteristics, and the amyloid precursor protein (APP) through the action of Aβ peptide can induce the expression of ODC. ODC gene expression and polyamine levels are increased in neocortical neurons and in the temporal cortex of AD patients. ODC protein expression increases in Purkinje cells of the cerebellum of AD patients and the protein is translocated from the nuclear compartment to the cytoplasm.

Also of interest, Lewandowski *et al.* recently revealed that the polyamine pathway contributes to the pathogenesis of Parkinson disease (PD).[92] In their work, they noted a disease-related decrease in the catabolic polyamine enzyme spermidine/spermine N^1-acetyltransferase 1 (SAT1). A causal link between polyamines and α-synuclein was revealed by showing that polyamines enhanced the toxicity of α-synuclein, thus suggesting that depletion of polyamines may be a useful strategy in PD.

While DFMO has been proven to be effective and safe at low doses in chemopreventive cancer trials as discussed earlier in this chapter, the fact that DFMO does not pass the blood–brain barrier easily[93] makes it unlikely that oral low-dose DFMO therapy for patients with advanced AD, PD or related disorders will become a reality in the future, unless efficacy can be defined at low regional brain concentrations. Clearly, this area of research is in its infancy, and more studies are warranted to further explore this avenue in order to verify whether polyamine inhibitors and analogs are useful to prevent, or at least retard, the onset of neurodegenerative diseases.

Acknowledgments

We would like to thank Dr Laurence Marton (Progen Pharmaceuticals) and Dr Giselle Sholler (Van Andel Institute) for comments and advice regarding this chapter.

References

1. P. S. Mamont, M. C. Duchesne, J. Grove and P. Bey, *Biochem. Biophys. Res. Commun.*, 1978, **81**, 58–66.
2. N. J. Prakash, P. J. Schechter, J. Grove and J. Koch-Weser, *Cancer Res.*, 1978, **38**, 3059–3062.
3. J. R. Fozard, M. L. Part, N. J. Prakash and J. Grove, *Eur. J. Pharmacol.*, 1980, **65**, 379–391.

4. E. W. Gerner, P. S. Mamont, A. Bernhardt and M. Siat, *Biochem. J.*, 1986, **239**, 379–386.
5. W. D. Heston, D. Kadmon and W. R. Fair, *Cancer Lett.*, 1982, **16**, 71–79.
6. J. Janne, H. Poso and A. Raina, *Biochim. Biophys. Acta*, 1978, **473**, 241–293.
7. C. W. Tabor and H. Tabor, *Annu. Rev. Biochem.*, 1984, **53**, 749–790.
8. H. M. Wallace, A. V. Fraser and A. Hughes, *Biochem. J.*, 2003, **376**, 1–14.
9. R. A. Casero, Jr and L. J. Marton, *Nat. Rev. Drug Discov.*, 2007, **6**, 373–390.
10. E. W. Gerner and F. L. Meyskens, Jr., *Nat. Rev. Cancer*, 2004, **4**, 781–792.
11. M. Auvinen, A. Paasinen, L. C. Andersson and E. Holtta, *Nature*, 1992, **360**, 355–358.
12. L. M. Shantz and V. A. Levin, *Amino Acids*, 2007, **33**, 213–223.
13. C. J. Bacchi, *Interdiscip. Perspect Infect Dis.*, 2009, **2009**, 1–5.
14. A. E. Pegg, *J. Biol. Chem.*, 2006, **281**, 14529–14532.
15. A. E. Pegg and P. P. McCann, *Am. J. Physiol.*, 1982, **243**, C212–C221.
16. B. W. Metcalf, P. Bey, C. Danzin, M. J. Jung, P. Casara and J. P. Vevert, *J. Am. Chem. Soc.*, 1978, **100**, 2551–2553.
17. N. Seiler, *Curr. Drug Targets*, 2003, **4**, 565–585.
18. N. Seiler, *Curr. Drug Targets*, 2003, **4**, 537–564.
19. M. A. Cornbleet, A. Kingsnorth, G. P. Tell, K. D. Haegele, A. M. Joder-Ohlenbusch and J. F. Smyth, *Cancer Chemother. Pharmacol.*, 1989, **23**, 348–352.
20. H. M. Wallace and K. Niiranen, *Amino Acids*, 2007, **33**, 261–265.
21. A. E. Pegg, *Cancer Res.*, 1988, **48**, 759–774.
22. J. Thiele and F. Dralle, *Ann. Chem.*, 1898, **302**, 275–299.
23. B. L. Freedlander and F. A. French, *Cancer Res.*, 1958, **18**, 1286–1289.
24. B. L. Freedlander and F. A. French, *Cancer Res.*, 1958, **18**, 360–363.
25. F. A. French, B. L. Freedlander and E. J. Blanz, Jr., *Cancer Res.*, 1961, **21**, 343–348.
26. H. G. Williams-Ashman and A. Schenone, *Biochem. Biophys. Res. Commun.*, 1972, **46**, 288–295.
27. K. Nishioka, *Polyamines in Cancer: Basic Mechanisms and Clinical Approaches*, R. G. Landes Company and Chapman & Hall, Austin, TX, 1996.
28. R. Paridaens, D. R. Uges, N. Barbet, L. Choi, M. Seeghers, van der W. T. Graaf, H. J. Groen, H. Dumez, I. V. Buuren, F. Muskiet, R. Capdeville, A. T. Oosterom and E. G. de Vries, *Br. J. Cancer*, 2000, **83**, 594–601.
29. L. L. Siu, E. K. Rowinsky, L. A. Hammond, G. R. Weiss, M. Hidalgo, G. M. Clark, J. Moczygemba, L. Choi, R. Linnartz, N. C. Barbet, I. T. Sklenar, R. Capdeville, G. Gan, C. W. Porter, D. D. Von Hoff and S. G. Eckhardt, *Clin. Cancer Res.*, 2002, **8**, 2157–2166.
30. M. Pless, K. Belhadj, H. D. Menssen, W. Kern, B. Coiffier, J. Wolf, R. Herrmann, E. Thiel, D. Bootle, I. Sklenar, C. Müller, L. Choi, C. Porter and R. Capdeville, *Clin. Cancer Res.*, 2004, **10**, 1299–1305.

31. M. J. Millward, A. Joshua, R. Kefford, S. Aamdal, D. Thomson, P. Hersey, G. Toner and K. Lynch, *Invest. New Drugs*, 2005, **23**, 253–256.
32. H. W. Herr, R. P. Warrel and J. H. Burchenal, *Urology*, 1986, **28**, 508–511.
33. J. Janne, L. Alhonen-Hongisto, P. Seppanen and M. Siimes, *Med. Biol.*, 1981, **59**, 448–457.
34. M. Siimes, P. Seppanen, L. Alhonen-Hongisto and J. Janne, *Int. J. Cancer*, 1981, **28**, 567–570.
35. T. A. Splinter and J. C. Romijn, *Eur. J. Cancer Clin. Oncol.*, 1986, **22**, 61–67.
36. F. L. Meyskens, E. M. Kingsley, T. Glattke, L. Loescher and A. Booth, *Invest. New Drugs*, 1986, **4**, 257–262.
37. M. Talpaz, C. Plager, J. Quesada, R. Benjamin, H. Kantarjian and J. Gutterman, *Eur. J. Cancer Clin. Oncol.*, 1986, **22**, 685–689.
38. V. Ganju, J. H. Edmonson and J. C. Buckner, *Invest. New Drugs*, 1994, **12**, 25–27.
39. M. D. Abeloff, S. T. Rosen, G. D. Luk, S. B. Baylin, M. Zeltzman and A. Sjoerdsma, *Cancer Treat. Rep.*, 1986, **70**, 843–845.
40. E. T. Creagan, D. J. Schaid, D. L. Ahmann and S. Frytak, *J. Invest. Dermatol.*, 1990, **95**(6 Suppl), 188S–192S.
41. O'J. A. Shaughnessy, L. M. Demers, S. E. Jones, J. Arseneau, P. Khandelwal, T. George, R. Gersh, D. Mauger and A. Manni, *Clin. Cancer Res.*, 1999, **5**, 3438–3444.
42. A. T. Vlastos, L. A. West, E. N. Atkinson, I. Boiko, A. Malpica, W. K. Hong and M. Follen, *Clin. Cancer Res.*, 2005, **11**, 390–396.
43. V. A. Levin, M. C. Chamberlain, M. D. Prados, A. K. Choucair, M. S. Berger, P. Silver, M. Seager, P. H. Gutin, R. L. Davis and C. B. Wilson, *Cancer Treat. Rep.*, 1987, **71**, 459–464.
44. M. Prados, L. Rodriguez, M. Chamberlain, P. Silver and V. Levin, *Neurosurgery*, 1989, **24**, 806–809.
45. V. A. Levin, K. R. Hess, A. Choucair, P. J. Flynn, K. A. Jaeckle, A. P. Kyritsis, W. K. Yung, M. D. Prados, J. M. Bruner, S. Ictech, M. J. Gleason and H. W. Kim, *Clin. Cancer Res.*, 2003, **9**, 981–990.
46. G. Scalabrino and M. E. Ferioli, *Cancer Detect. Prev.*, 1985, **8**, 11–16.
47. R. I. Ernestus, G. Rohn, R. Schroder, T. Els, A. Klekner, W. Paschen and N. Klug, *J. Neurol. Neurosurg. Psychiatry*, 2001, **71**, 88–92.
48. V. A. Levin, J. L. Jochec, L. M. Shantz, P. E. Koch and A. E. Pegg, *J. Histochem. Cytochem.*, 2004, **52**, 1467–1474.
49. V. A. Levin, J. L. Jochec, L. M. Shantz and K. D. Aldape, *Int. J. Cancer*, 2007, **121**(10), 2279–2283.
50. A. S. Bachmann, *Hawaii Med. J.*, 2004, **63**, 371–374.
51. D. L. Koomoa, T. Borsics, D. J. Feith, C. C. Coleman, C. J. Wallick, I. Gamper, A. E. Pegg and A. S. Bachmann, *Mol. Cancer Ther.*, 2009, **8**, 2067–2075.
52. D. L. Koomoa, R. C. Go, K. Wester and A. S. Bachmann, *Neurosci. Lett.*, 2008, **436**, 171–176.

53. D. L. Koomoa, L. P. Yco, T. Borsics, C. J. Wallick and A. S. Bachmann, *Cancer Res.*, 2008, **68**, 9825–9831.
54. C. J. Wallick, I. Gamper, M. Thorne, D. J. Feith, K. Y. Takasaki, S. M. Wilson, J. A. Seki, A. E. Pegg, C. V. Byus and A. S. Bachmann, *Oncogene*, 2005, **24**, 5606–5618.
55. D. Geerts, J. Koster, D. Albert, D. L. Koomoa, D. J. Feith, A. E. Pegg, R. Volckmann, H. Caron, R. Versteeg and A. S. Bachmann, *Int. J. Cancer*, 2010, **126**, 2012–2024.
56. G. L. Sholler, E. Currier, D. L. Koomoa and A. S. Bachmann, in *Proceedings of the 101st Annual Meeting of the American Association for Cancer Research: 2010*; Washington, DC: AACR; 2010: Abstract nr, P. O.T74.
57. M. D. Hogarty, M. D. Norris, K. Davis, X. Liu, N. F. Evageliou, C. S. Hayes, B. Pawel, R. Guo, H. Zhao, E. Sekyere, J. Keating, W. Thomas, N. C. Cheng, J. Murray, J. Smith, R. Sutton, N. Venn, W. B. London, A. Buxton, S. K. Gilmour, G. M. Marshall and M. Haber, *Cancer Res.*, 2008, **68**, 9735–9745.
58. R. J. Rounbehler, W. Li, M. A. Hall, C. Yang, M. Fallahi and J. L. Cleveland, *Cancer Res.*, 2009, **69**, 547–553.
59. N. F. Evageliou and M. D. Hogarty, *Clin. Cancer Res.*, 2009, **15**, 5956–5961.
60. F. Checchi, P. Piola, H. Ayikoru, F. Thomas, D. Legros and G. Priotto, *PLoS Negl. Trop. Dis.*, 2007, **1**, e64.
61. G. Priotto, C. Fogg, M. Balasegaram, O. Erphas, A. Louga, F. Checchi, S. Ghabri and P. Piola, *PLoS Clin. Trials*, 2006, **1**, e39.
62. F. Chappuis, *Clin. Infect. Dis.*, 2007, **45**, 1443–1445.
63. G. Priotto, S. Kasparian, D. Ngouama, S. Ghorashian, U. Arnold, S. Ghabri and U. Karunakara, *Clin. Infect. Dis.*, 2007, **45**, 1435–1442.
64. Saulnier G. L. Sholler, S. Kalkunte, C. Greenlaw, K. McCarten and E. Forman, *J. Pediatr. Hematol. Oncol.*, 2006, **28**, 693–695.
65. P. P. Carbone, J. A. Douglas, P. O. Larson, A. K. Verma, I. A. Blair, M. Pomplun and K. D. Tutsch, *Cancer Epidemiol. Biomark. Prev.*, 1998, **7**, 907–912.
66. D. S. Alberts, R. T. Dorr, J. G. Einspahr, M. Aickin, K. Saboda, M. J. Xu, Y. M. Peng, R. Goldman, J. A. Foote, J. A. Warneke, S. Salasche, D. J. Roe and G. T. Bowden, *Cancer Epidemiol. Biomarkers Prev.*, 2000, **9**, 1281–1286.
67. J. G. Einspahr, M. A. Nelson, K. Saboda, J. Warneke, G. T. Bowden and D. S. Alberts, *Clin. Cancer Res.*, 2002, **8**, 149–155.
68. J. G. Einspahr, G. T. Bowden and D. S. Alberts, *Recent Results Cancer Res.*, 2003, **163**, 151–164.
69. S. M. Fischer, C. J. Conti, J. Viner, C. M. Aldaz and R. A. Lubet, *Carcinogenesis*, 2003, **24**, 945–952.
70. A. R. Simoneau, E. W. Gerner, R. Nagle, A. Ziogas, S. Fujikawa-Brooks, H. Yerushalmi, T. E. Ahlering, R. Lieberman, C. E. McLaren, H. Anton-Culver and F. L. Meyskens, Jr, *Cancer Epidemiol. Biomark. Prev.*, 2008, **17**, 292–299.

71. F. L. Meyskens, Jr, C. E. McLaren, D. Pelot, S. Fujikawa-Brooks, P. M. Carpenter, E. Hawk, G. Kelloff, M. J. Lawson, J. Kidao, J. McCracken, C. G. Albers, D. J. Ahnen, D. K. Turgeon, S. Goldschmid, P. Lance, C. H. Hagedorn, D. L. Gillen and E. W. Gerner, *Cancer Prev. Res. (Phil. PA)*, 2008, **1**, 32–38.
72. E. W. Gerner, F. L. Meyskens, Jr, S. Goldschmid, P. Lance and D. Pelot, *Amino Acids*, 2007, **33**, 189–195.
73. A. R. Simoneau, E. W. Gerner, M. Phung, C. E. McLaren and F. L. Meyskens, Jr, *J. Natl. Cancer Inst.*, 2001, **93**, 57–59.
74. F. L. Meyskens, Jr and E. W. Gerner, *Clin. Cancer Res.*, 1999, **5**, 945–951.
75. F. L. Meyskens, Jr and E. W. Gerner, *J. Cell. Biochem. Suppl.*, 1995, **22**, 126–131.
76. A. M. Kamat and D. L. Lamm, *Urol. Clin. North Am.*, 2002, **29**, 157–168.
77. R. R. Love, P. P. Carbone, A. K. Verma, D. Gilmore, P. Carey, K. D. Tutsch, M. Pomplun and G. Wilding, *J. Natl. Cancer Inst.*, 1993, **85**, 732–737.
78. C. L. Loprinzi and E. M. Messing, *J. Cell. Biochem. Suppl.*, 1992, **16I**, 153–155.
79. E. W. Gerner and F. L. Meyskens, Jr, *Clin. Cancer Res.*, 2009, **15**, 758–761.
80. H. S. Garewal, R. E. Sampliner and M. B. Fennerty, *J. Natl. Cancer Inst. Monogr.*, 1992, **13**, 51–54.
81. R. A. Casero, Jr, S. J. Ervin, P. Celano, S. B. Baylin and R. J. Bergeron, *Cancer Res.*, 1989, **49**, 639–643.
82. G. Wilding, D. King, K. Tutsch, M. Pomplun, C. Feierabend, D. Alberti and R. Arzoomanian, *Invest. New Drugs*, 2004, **22**, 131–138.
83. H. A. Hahm, D. S. Ettinger, K. Bowling, B. Hoker, T. L. Chen, Y. Zabelina and R. A. Casero, Jr, *Clin. Cancer Res.*, 2002, **8**, 684–690.
84. R. R. Streiff and J. F. Bender, *Invest. New Drugs*, 2001, **19**, 29–39.
85. A. C. Wolff, D. K. Armstrong, J. H. Fetting, M. K. Carducci, C. D. Riley, J. F. Bender, R. A. Casero, Jr and N. E. Davidson, *Clin. Cancer Res.*, 2003, **9**, 5922–5928.
86. J. S. Carew, S. T. Nawrocki, V. K. Reddy, D. Bush, J. E. Rehg, A. Goodwin, J. A. Houghton, R. A. Casero, Jr, L. J. Marton and J. L. Cleveland, *Cancer Res.*, 2008, **68**, 4783–4790.
87. M. R. Burns, G. F. Graminski, R. S. Weeks, Y. Chen and O'T. G. Brien, *J. Med. Chem.*, 2009, **52**, 1983–1993.
88. O. Heby, L. Persson and M. Rentala, *Amino Acids*, 2007, **33**, 359–366.
89. L. M. Fiori and G. Turecki, *J. Psychiatry Neurosci.*, 2008, **33**, 102–110.
90. H. G. Bernstein and M. Muller, *Prog. Neurobiol.*, 1999, **57**, 485–505.
91. S. M. Yatin, M. Yatin, S. Varadarajan, K. B. Ain and D. A. Butterfield, *J. Neurosci. Res.*, 2001, **63**, 395–401.
92. N. M. Lewandowski, S. Ju, M. Verbitsky, B. Ross, M. L. Geddie, E. Rockenstein, A. Adame, A. Muhammad, J. P. Vonsattel, D. Ringe, L. Cote, S. Lindquist, E. Masliah, G. A. Petsko, K. Marder, L. N. Clark and S. A. Small, *Proc. Natl. Acad. Sci. U. S. A.*, 2010, **107**, 16970–16975.
93. V. A. Levin, J. Csejtey and D. J. Byrd, *Cancer Chemother. Pharmacol.*, 1983, **10**, 196–199.

94. J. A. Gastaut, G. Tell, P. J. Schechter, D. Maraninchi, B. Mascret and Y. Carcassonne, *Cancer Chemother. Pharmacol.*, 1987, **20**, 344–348.
95. M. Mitchell, G. Tortolero-Luna, J. Lee, W. Hittelman, R. Lotan, J. Wharton, W. Hong and K. Nishioka, *Clin. Cancer Res.*, 1998, **4**, 303–310.
96. E. T. Creagan, H. J. Long, D. L. Ahmann and D. J. Schaid, *Am. J. Clin. Oncol.*, 1990, **13**, 218–220.
97. M. K. Croghan, A. Booth and F. L. Meyskens, Jr, *J. Biol. Response Mod.*, 1988, **7**, 409–415.
98. V. A. Levin, M. D. Prados, W. K. Yung, M. J. Gleason, S. Ictech and M. Malec, *J. Natl. Cancer Inst.*, 1992, **84**, 1432–1437.
99. R. M. Bukowski, T. R. Fleming, J. S. Macdonald, N. Oishi, S. A. Taylor and L. H. Baker, *Cancer*, 1993, **71**, 322–325.
100. J. N. Winter, P. S. Ritch, S. T. Rosen, M. M. Oken, J. M. Wolter, P. H. Wiernik and M. J. O'Connell, *Cancer Invest.*, 1990, **8**, 143–146.
101. M. S. Simon, J. Eckenrode and R. B. Natale, *Invest. New Drugs*, 1990, **8**, S79–S81.
102. J. T. Barrett, B. Orofiamma, J. D. Khandekar, P. P. Carbone, R. L. Comis and T. E. Davis, *Cancer*, 1989, **64**, 2445–2447.
103. H. O. Douglass, Jr, M. Lefkopoulou, H. L. Davis, S. G. Taylor, G. Falkson, A. Mittelman, J. MacIntyre and P. F. Engstrom, *Am. J. Clin. Oncol.*, 1988, **11**, 646–649.
104. W. A. Knight, 3rd, D. M. Loesch, L. P. Leichman, C. Fabian and O'R. M. Bryan, *Cancer Treat. Rep.*, 1982, **66**, 2099–2100.
105. P. H. Wiernik, L. I. Gordon, M. M. Oken, J. E. Harris and O'M. J. Connell, *Leuk. Lymphoma*, 1999, **35**, 375–377.
106. W. A. Knight, 3rd, C. Fabian, J. J. Costanzi, S. E. Jones and C. A. Coltman, Jr, *Invest. New Drugs*, 1983, **1**, 235–237.
107. A. M. Levine, A. Tulpule, D. Tessman, L. Kaplan, F. Giles, B. D. Luskey, D. T. Scadden, D. W. Northfelt, I. Silverberg, J. Wernz, B. Espina and D. Von Hoff, *J. Clin. Oncol.*, 1997, **15**, 1094–1103.
108. H. I. Scher, A. Yagoda, T. Ahmed and R. C. Watson, *J. Clin. Oncol.*, 1985, **3**, 224–228.
109. W. A. Knight, 3rd, A. Drelichman, C. Fabian and R. M. Bukowski, *Cancer Treat. Rep.*, 1983, **67**, 1139–1140.
110. V. A. Levin, M. C. Chamberlain, M. D. Prados, A. K. Choucair, M. S. Berger, P. Silver, M. Seager, P. H. Gutin, R. L. Davis and C. B. Wilson, *Cancer Treat. Rep.*, 1987, **71**, 459–464.
111. M. D. Bregman and F. L. Meyskens, Jr, *Int. J. Cancer*, 1986, **37**, 101–107.
112. V. A. Levin, M. D. Prados, W. K. Yung, M. J. Gleason, S. Ictech and M. Malec, *J. Natl. Cancer Inst.*, 1992, **84**, 1432–1437.
113. M. Prados, L. Rodriguez, M. Chamberlain, P. Silver and V. A. Levin, *Neurosurgery*, 1989, **24**, 806–809.
114. J. C. Buckner, P. A. Burch, T. L. Cascino, O'J. R. Fallon and B. W. Scheithauer, *J. Neurooncol.*, 1998, **36**, 65–70.

Subject Index